Student Solutions Manual

for

McKeague/Turner's

Trigonometry

Fifth Edition

Judy Barclay
Cuesta College

THOMSON
BROOKS/COLE

Australia • Canada • Mexico • Singapore • Spain • United Kingdom • United States

Printed in the United States of America
   3 4 5 6 7 07

Printer: Thomson/West

ISBN-13: 978-0-534-40395-9
ISBN-10: 0-534-40395-6

For more information about our products,
contact us at:
Thomson Learning Academic Resource Center
1-800-423-0563

For permission to use material from this text,
contact us by:
Phone: 1-800-730-2214
Fax: 1-800-730-2215
Web: http://www.thomsonrights.com

Brooks/Cole—Thomson Learning
10 Davis Drive
Belmont, CA 94002-3098
USA

Asia
Thomson Learning
5 Shenton Way #01-01
UIC Building
Singapore 068808

Australia/New Zealand
Thomson Learning
102 Dodds Street
Southbank, Victoria 3006
Australia

Canada
Nelson
1120 Birchmount Road
Toronto, Ontario M1K 5G4
Canada

Europe/Middle East/South Africa
Thomson Learning
High Holborn House
50/51 Bedford Row
London WC1R 4LR
United Kingdom

Latin America
Thomson Learning
Seneca, 53
Colonia Polanco
11560 Mexico D.F.
Mexico

Spain/Portugal
Paraninfo
Calle/Magallanes, 25
28015 Madrid, Spain

Preface to the Student

This manual contains detailed explanations to the odd exercises from the problem sets and every exercise from the chapter tests. There is no substitute for working through a problem on your own, but I hope that the manual will end some of the frustration involved in problem solving.

I suggest that you close this manual while doing your homework. Only look up those exercises that you cannot do on your own. It will also be helpful to study one problem and then use the same technique to solve an even problem whose solution is not included in this manual. The solution of a problem is the important part, not the answer.

When using a calculator to compute the answer to an exercise, do not round until the end of the calculation. In this manual intermediate steps may be shown to three or four decimal places for you to check your work to that point. However, the calculations have been performed on a calculator without rounding until the last step. In other words, the calculations only appear to be rounded, they are not.

You would not think about becoming a good tennis player or golfer without a great deal of practice. The same is true in mathematics. I hope this manual helps you to be successful in your trigonometry class and I wish you good luck in your problem solving!

Acknowledgments

I would like to thank all those who assisted me in the development and completion of this manual: the many students I have had in classes over the past thirty years, Pat McKeague and Mark Turner, Lisa Chow at Thomson Learning, and Renoda Campbell-Monza, my assistant and typist. Without Renoda's help, I would not have completed this project on time.

Special thanks to my daughters, Stephanie and Erika, and to my husband, Ken, for their encouragement and loving support always.

About the Author

Judy Barclay received her bachelor's degree from the State University of New York College at Cortland and her master's degree from the University of Massachusetts. She has been teaching for 30 years at the secondary and community college levels. She has taught trigonometry many times. She is currently an instructor and the chair in the Mathematics division at Cuesta College in San Luis Obispo, California.

Contents

CHAPTER 1   The Six Trigonometric Functions

Problem Set 1.1

1. $10°$ is an acute angle.

The complement of $10°$ is $80°$ because $10° + 80° = 90°$.

The supplement of $10°$ is $170°$ because $10° + 170° = 180°$.

3. $45°$ is an acute angle.

The complement of $45°$ is $45°$ because $45° + 45° = 90°$.

The supplement of $45°$ is $135°$ because $45° + 135° = 180°$.

5. $120°$ is an obtuse angle.

The complement of $120°$ is $-30°$ because $120° + (-30)° = 90°$.

The supplement of $120°$ is $60°$ because $120° + 60° = 180°$.

7.   We can't tell if $x°$ is acute or obtuse (or neither).

The complement of $x°$ is $90° - x$ because $x° + (90° - x°) = 90°$.

The supplement of $x°$ is $180° - x°$ because $x° + (180° - x°) = 180°$.

9.

$$\alpha = 180° - (\angle A + \angle D)$$ — The sum of the angles of a triangle is $180°$

$$= 180° - (30° + 90°)$$ — Substitute given values

$$= 180° - 120°$$ — Simplify

$$= 60°$$

11.

$$\alpha = 180° - (\angle A + \angle D)$$ — The sum of the angles of a triangle is $180°$

$$= 180° - (\alpha + 90°)$$ — Substitute the given values

$$= 90° - \alpha$$ — Simplify right side

$$2\alpha = 90°$$ — Add α to both sides

$$\alpha = 45°$$ — Divide both sides by 2

13.

$$\angle A = 180° - (\alpha + \beta + \angle B)$$ — The sum of the angles of a triangle is $180°$

$$= 180° - (100° + 30°)$$ — Substitute given values

$$= 180° - 130°$$ — Simplify

$$= 50°$$

15. Angles α and β are complementary because
$$\alpha + \beta + 90° = 180°$$
$$\alpha + \beta = 90°$$

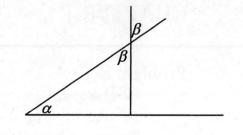

17.

$\alpha + \beta = 90°$	α and β are complementary
$\beta = 90° - \alpha$	Subtract α from both sides
$= 90° - 25°$	Substitute given value
$= 65°$	Simplify

19. One complete revolution equals $360°$.

Therefore, it rotates $360°$ in 2 seconds and $180°$ in 1 second.

21. One complete revolution equals $360°$.

In 4 hours, the hour hand revolves $\dfrac{4}{12}$ or $\dfrac{1}{3}$ of a revolution.

$\dfrac{1}{3}$ of $360° = 120°$.

23. Let $\alpha =$ the degree measure of each angle

Then
$$\alpha + \alpha + \alpha = 180°$$
$$3\alpha = 180°$$
$$\alpha = 60°$$

Therefore, each angle of an equilateral triangle is $60°$

25.

$c^2 = a^2 + b^2$	Pythagorean Theorem
$= 4^2 + 3^2$	Substitute given values
$= 16 + 9$	Simplify
$= 25$	

Therefore, $c = \pm 5$. Our only solution is $c = 5$, because we cannot use $c = -5$.

27.

$a^2 + b^2 = c^2$	Pythagorean Theorem
$b^2 = c^2 - a^2$	Subtract a^2 from both sides
$= 17^2 - 8^2$	Substitute given values
$= 289 - 64$	Simplify
$= 225$	

Therefore, $b = \pm 15$. Our only solution is $b = 15$, because we cannot use $b = -15$.

29.

$$a^2 + b^2 = c^2 \qquad \text{Pythagorean Theorem}$$
$$a^2 = c^2 - b^2 \qquad \text{Subtract } b^2 \text{ from both sides}$$
$$= 13^2 - 12^2 \qquad \text{Substitute given values}$$
$$= 169 - 144 \qquad \text{Simplify}$$
$$= 25$$

Therefore, $a = \pm 5$. Our only solution is $a = 5$, because we **cannot** use $a = -5$.

31.

$$x^2 = 3^2 + 3^2 \qquad \text{Pythagorean Theorem}$$
$$= 9 + 9 \qquad \text{Simplify}$$
$$= 18$$

Therefore, $x = \pm\sqrt{18} = \pm 3\sqrt{2}$. Our only solution is $x = 3\sqrt{2}$ because we **cannot** use $x = -3\sqrt{2}$.
Note: This must be a $45° - 45° - 90°$ triangle.

33.

$$x^2 = (2)^2 + \left(2\sqrt{3}\right)^2 \qquad \text{Pythagorean Theorem}$$
$$= 4 + 12 \qquad \text{Simplify}$$
$$= 16$$

Therefore $x = \pm 4$. Our only solution is $x = 4$, because we **cannot** use $x = -4$.

35.

$$\left(\sqrt{10}\right)^2 = x^2 + (x+2)^2 \qquad \text{Pythagorean Theorem}$$
$$10 = x^2 + x^2 + 4x + 4 \qquad \text{Simplify}$$
$$10 = 2x^2 + 4x + 4 \qquad \text{Combine like terms}$$
$$0 = 2x^2 + 4x - 6 \qquad \text{Subtract 10 from both sides}$$
$$0 = x^2 + 2x - 3 \qquad \text{Divide both sides by 2}$$
$$0 = (x+3)(x-1) \qquad \text{Factor}$$
$$x+3 = 0 \text{ or } x-1 = 0 \qquad \text{Set each factor equal to zero}$$
$$x = -3 \qquad x = 1$$

Therefore, $x = 1$ because $x = -3$ is not possible.

37.

$$(BD)^2 = (CD)^2 + (BC)^2 \qquad \text{Pythagorean Theorem}$$
$$5^2 = (CD)^2 + (4)^2 \qquad \text{Substitute given values}$$
$$25 = (CD)^2 + 16 \qquad \text{Simplify}$$
$$9 = (CD)^2 \qquad \text{Subtract 16 from both sides}$$

$$CD = 3 \text{ or } CD = -3 \qquad \text{Take square root of both sides}$$
$$CD = 3 \qquad \text{Eliminate negative solution}$$

Therefore, $AC = 2 + 3 = 5 \qquad AC = AD + DC$

This problem is continued on the next page.

$$(AB)^2 = (AC)^2 + (BC)^2 \qquad \text{Pythagorean Theorem}$$
$$= 5^2 + 4^2 \qquad \text{Substitute given values}$$
$$= 25 + 16 \qquad \text{Simplify}$$
$$= \sqrt{41}$$
$$AB = \sqrt{41} \text{ or } AB = -\sqrt{41} \qquad \text{Take square root of both sides}$$
$$AB = \sqrt{41} \qquad \text{Eliminate negative solution}$$

39.
$$(AB + BC)^2 = (CD)^2 + (AD)^2 \qquad \text{Pythagorean Theorem}$$
$$(4 + r)^2 = r^2 + 8^2 \qquad \text{Substitute given values}$$
$$16 + 8r + r^2 = r^2 + 64 \qquad \text{Simplify}$$
$$16 + 8r = 64 \qquad \text{Subtract } r^2 \text{ from both sides}$$
$$8r = 48 \qquad \text{Subtract 16 from both sides}$$
$$r = 6 \qquad \text{Divide both sides by 8}$$

41. This is an isosceles triangle. Therefore, the altitude must bisect the base
$$x^2 = (18)^2 + (13.5)^2 \qquad \text{Pythagorean Theorem}$$
$$= 324 + 182.5 \qquad \text{Simplify}$$
$$= 506.25$$
$$x = 22.5 \text{ or } x = -22.5 \qquad \text{Take square root of both sides}$$
$$x = 22.5 \text{ ft} \qquad \text{Eliminate negative solution}$$

43. The shortest side is 1.
The longest side is twice the shortest side. Therefore, it is 2.
The side opposite the $60°$ angle is $1\sqrt{3}$ or $\sqrt{3}$.

45. The longest side is 8 which is twice the shortest side.
Therefore, the shortest side is 4.
The side opposite the $60°$ angle is $4\sqrt{3}$.

47. Let t = the shortest side, $2t$ = the longest side, and $t\sqrt{3}$ = the side opposite $60°$

Therefore, $t\sqrt{3} = 6$ \qquad Side opposite $60°$ is 6

$$t = \frac{6}{\sqrt{3}} \qquad \text{Divide both sides by } \sqrt{3}$$

$$= \frac{6}{\sqrt{3}} \cdot \frac{\sqrt{3}}{\sqrt{3}} \qquad \text{Rationalize the denominator}$$

$$= \frac{6\sqrt{3}}{3} = 2\sqrt{3}$$

Since $t = 2\sqrt{3}$, $2t = 2(2\sqrt{3}) = 4\sqrt{3}$. \qquad The shortest side is $2\sqrt{3}$ and the longest side is $4\sqrt{3}$.

49. The shortest side is 20 feet.
The longest side is twice the shortest side
Therefore, $x = 2(20)$
$\qquad x = 40$ feet

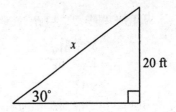

51. The tent is made up of 3 congruent rectangles and 2 congruent triangles.
First we'll find the sides of the $30° - 60° - 90°$ triangle.
The side opposite $60°$ is 4 ft. Let t = the shortest side.

$t\sqrt{3} = 4$

$\qquad t = \dfrac{4}{\sqrt{3}}$

$\qquad t = \dfrac{4}{\sqrt{3}} \cdot \dfrac{\sqrt{3}}{\sqrt{3}} = \dfrac{4\sqrt{3}}{3}$

The shortest side is $\dfrac{4\sqrt{3}}{3}$. The hypotenuse is $2\left(\dfrac{4\sqrt{3}}{3}\right) = \dfrac{8\sqrt{3}}{3}$.

Also the base of the triangular sides are $2\left(\dfrac{4\sqrt{3}}{3}\right) = \dfrac{8\sqrt{3}}{3}$.

Note: This is an equilateral triangle.

Area of rectangles = length · width

Area of rectangles $= 6\left(\dfrac{8\sqrt{3}}{3}\right)$

$\qquad\qquad\qquad = 16\sqrt{3}$ ft^2

Area of triangles $= \dfrac{1}{2} \cdot$ base · height

$\qquad\qquad\quad = \dfrac{1}{2}\left(\dfrac{8\sqrt{3}}{3}\right)(4)$

$\qquad\qquad\quad = \dfrac{16\sqrt{3}}{3}$ ft^2

Area of tent = 3 rectangles + 2 triangles

$\qquad\qquad = 3\left(16\sqrt{3}\right) + 2\left(\dfrac{16\sqrt{3}}{3}\right)$

$\qquad\qquad = 48\sqrt{3} + \dfrac{32\sqrt{3}}{3}$

$\qquad\qquad \approx 101.6$ ft^2

53. $\text{hypotenuse} = \dfrac{4}{5} \cdot \sqrt{2}$ Hypotenuse is $t \cdot \sqrt{2}$

$\qquad\qquad\quad = \dfrac{4\sqrt{2}}{5}$ Simplify

55. $\text{hypotenuse} = t\sqrt{2}$ t is the shorter side

$\qquad\quad 8\sqrt{2} = t\sqrt{2}$ Substitute given value

$\qquad\qquad 8 = t$ Divide both sides by $\sqrt{2}$

57. $\text{hypotenuse} = t\sqrt{2}$ t is the shorter side

$\qquad\qquad 4 = t\sqrt{2}$ Substitute given value

$\qquad\quad \dfrac{4}{\sqrt{2}} = t$ Divide both sides by $\sqrt{2}$

$\qquad\quad t = 2\sqrt{2}$ Rationalize denominator by multiplying numerator and denominator by $\sqrt{2}$

59. We are looking for the hypotenuse of a $45° - 45° - 90°$ triangle where the shorter sides are 1000 feet.

$\text{hypotenuse} = 1000\sqrt{2}$ Hypotenuse is $t\sqrt{2}$

$\qquad\qquad\quad \approx 1414 \text{ ft}$ Rounded to the nearest foot

The bullet travels 1414 feet.

61. (a) $\text{hypotenuse} = t\sqrt{2}$ t is the edge of the cube

$\qquad\qquad\qquad = 1\sqrt{2}$ Substitute given value

$\qquad\qquad\qquad = \sqrt{2}$ Simplify

Therefore, $CH = \sqrt{2}$ inches

(b) $(CF)^2 = (CH)^2 + (FH)^2$ Pythagorean Theorem

$\qquad\quad = \left(\sqrt{2}\right)^2 + (1)^2$ Substitute given values

$\qquad\quad = 2 + 1$ Simplify

$\qquad\quad = 3$

$CF = \pm\sqrt{3}$ Take easy square root of both sides

$CF = \sqrt{3}$ inches   Eliminate the negative solution

63. (a) hypotenuse $= x\sqrt{2}$ $\qquad$ x is the edge of the cube

The length of the diagonal of any face of the cube will be $x\sqrt{2}$.

(b) $(CF)^2 = (CH)^2 + (FH)^2$ $\qquad$ Pythagorean Theorem

$\qquad = \left(x\sqrt{2}\right)^2 + (x)^2$ $\qquad$ Substitute given values

$\qquad = 2x^2 + x^2$ $\qquad$ Simplify

$\qquad = 3x^2$

$CF = \sqrt{3x^2}$ $\qquad$ or $\qquad$ $CF = -\sqrt{3x^2}$

$\quad\; = x\sqrt{3}$ $\qquad\qquad\qquad\quad$ This is impossible.

The length of any diagonal that passes through the center of the cube will be $x\sqrt{3}$.

65. $(CF)^2 = (CH)^2 + (FH)^2$ $\qquad$ Pythagorean Theorem

$\qquad = \left(t\sqrt{2}\right)^2 + (t)^2$ $\qquad$ From Problem 61a above

$\qquad = 2t^2 + t^2$ $\qquad$ Simplify

$\qquad = 3t^2$

$(3)^2 = 3t^2$ $\qquad$ Substitute given value

$3 = t^2$ $\qquad$ Divide both sides by 3

$t = \pm\sqrt{3}$ $\qquad$ Take square root of both sides

$t = \sqrt{3}$ ft $\qquad$ Eliminate the negative solution.

Problem Set 1.2

13. If we let $x = 0$, the equation $3x + 2y = 6$ becomes: $\quad 3(0) + 2y = 6$

$$2y = 6$$

$$y = 3$$

This gives us $(0, 3)$ as one solution to $3x + 2y = 6$.

If we let $y = 0$, the equation $3x + 2y = 6$ becomes: $\quad 3x + 2(0) = 6$

$$3x = 6$$

$$x = 2$$

It gives us $(2, 0)$ as a second solution to $3x + 2y = 6$.

Graphing the points $(0, 3)$ and $(2, 0)$ and then drawing a line through them, we have the graph of $3x + 2y = 6$.

15. If we let $x = 0$, the equation $y = \frac{1}{2}x$ becomes: $y = \frac{1}{2}(0)$

$$y = 0$$

This gives us (0, 0) as one solution to $y = \frac{1}{2}x$.

If we let $x = 4$, the equation $y = \frac{1}{2}x$ becomes: $y = \frac{1}{2}(4)$

$$y = 2$$

This gives us (4, 2) as a second solution to $y = \frac{1}{2}x$.

Graphing the points (0, 0) and (4, 2) and then drawing a line through them, we have the graph of $y = \frac{1}{2}x$.

17. The vertex of this parabola is at $(0, -4)$.

If we let $x = -2$, the equation $y = x^2 - 4$ becomes $y = (-2)^2 - 4$

$$= 4 - 4$$
$$= 0$$

This gives us $(-2, 0)$ as a point of the curve.

If we let $x = -1$, the equation $y = x^2 - 4$ becomes $y = (-1)^2 - 4$

$$= 1 - 4$$
$$= -3$$

This gives us $(-1, -3)$ as a point on the curve.

Using the symmetry of a parabola, the points $(2, 0)$ and $(1, -3)$ will also be points on the curve.

Graphing the points $(-2, 0), (-1, -3), (0, -4), (1, -3)$, and $(2, 0)$, and then drawing a smooth curve through them, we have the graph of the parabola $y = x^2 - 4$.

19. The vertex of this parabola is $(-2, 4)$

If we let $x = -1$, the equation $y = (x + 2)^2 + 4$ becomes

$$y = (-1 + 2)^2 + 4$$
$$= 1^2 + 4$$
$$= 1 + 4$$
$$= 5$$

This problem is continued on the next page

This gives us $(-1, 5)$ as a point on the curve.

If we let $x = 0$, the equation $y = (x + 2)^2 + 4$ becomes
$$y = (0 + 2)^2 + 4$$
$$= 2^2 + 4$$
$$= 4 + 4$$
$$= 8$$

This gives us $(0, 8)$ as a point on the curve.

Using the symmetry of a parabola, the points $(-3, 5)$ and $(-4, 8)$ will also be points on the curve.

Graphing the points $(0, 8), (-1, 5), (-2, 4), (-3, 5)$ and $(-4, 8)$ and then drawing a smooth curve through them, we have the graph of the parabola $y = (x + 2)^2 + 4$.

25. The center of this circle is $(0, 0)$ and the radius is 5.

27. The center of the circle is $(0, 0)$ and the radius is $\sqrt{5}$.

35. From the graph $x^2 + y^2 = 25$, we can see that $(0, 5)$ and $(5, 0)$ are the points at which the line $x + y = 5$ will intersect the circle.

37. Solving this system of equations by substitution,, we get:

$$x^2 + y^2 = 1 \qquad\qquad x^2 + x^2 = 1$$
$$y = x \qquad\qquad\qquad 2x^2 = 1$$
$$x^2 = \frac{1}{2}$$
$$x = \pm\sqrt{\frac{1}{2}}$$
$$x = \pm\frac{\sqrt{2}}{2}$$

If $x = \frac{\sqrt{2}}{2}, y = \frac{\sqrt{2}}{2}$

If $x = -\frac{\sqrt{2}}{2}, y = -\frac{\sqrt{2}}{2}$

The solution is $\left(\frac{\sqrt{2}}{2}, \frac{\sqrt{2}}{2}\right)$ and $\left(-\frac{\sqrt{2}}{2}, -\frac{\sqrt{2}}{2}\right)$.

39.

$$r = \sqrt{(x_2 - x_1)^2 + (y_2 - y_1)^2}$$ Distance formula

$$= \sqrt{(3-6)^2 + (7-3)^2}$$ Substitute given values

$$= \sqrt{(-3)^2 + 4^2}$$ Simplify

$$= \sqrt{9+16}$$

$$= \sqrt{25} = 5$$

41.

$$r = \sqrt{(x_2 - x_1)^2 + (y_2 - y_1)^2}$$ Distance formula

$$= \sqrt{(0-5)^2 + (12-0)^2}$$ Substitute given values

$$= \sqrt{(-5)^2 + 12^2}$$ Simplify

$$= \sqrt{25+144}$$

$$= \sqrt{169} = 13$$

43.

$$r = \sqrt{(x_2 - x_1)^2 + (y_2 - y_1)^2}$$ Distance formula

$$= \sqrt{[-1-(-10)]^2 + (-2-5)^2}$$ Substitute given values

$$= \sqrt{9^2 + (-7)^2}$$ Simplify

$$= \sqrt{81+49}$$

$$= \sqrt{130}$$

45.

$$r = \sqrt{(x_2 - x_1)^2 + (y_2 - y_1)^2}$$ Distance formula

$$= \sqrt{(3-0)^2 + (-4-0)^2}$$ Substitute given values

$$= \sqrt{(3)^2 + (-4)^2}$$ Simplify

$$= \sqrt{9+16}$$

$$= \sqrt{25} = 5$$

47.

$$r = \sqrt{(x_2 - x_1)^2 + (y_2 - y_1)^2} \qquad \text{Distance formula}$$

$$\sqrt{13} = \sqrt{(x-1)^2 + (2-5)^2} \qquad \text{Substitute given values}$$

$$\sqrt{13} = \sqrt{(x-1)^2 + (-3)^2} \qquad \text{Simplify}$$

$$\sqrt{13} = \sqrt{(x-1)^2 + 9}$$

$$13 = (x-1)^2 + 9 \qquad \text{Square both sides}$$

$$4 = (x-1)^2 \qquad \text{Subtract 9 from both sides}$$

$$\pm 2 = x - 1 \qquad \text{Use the square root method to solve}$$

$$x - 1 = 2 \text{ or } x - 1 = -2$$

$$x = 3 \text{ or } \quad x = -1$$

49. First, we convert 1.2 miles to feet: 1.2 mi = 1.2(5,280) ft = 6,336 ft

$$c^2 = a^2 + b^2 \qquad \text{Pythagorean Theorem}$$

$$= (2,640)^2 + (6,336)^2 \qquad \text{Substitute given values}$$

$$= 6,969,600 + 40,144,896 \qquad \text{Simplify}$$

$$= 47,114,496$$

$$c = 6,864 \text{ ft (or 1.3 mi)} \qquad \text{Take square root of both sides}$$

51. The x-axis goes from home plate to first base a distance of 60 feet. Therefore, home plate is (0, 0) and first base is (60, 0).

The y-axis goes from home plate to third base. Therefore, third base is (0, 60). Second base will be at (60, 60).

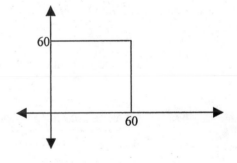

53. Quadrants II and III lie to the left of the y-axis. Therefore, all points in these two quadrants have negative x-coordinates.

55. In quadrant III, x and y are always negative. Therefore the ratio $\dfrac{x}{y}$ will always be positive.

57. The cannonball is at ground ($y = 0$) when $x = 0$ and when $x = 160$. The x-coordinate of the vertex (the maximum) will be at $\dfrac{1}{2}(160)$ or 80, and the y-coordinate will be 60.

We can now sketch the parabola through the points (0, 0), (80, 60), and (160, 0).

This problem is continued on the next page

The equation will be in the form $y = a(x-80)^2 + 60$. We will use the point (160, 0) to find a.
Let $x = 160$ and $y = 0$:

$$y = a(x-80)^2 + 60$$
$$0 = a(160-80)^2 + 60$$
$$-60 = a(80)^2$$
$$-60 = 6400a$$
$$a = -\frac{3}{320}$$

Therefore, the equation is $y = -\frac{3}{320}(x-80)^2 + 60$.

59. The complement of a $45°$ angle is $90° - 45° = 45°$.

61. The complement of a $60°$ angle is $90° - 60° = 30°$.

63. The supplement of a $120°$ angle is $180° - 120° = 60°$.

65. The supplement of a $90°$ angle is $180° - 90° = 90°$.

67. An angle coterminal with an angle of $-60°$ is $-60° + 360° = 300°$.

69. An angle coterminal with an angle of $-210°$ is $-210° + 360° = 150°$.

71. (a) If we draw $135°$ in standard position, we see that the terminal side is along the line $y = -x$. Since the terminal side lies in the second quadrant, x is negative and y is positive. Some of points on the terminal side are $(-1,1), (-3,3),$ and $\left(-\sqrt{2}, \sqrt{2}\right)$.

(b) To find the distance from $(0, 0)$ to $(-3,3)$, we use the distance formula:

$$r = \sqrt{(-3-0)^2 + (3-0)^2}$$
$$= \sqrt{(-3)^2 + (3)^2}$$
$$= \sqrt{9+9}$$
$$= \sqrt{18} = 3\sqrt{2}$$

(c) The angle between $0°$ and $-360°$ that is coterminal with $135°$ is $135° - 360° = -225°$.

73.   (a) If we draw $225°$ in standard position, we see that the terminal side is along the line $y = x$. Since the terminal side lies in the third quadrant, x and y are both negative. Some of points on the terminal side are $(-1,-1), (-3,-3)$, and $\left(-\sqrt{2}, -\sqrt{2}\right)$.

    (b) To find the distance from $(0, 0)$ to $(-3, -3)$, we use the distance formula:

$$r = \sqrt{(-3-0)^2 + (-3-0)^2}$$
$$= \sqrt{(-3)^2 + (-3)^2}$$
$$= \sqrt{9+9}$$
$$= \sqrt{18} = 3\sqrt{2}$$

    (c) The angle between $0°$ and $-360°$ that is coterminal with $225°$ is $225° - 360° = -135°$.

75.   (a) If we draw $90°$ in standard position, we see that the terminal side is the positive y-axis. Some of the points on the terminal side are $(0,1), (0,2)$, and $(0,3)$.

    (b) The distance between $(0, 0)$ and $(0, 3)$ is 3 units.

    (c) The angle between $0°$ and $-360°$ that is coterminal with $90°$ is $90° - 360° = -270°$.

77.   (a) If we draw $-45°$ in standard position, we see that the terminal side is along the line $y = -x$. Since the terminal side lies in the fourth quadrant, x is positive and y is negative. Some of points on the terminal side are $(1,-1), (3,-3)$, and $\left(\sqrt{2}, -\sqrt{2}\right)$.

    (b) To find the distance from $(0, 0)$ to $(3, -3)$, we use the distance formula:

$$r = \sqrt{(3-0)^2 + (-3-0)^2}$$
$$= \sqrt{(3)^2 + (-3)^2}$$
$$= \sqrt{9+9}$$
$$= \sqrt{18} = 3\sqrt{2}$$

    (c) The angle between $0°$ and $360°$ that is coterminal with $-45°$ is $-45° + 360° = 315°$.

79.   The side opposite $30°$ is 1.
    The side opposite $60°$ is $1\sqrt{3}$ or $\sqrt{3}$.
    Therefore, the point is $\left(\sqrt{3},1\right)$.

81.
$$r = \sqrt{(3-0)^2 + (-2-0)^2}$$
$$= \sqrt{(3)^2 + (-2)^2}$$
$$= \sqrt{9+4}$$
$$= \sqrt{13}$$

83. We will find the lengths of the three sides:
From (0, 0) to (5, 0) is 5 units.
From (5, 0) to (5, 12) is 12 units.
From (0. 0) to (5, 12), we use the distance formula:

$$r = \sqrt{(5-0)^2 + (12-0)^2}$$
$$= \sqrt{5^2 + 12^2}$$
$$= \sqrt{25+144}$$
$$= \sqrt{169} = 13$$

If this is a right triangle, then $c^2 = a^2 + b^2$.
We check: $13^2 = 5^2 + 12^2$
$$169 = 25 + 144$$
$$169 = 169 \quad \text{It checks.}$$

Problem Set 1.3

1. $(x,y) = (3,4)$

$\sin\theta = \dfrac{y}{r} = \dfrac{4}{5}$ $\qquad$ $\cot\theta = \dfrac{x}{y} = \dfrac{3}{4}$

$x = 3$ and $y = 4$

$\cos\theta = \dfrac{x}{r} = \dfrac{3}{5}$ $\qquad$ $\sec\theta = \dfrac{r}{x} = \dfrac{5}{3}$

$r = \sqrt{3^2 + 4^2}$

$\tan\theta = \dfrac{y}{x} = \dfrac{4}{3}$ $\qquad$ $\csc\theta = \dfrac{r}{y} = \dfrac{5}{4}$

$ = \sqrt{9+16}$

$ = \sqrt{25} = 5$

3. $(x,y) = (-3,4)$

$\sin\theta = \dfrac{y}{r} = \dfrac{4}{5}$ $\qquad$ $\cot\theta = \dfrac{x}{y} = -\dfrac{3}{4}$

$x = -3$ and $y = 4$

$\cos\theta = \dfrac{x}{r} = -\dfrac{3}{5}$ $\qquad$ $\sec\theta = \dfrac{r}{x} = -\dfrac{5}{3}$

$r = \sqrt{(-3)^2 + 4^2}$

$\tan\theta = \dfrac{y}{x} = -\dfrac{4}{3}$ $\qquad$ $\csc\theta = \dfrac{r}{y} = \dfrac{5}{4}$

$ = \sqrt{9+16}$

$ = \sqrt{25} = 5$

5.　　$(x, y) = (-5, 12)$

$x = -5$ and $y = 12$

$r = \sqrt{(-5)^2 + 12^2}$

$\quad = \sqrt{25 + 144}$

$\quad = \sqrt{169} = 13$

$\sin\theta = \dfrac{y}{r} = \dfrac{12}{13}$

$\cos\theta = \dfrac{x}{r} = -\dfrac{5}{13}$

$\tan\theta = \dfrac{y}{x} = -\dfrac{12}{5}$

$\cot\theta = \dfrac{x}{y} = -\dfrac{5}{12}$

$\sec\theta = \dfrac{r}{x} = -\dfrac{13}{5}$

$\csc\theta = \dfrac{r}{y} = \dfrac{13}{12}$

7.　　$(x, y) = (-1, -2)$

$x = -1$ and $y = -2$

$r = \sqrt{(-1)^2 + (-2)^2}$

$\quad = \sqrt{1 + 4}$

$\quad = \sqrt{5}$

$\sin\theta = \dfrac{y}{r} = -\dfrac{2}{\sqrt{5}}$

$\cos\theta = \dfrac{x}{r} = -\dfrac{1}{\sqrt{5}}$

$\tan\theta = \dfrac{y}{x} = \dfrac{-2}{-1} = 2$

$\cot\theta = \dfrac{x}{y} = \dfrac{-1}{-2} = \dfrac{1}{2}$

$\sec\theta = \dfrac{r}{x} = \dfrac{\sqrt{5}}{-1} = -\sqrt{5}$

$\csc\theta = \dfrac{r}{y} = -\dfrac{\sqrt{5}}{2}$

9.　　$(x, y) = (a, b)$

$x = a$ and $y = b$

$r = \sqrt{a^2 + b^2}$

$\sin\theta = \dfrac{y}{r} = \dfrac{b}{\sqrt{a^2 + b^2}}$

$\cos\theta = \dfrac{x}{r} = \dfrac{a}{\sqrt{a^2 + b^2}}$

$\tan\theta = \dfrac{y}{x} = \dfrac{b}{a}$

$\cot\theta = \dfrac{x}{y} = \dfrac{a}{b}$

$\sec\theta = \dfrac{r}{x} = \dfrac{\sqrt{a^2 + b^2}}{a}$

$\csc\theta = \dfrac{r}{y} = \dfrac{\sqrt{a^2 + b^2}}{b}$

11.　　$(x, y) = (-3, 0)$

$x = -3$ and $y = 0$

$r = \sqrt{(-3)^2 + 0^2}$

$\quad = \sqrt{9}$

$\quad = 3$

$\sin\theta = \dfrac{y}{r} = \dfrac{0}{3} = 0$

$\cos\theta = \dfrac{x}{r} = \dfrac{-3}{3} = -1$

$\tan\theta = \dfrac{y}{x} = \dfrac{0}{-3} = 0$

$\cot\theta = \dfrac{x}{y} = \dfrac{-3}{0}$ is undefined

$\sec\theta = \dfrac{r}{x} = \dfrac{3}{-3} = -1$

$\csc\theta = \dfrac{r}{y} = \dfrac{3}{0}$ is undefined

13. $(x,y)=\left(\sqrt{3},-1\right)$

$x=\sqrt{3}$ and $y=-1$

$r=\sqrt{\left(\sqrt{3}\right)^2+(-1)^2}$

$\quad =\sqrt{3+1}$

$\quad =\sqrt{4}$

$\quad =2$

$\sin\theta=\dfrac{y}{r}=-\dfrac{1}{2}$

$\cos\theta=\dfrac{x}{r}=\dfrac{\sqrt{3}}{2}$

$\tan\theta=\dfrac{y}{x}=-\dfrac{1}{\sqrt{3}}$

$\cot\theta=\dfrac{x}{y}=-\dfrac{\sqrt{3}}{1}=-\sqrt{3}$

$\sec\theta=\dfrac{r}{x}=\dfrac{2}{\sqrt{3}}$

$\csc\theta=\dfrac{r}{y}=-\dfrac{2}{1}=-2$

15. $(x,y)=\left(-\sqrt{5},2\right)$

$x=-\sqrt{5}$ and $y=2$

$r=\sqrt{\left(-\sqrt{5}\right)^2+2^2}$

$\quad =\sqrt{5+4}$

$\quad =\sqrt{9}$

$\quad =3$

$\sin\theta=\dfrac{y}{r}=\dfrac{2}{3}$

$\cos\theta=\dfrac{x}{r}=-\dfrac{\sqrt{5}}{3}$

$\tan\theta=\dfrac{y}{x}=-\dfrac{2}{\sqrt{5}}$

$\cot\theta=\dfrac{x}{y}=-\dfrac{\sqrt{5}}{2}$

$\sec\theta=\dfrac{r}{x}=-\dfrac{3}{\sqrt{5}}$

$\csc\theta=\dfrac{r}{y}=\dfrac{3}{2}$

17. $(x,y)=(60,80)$

$x=60$ and $y=80$

$r=\sqrt{(60)^2+(80)^2}$

$\quad =\sqrt{3600+6400}$

$\quad =\sqrt{10,000}$

$\quad =100$

$\sin\theta=\dfrac{y}{r}=\dfrac{80}{100}=\dfrac{4}{5}$

$\cos\theta=\dfrac{x}{r}=\dfrac{60}{100}=\dfrac{3}{5}$

$\tan\theta=\dfrac{y}{x}=\dfrac{80}{60}=\dfrac{4}{3}$

$\cot\theta=\dfrac{x}{y}=\dfrac{60}{80}=\dfrac{3}{4}$

$\sec\theta=\dfrac{r}{x}=\dfrac{100}{60}=\dfrac{5}{3}$

$\csc\theta=\dfrac{r}{y}=\dfrac{100}{80}=\dfrac{5}{4}$

19. $(x,y)=(5a,-12a)$

$x=5a$ and $y=-12a$

$r=\sqrt{(5a)^2+(-12a)^2}$

$\quad =\sqrt{25a^2+144a^2}$

$\quad =\sqrt{169a^2}=13a$

$\sin\theta=\dfrac{y}{r}=\dfrac{-12a}{13a}=-\dfrac{12}{13}$

$\cos\theta=\dfrac{x}{r}=\dfrac{5a}{13a}=\dfrac{5}{13}$

$\tan\theta=\dfrac{y}{x}=\dfrac{-12a}{5a}=-\dfrac{12}{5}$

$\cot\theta=\dfrac{x}{y}=\dfrac{5a}{-12a}=-\dfrac{5}{12}$

$\sec\theta=\dfrac{r}{x}=\dfrac{13a}{5a}=\dfrac{13}{5}$

$\csc\theta=\dfrac{r}{y}=\dfrac{13a}{-12a}=-\dfrac{13}{12}$

21.   (3, 4) lies on the terminal side of θ. Therefore,

$x = 3, y = 4,$ and $r = \sqrt{3^2 + 4^2} = \sqrt{9 + 16} = \sqrt{25} = 5$

$\sin\theta = \dfrac{y}{r} = \dfrac{4}{5}$ $\qquad\qquad$ $\cos\theta = \dfrac{x}{r} = \dfrac{3}{5}$ $\qquad\qquad$ $\tan\theta = \dfrac{y}{x} = \dfrac{4}{3}$

23.   (3, 5) lies on the terminal side of θ. Therefore,

$x = 3, y = 5,$ and $r = \sqrt{3^2 + 5^2} = \sqrt{9 + 25} = \sqrt{34}$.

$\sin\theta = \dfrac{y}{r} = \dfrac{5}{\sqrt{34}}$ $\qquad\qquad$ $\cos\theta = \dfrac{x}{r} = \dfrac{3}{\sqrt{34}}$ $\qquad\qquad$ $\tan\theta = \dfrac{y}{x} = \dfrac{5}{3}$

25. $(x, y) = (9.36, 7.02)$ $\qquad\qquad$ $\sin\theta = \dfrac{y}{r} = \dfrac{7.02}{11.7} = 0.6$

$x = 9.36$ and $y = 7.02$ $\qquad\qquad$ $\cos\theta = \dfrac{x}{r} = \dfrac{9.36}{11.7} = 0.8$

$r = \sqrt{(9.36)^2 + (7.02)}$

$\quad = \sqrt{87.6096 + 49.2804}$

$\quad = \sqrt{136.89}$

$\quad = 11.7$

27.   The terminal side of 135° in standard position lies on the line $y = -x$. A point on this line in quadrant II is $(-1, 1)$.

$(x, y) = (1, -1)$

$x = -1, y = 1$ and $r = \sqrt{(-1)^2 + 1^2} = \sqrt{1 + 1} = \sqrt{2}$.

$\sin\theta = \dfrac{y}{r} = \dfrac{1}{\sqrt{2}}$ $\qquad\qquad$ $\cos\theta = \dfrac{x}{r} = -\dfrac{1}{\sqrt{2}}$ $\qquad\qquad$ $\tan\theta = \dfrac{y}{x} = \dfrac{1}{-1} = -1$

29.   A point on the terminal side of an angle of 90° is $(0, 1)$.

$(x, y) = (0, 1)$

$x = 0, y = 1$ and $r = \sqrt{0^2 + 1^2} = \sqrt{1} = 1$.

$\sin\theta = \dfrac{y}{r} = \dfrac{1}{1} = 1$ $\qquad\qquad$ $\cos\theta = \dfrac{x}{r} = \dfrac{0}{1} = 0$ $\qquad\qquad$ $\tan\theta = \dfrac{y}{x} = \dfrac{1}{0}$ (undefined)

31. The terminal side of $-45°$ in standard position lies on the line $y = -x$. A point on this line in quadrant IV is $(1, -1)$.

$(x, y) = (1, -1)$

$x = 1$ and $y = -1$

$r = \sqrt{1^2 + (-1)^2} = \sqrt{1 + 1} = \sqrt{2}.$

$\sin\theta = \dfrac{y}{r} = -\dfrac{1}{\sqrt{2}}$ $\qquad$ $\cos\theta = \dfrac{x}{r} = \dfrac{1}{\sqrt{2}}$ $\qquad$ $\tan\theta = \dfrac{y}{x} = \dfrac{-1}{1} = -1$

33. A point on the terminal side of an angle of $0°$ is $(1, 0)$.

$(x, y) = (1, 0)$

$x = 1$ and $y = 0$

$r = \sqrt{1^2 + 0^2} = \sqrt{1} = 1.$

$\sin\theta = \dfrac{y}{r} = \dfrac{0}{1} = 0$ $\qquad$ $\cos\theta = \dfrac{x}{r} = \dfrac{1}{1} = 1$ $\qquad$ $\tan\theta = \dfrac{y}{x} = \dfrac{0}{1} = 0$

35. $\cos\theta = \dfrac{x}{r}$ (r is always positive)

If $\cos\theta$ is positive, then x must be positive. x is positive in quadrants I and IV. Therefore, $\cos\theta$ is positive in QI and QIV.

37. $\sin\theta = \dfrac{y}{r}$ (r is always positive)

If $\sin\theta$ is negative, then y must be negative. y is negative in quadrants III and IV. Therefore, $\sin\theta$ is negative in QIII and QIV.

39. The sine function is negative in quadrants III and IV.
If tangent function is positive in quadrants I and III.
Therefore, $\sin\theta$ is negative and $\tan\theta$ is positive in QIII.

41. The sine function is positive in quadrants I and II.
If tangent function is positive in quadrants I and III.
Therefore, both functions are positive in quadrant I.

The sine function is negative in quadrants III and IV.
If tangent function is negative in quadrants II and IV.
Therefore, both functions are negaitive in quadrant IV.

43. $\sin\theta = \dfrac{y}{r} = \dfrac{12}{13}$ and θ terminates in Q1. $\cos\theta = \dfrac{x}{r} = \dfrac{5}{13}$ $\sec\theta = \dfrac{r}{x} = \dfrac{13}{5}$

$y = 12$ and $r = 13$ $\tan\theta = \dfrac{y}{x} = \dfrac{12}{5}$ $\csc\theta = \dfrac{r}{y} = \dfrac{13}{12}$

$x^2 + y^2 = r^2$ $\cot\theta = \dfrac{x}{y} = \dfrac{5}{12}$

$x^2 + (12)^2 = (13)^2$

$x^2 + 144 = 169$

$x^2 = 25$

$x = \pm 5$

Since θ terminates in Q1, x must equal 5

45. θ terminates in QIV. Therefore, x is positive and y is negative.

$\cos\theta = \dfrac{x}{r} = \dfrac{24}{25}$

Therefore, $x = 24$ and $r = 25$

$x^2 + y^2 = r^2$ $\sin\theta = \dfrac{y}{r} = -\dfrac{7}{25}$ $\sec\theta = \dfrac{r}{x} = \dfrac{25}{24}$

$24^2 + (y)^2 = (25)^2$ $\tan\theta = \dfrac{y}{x} = -\dfrac{7}{24}$ $\csc\theta = \dfrac{r}{y} = -\dfrac{25}{7}$

$576 + y^2 = 625$ $\cot\theta = \dfrac{x}{y} = -\dfrac{24}{7}$

$y^2 = 49$

$y = \pm 7$

Therefore, $y = -7$ because θ is in QIV.

47. θ terminates in QI. Therefore, both x and y are positive.

$\tan\theta = \dfrac{y}{x} = \dfrac{3}{4}$

Therefore $x = 4$ and $y = 3$

$r = \sqrt{x^2 + y^2}$ $\sin\theta = \dfrac{y}{r} = \dfrac{3}{5}$ $\cot\theta = \dfrac{x}{y} = \dfrac{4}{3}$

$= \sqrt{4^2 + 3^2}$ $\cos\theta = \dfrac{x}{r} = \dfrac{4}{5}$ $\sec\theta = \dfrac{r}{x} = \dfrac{5}{4}$

$= \sqrt{16 + 9}$ $\csc\theta = \dfrac{r}{y} = \dfrac{5}{3}$

$= \sqrt{25} = 5$

49. θ terminates in QIII. Therefore, both x and y are negative.

$$\sin\theta = \frac{y}{r} = \frac{-20}{29}$$

Therefore $y = -20$ and $r = 29$

$$x^2 + y^2 = r^2 \qquad \cos\theta = \frac{x}{r} = -\frac{21}{29} \qquad \sec\theta = \frac{r}{x} = -\frac{29}{21}$$

$$x^2 + (-20)^2 = (29)^2 \qquad \tan\theta = \frac{y}{x} = \frac{-20}{-21} = \frac{20}{21} \qquad \csc\theta = \frac{r}{y} = -\frac{29}{20}$$

$$x^2 + 400 = 841 \qquad \cot\theta = \frac{x}{y} = \frac{-21}{-20} = \frac{21}{20}$$

$$x^2 = 441$$
$$x = \pm 21$$

Therefore, $x = -21$.

51. The secant function is positive in QI and QIV. The sine function is negative in QIII and QIV. Therefore, θ is in QIV, where x is positive and y is negative.

$$\sec\theta = \frac{r}{x} = \frac{13}{5}$$

Therefore $x = 5$ and $r = 13$

$$x^2 + y^2 = r^2 \qquad \sin\theta = \frac{y}{r} = -\frac{12}{13} \qquad \cot\theta = \frac{x}{y} = -\frac{5}{12}$$

$$5^2 + y^2 = 13^2 \qquad \cos\theta = \frac{x}{r} = \frac{5}{13} \qquad \csc\theta = \frac{r}{y} = -\frac{13}{12}$$

$$25 + y^2 = 169 \qquad \tan\theta = \frac{y}{x} = -\frac{12}{5}$$

$$y^2 = 144$$
$$y = \pm 12$$

Therefore, $y = -12$.

53. $\cot\theta$ is positive in QI and QIII. $\cos\theta$ is positive in QI and QIV. Therefore, θ is in QI and both x and y are positive.

$$\cot\theta = \frac{x}{y} = \frac{1}{2}$$

Therefore, $x = 1$ and $y = 2$.

$$x^2 + y^2 = r^2 \qquad \sin\theta = \frac{y}{r} = \frac{2}{\sqrt{5}} \qquad \sec\theta = \frac{r}{x} = \frac{\sqrt{5}}{1} = \sqrt{5}$$

$$1^2 + 2^2 = r^2 \qquad \cos\theta = \frac{x}{r} = \frac{1}{\sqrt{5}} \qquad \csc\theta = \frac{r}{y} = \frac{\sqrt{5}}{2}$$

$$1 + 4 = r^2 \qquad \tan\theta = \frac{y}{x} = \frac{2}{1} = 2$$

$$r^2 = 5$$

$$r = \sqrt{5} \text{ because } r \text{ is always positive.}$$

55. θ terminates in QIV. Therefore x is positive and y is negative.

$$\cos\theta = \frac{x}{r} = \frac{\sqrt{3}}{2}$$

Therefore $x = \sqrt{3}$ and $r = 2$

$$x^2 + y^2 = r^2 \qquad\qquad \sin\theta = \frac{y}{r} = -\frac{1}{2} \qquad\qquad \sec\theta = \frac{r}{x} = \frac{2}{\sqrt{3}}$$

$$\left(\sqrt{3}\right)^2 + \left(y\right)^2 = \left(2\right)^2 \qquad\qquad \tan\theta = \frac{y}{x} = -\frac{1}{\sqrt{3}} \qquad\qquad \csc\theta = \frac{r}{y} = -\frac{2}{1} = -2$$

$$3 + y^2 = 4 \qquad\qquad \cot\theta = \frac{x}{y} = -\frac{\sqrt{3}}{1} = -\sqrt{3}$$

$$y^2 = 1$$

Therefore, $y = -1$.

57. In QII, x is negative and y is positive.

$$\cot\theta = \frac{x}{y} = \frac{-2}{1}. \qquad\qquad \sin\theta = \frac{y}{r} = \frac{1}{\sqrt{5}} \qquad\qquad \sec\theta = \frac{r}{x} = -\frac{\sqrt{5}}{2}$$

Therefore, $x = -2$ and $y = 1$. $\qquad \cos\theta = \frac{x}{r} = -\frac{2}{\sqrt{5}} \qquad\qquad \csc\theta = \frac{r}{y} = \frac{\sqrt{5}}{1} = \sqrt{5}$

From Problem 53, we found that $r = \sqrt{5}$ $\qquad \tan\theta = \frac{y}{x} = -\frac{1}{2}$

59. $\tan\theta = \frac{a}{b}$ where a and b are both positive

$$\tan\theta = \frac{y}{x} = \frac{a}{b} \qquad\qquad \sin\theta = \frac{y}{r} = \frac{a}{\sqrt{a^2 + b^2}} \qquad\qquad \cot\theta = \frac{x}{y} = \frac{b}{a}$$

Therefore, $x = b$ and $y = a$. $\qquad \cos\theta = \frac{x}{r} = \frac{b}{\sqrt{a^2 + b^2}} \qquad\qquad \sec\theta = \frac{r}{x} = \frac{\sqrt{a^2 + b^2}}{b}$

$$r = \sqrt{a^2 + b^2} \qquad\qquad\qquad\qquad\qquad\qquad\qquad\qquad\qquad\quad \csc\theta = \frac{r}{y} = \frac{\sqrt{a^2 + b^2}}{a}$$

61. $\sin\theta = \frac{y}{r} = \frac{1}{1}$

$$x^2 + y^2 = r^2$$
$$x^2 + 1^2 = 1^2$$
$$x^2 + 1 = 1$$
$$x^2 = 0$$
$$x = 0 \qquad\qquad \text{Therefore, the point } (0, 1) \text{ lies on the terminal side of } \theta \text{ and } \theta \text{ must be } 90°.$$

63. In QIII, both x and y are negative.

$$\tan\theta = \frac{y}{x} = \frac{-1}{-1}$$

Therefore, a point on the terminal side is $(-1,-1)$. This point lies on the line $y = x$ in quadrant III. Therefore, the angle must be $225°$.

65. The terminal side of θ lies in QI. The point $(1,2)$ lies on $y = 2x$ in QI.

$(x,y) = (1,2)$

$x = 1$ and $y = 2$

$$r = \sqrt{x^2 + y^2} \qquad\qquad \sin\theta = \frac{y}{r} = \frac{2}{\sqrt{5}}$$

$$= \sqrt{1^2 + 2^2} \qquad\qquad \cos\theta = \frac{x}{r} = \frac{1}{\sqrt{5}}$$

$$= \sqrt{1+4} = \sqrt{5}$$

67. The terminal side of θ lies in QII. The point $(-1,3)$ lies on $y = -3x$ in QII.

$(x,y) = (-1,3)$

$x = -1$ and $y = 3$

$$r = \sqrt{x^2 + y^2} \qquad\qquad \sin\theta = \frac{y}{r} = \frac{3}{\sqrt{10}}$$

$$= \sqrt{(-1)^2 + 3^2} \qquad\qquad \tan\theta = \frac{y}{x} = \frac{3}{-1} = -3$$

$$= \sqrt{1+9} = \sqrt{10}$$

69. The point $(1,1)$ lies on the terminal side of $45°$ in QI. Therefore, $x = 1$ and $y = 1$.

$$r = \sqrt{x^2 + y^2}$$

$$= \sqrt{1^2 + 1^2}$$

$$= \sqrt{2}$$

$$\cos 45° = \frac{x}{r} = \frac{1}{\sqrt{2}}$$

The point $(1,-1)$ lies on the terminal side of $-45°$ in QIV.

Therefore, $x = 1$ and $y = -1$.

$$r = \sqrt{1^2 + (-1)^2}$$

$$= \sqrt{1+1}$$

$$= \sqrt{2}$$

$$\cos(-45°) = \frac{x}{r} = \frac{1}{\sqrt{2}}$$

71.
$$\sin\theta = \frac{y}{r} = \frac{-3}{5}$$
$$y = -3 \text{ and } r = 5$$
$$x^2 + y^2 = r^2$$
$$x^2 + (-3)^2 = 5^2$$
$$x^2 + 9 = 25$$
$$x^2 = 16$$
$$x = \pm 4$$

Problem Set 1.4

1. $\dfrac{1}{7}$

3. $\dfrac{1}{-\dfrac{2}{3}} = 1 \cdot -\dfrac{3}{2} = -\dfrac{3}{2}$

5. $\dfrac{1}{-\dfrac{1}{\sqrt{2}}} = 1 \cdot \dfrac{\sqrt{2}}{-1} = -\sqrt{2}$

7. $\dfrac{1}{x}, x \neq 0$

9.
$$\csc\theta = \frac{1}{\sin\theta}$$
$$= \frac{1}{\dfrac{4}{5}}$$
$$= \frac{5}{4}$$

11.
$$\cos\theta = \frac{1}{\sec\theta}$$
$$= \frac{1}{-2}$$
$$= -\frac{1}{2}$$

13.
$$\cot\theta = \frac{1}{\tan\theta}$$
$$= \frac{1}{a}, a \neq 0$$

15.
$$\tan\theta = \frac{\sin\theta}{\cos\theta}$$
$$= \frac{\dfrac{3}{5}}{-\dfrac{4}{5}}$$
$$= -\frac{3}{4}$$

17. $\cot\theta = \dfrac{\cos\theta}{\sin\theta}$

$ = \dfrac{-12/13}{-5/13}$

$ = \dfrac{12}{5}$

19. $\sin^2\theta = (\sin\theta)^2$

$ = \left(\dfrac{1}{\sqrt{2}}\right)^2$

$ = \dfrac{1}{2}$

21. $\tan^3\theta = (\tan\theta)^3$

$ = (2)^3$

$ = 8$

23. $\tan\theta = \dfrac{\sin\theta}{\cos\theta}$

$ = \dfrac{-\dfrac{12}{13}}{-\dfrac{5}{13}}$

$ = \dfrac{12}{5}$

25. $\sec\theta = \dfrac{1}{\cos\theta}$

$ = \dfrac{1}{-\dfrac{5}{13}}$

$ = -\dfrac{13}{5}$

27. $\sin\theta = \pm\sqrt{1 - \cos^2\theta}$

$ = \pm\sqrt{1 - \left(\dfrac{3}{5}\right)^2}$

$ = \pm\sqrt{1 - \dfrac{9}{25}}$

$ = \pm\sqrt{\dfrac{16}{25}}$

$ = \pm\dfrac{4}{5}$

Since θ terminates in QI, $\sin\theta$ is positive.

Therefore, $\sin\theta = \dfrac{4}{5}$

29. $\cos\theta = \pm\sqrt{1 - \sin^2\theta}$

$ = \pm\sqrt{1 - \left(\dfrac{1}{3}\right)^2}$

$ = \pm\sqrt{1 - \dfrac{1}{9}}$

$ = \pm\sqrt{\dfrac{8}{9}} = \pm\dfrac{2\sqrt{2}}{3}$

Since θ terminates in QII, $\cos\theta$ is negative.

Therefore, $\cos\theta = -\dfrac{2\sqrt{2}}{3}$

31.

$$\cos\theta = \pm\sqrt{1-\sin^2\theta}$$

$$= \pm\sqrt{1-\left(-\frac{4}{5}\right)^2}$$

$$= \pm\sqrt{1-\frac{16}{25}}$$

$$= \pm\sqrt{\frac{9}{25}}$$

$$= \pm\frac{3}{5}$$

Since θ terminates in QIII, $\cos\theta$ is negative.

Therefore, $\cos\theta = -\frac{3}{5}$

33.

$$\sin\theta = \pm\sqrt{1-\cos^2\theta}$$

$$= \pm\sqrt{1-\left(\frac{\sqrt{3}}{2}\right)^2}$$

$$= \pm\sqrt{1-\frac{3}{4}}$$

$$= \pm\sqrt{\frac{1}{4}}$$

$$= \pm\frac{1}{2}$$

Since θ terminates in QI, $\sin\theta$ is positive.

Therefore, $\sin\theta = \frac{1}{2}$

35.

$$\cos\theta = \pm\sqrt{1-\sin^2\theta}$$

$$= \pm\sqrt{1-\left(\frac{1}{\sqrt{5}}\right)^2}$$

$$= \pm\sqrt{1-\frac{1}{5}}$$

$$= \pm\sqrt{\frac{4}{5}}$$

$$= \pm\frac{2}{\sqrt{5}}$$

Since θ terminates in QII, $\cos\theta$ is negative.

Therefore, $\cos\theta = -\frac{2}{\sqrt{5}}$

37.

$$\cos\theta = \pm\sqrt{1-\sin^2\theta}$$

$$= \pm\sqrt{1-\left(\frac{1}{3}\right)^2}$$

$$= \pm\sqrt{1-\frac{1}{9}}$$

$$= \pm\sqrt{\frac{8}{9}}$$

$$= \pm\frac{2\sqrt{2}}{3}$$

Since θ terminates in QI, $\cos\theta$ is positive.

Therefore, $\cos\theta = \frac{2\sqrt{2}}{3}$

$$\tan\theta = \frac{\sin\theta}{\cos\theta}$$

$$= \frac{\dfrac{1}{3}}{\dfrac{2\sqrt{2}}{3}}$$

$$= \frac{1}{2\sqrt{2}}$$

39.
$$\sec^2\theta = \tan^2\theta + 1$$
$$\sec\theta = \pm\sqrt{\tan^2\theta + 1}$$
$$= \pm\sqrt{\left(\frac{8}{15}\right)^2 + 1}$$
$$= \sqrt{\frac{64}{225} + 1}$$
$$= \pm\sqrt{\frac{289}{225}} = \pm\frac{17}{15}$$

Since θ terminates in QIII, $\sec\theta$ is negative. Therefore, $\sec\theta = -\frac{17}{15}$

41.
$$\csc\theta = \pm\sqrt{1 + \cot^2\theta} \qquad \text{Pythagorean identity}$$
$$= \pm\sqrt{1 + \left(-\frac{21}{20}\right)^2} \qquad \text{Substitute given value}$$
$$= \pm\sqrt{1 + \frac{441}{400}} \qquad \text{Simplify}$$
$$= \pm\sqrt{\frac{841}{400}}$$
$$= \pm\frac{29}{20}$$

Since $\cot\theta$ is negative and $\sin\theta$ is positive, θ must lie in QII. Therefore, $\csc\theta = \frac{29}{20}$.

43. θ terminates in QI. Therefore, all functions are positive.

$$\sin\theta = \sqrt{1 - \cos^2\theta} \qquad\qquad \tan\theta = \frac{\sin\theta}{\cos\theta}$$
$$= \sqrt{1 - \left(\frac{12}{13}\right)^2} \qquad\qquad = \frac{\frac{5}{13}}{\frac{12}{13}}$$
$$= \sqrt{1 - \frac{144}{169}} \qquad\qquad = \frac{5}{12}$$
$$= \sqrt{\frac{25}{169}}$$
$$= \frac{5}{13}$$

This problem is continued on the next page

$$\sec\theta = \frac{1}{\cos\theta} \qquad\qquad \cot\theta = \frac{1}{\tan\theta}$$

$$= \frac{1}{\frac{12}{13}} \qquad\qquad = \frac{1}{\frac{5}{12}}$$

$$= \frac{13}{12} \qquad\qquad = \frac{12}{5}$$

$$\csc\theta = \frac{1}{\sin\theta} \qquad\qquad \text{All six ratios are:}$$

$$= \frac{1}{\frac{5}{13}} \qquad\qquad \sin\theta = \frac{5}{13} \qquad\qquad \cot\theta = \frac{12}{5}$$

$$= \frac{13}{5} \qquad\qquad \cos\theta = \frac{12}{13} \qquad\qquad \sec\theta = \frac{13}{12}$$

$$\tan\theta = \frac{5}{12} \qquad\qquad \csc\theta = \frac{13}{5}$$

45. In QIII and QIV, $\sin\theta$ is negative, but θ is not in QIII. Therefore, θ must be in QIV.

$$\csc\theta = \frac{1}{\sin\theta} \qquad\qquad \tan\theta = \frac{\sin\theta}{\cos\theta}$$

$$= \frac{1}{-1/2} \qquad\qquad = \frac{-1/2}{\sqrt{3}/2}$$

$$= -2 \qquad\qquad = -\frac{1}{\sqrt{3}}$$

$$\cos\theta = \pm\sqrt{1 - \sin^2\theta} \qquad\qquad \cot\theta = \frac{1}{\tan\theta}$$

$$= \pm\sqrt{1 - \left(-\frac{1}{2}\right)^2} \qquad\qquad = \frac{1}{-1/\sqrt{3}}$$

$$= \pm\sqrt{1 - \frac{1}{4}} \qquad\qquad = -\sqrt{3}$$

$$= \pm\sqrt{\frac{3}{4}} \qquad\qquad \sec\theta = \frac{1}{\cos\theta}$$

$$= \pm\frac{\sqrt{3}}{2} \qquad\qquad = \frac{1}{\sqrt{3}/2}$$

$$\cos\theta = \frac{\sqrt{3}}{2} \text{ because } \cos\theta \text{ is positive in QIV.} \qquad\qquad = \frac{2}{\sqrt{3}}$$

This problem is continued on the next page

All six ratios are:

$$\sin\theta = -\frac{1}{2}$$

$$\cot\theta = -\sqrt{3}$$

$$\cos\theta = \frac{\sqrt{3}}{2}$$

$$\sec\theta = \frac{2}{\sqrt{3}}$$

$$\tan\theta = -\frac{1}{\sqrt{3}}$$

$$\csc\theta = -2$$

47. The secant and sine functions are both positive in QI. Therefore all functions must be positive.

$$\cos\theta = \frac{1}{\sec\theta}$$

$$\tan\theta = \frac{\sin\theta}{\cos\theta}$$

$$=\frac{1}{2}$$

$$=\frac{\frac{\sqrt{3}}{2}}{\frac{1}{2}} = \sqrt{3}$$

$$\sin\theta = \pm\sqrt{1-\cos^2\theta}$$

$$\cot\theta = \frac{1}{\tan\theta}$$

$$=\sqrt{1-\left(\frac{1}{2}\right)^2}$$

$$=\frac{1}{\sqrt{3}}$$

$$=\sqrt{1-\frac{1}{4}}$$

$$\csc\theta = \frac{1}{\sin\theta}$$

$$=\sqrt{\frac{3}{4}}$$

$$=\frac{1}{\sqrt{3}/2}$$

$$=\frac{\sqrt{3}}{2}$$

$$=\frac{2}{\sqrt{3}}$$

All six ratios are:

$$\sin\theta = \frac{\sqrt{3}}{2}$$

$$\cot\theta = \frac{1}{\sqrt{3}}$$

$$\cos\theta = \frac{1}{2}$$

$$\sec\theta = 2$$

$$\tan\theta = \sqrt{3}$$

$$\csc\theta = \frac{2}{\sqrt{3}}$$

49.

$$\sin\theta = \pm\sqrt{1-\cos^2\theta}$$

$$= \pm\sqrt{1-\left(\frac{2}{\sqrt{13}}\right)^2}$$

$$= \pm\sqrt{1-\frac{4}{13}}$$

$$= \pm\sqrt{\frac{9}{13}}$$

$$= \pm\frac{3}{\sqrt{13}}$$

$\sin\theta = -\dfrac{3}{\sqrt{13}}$ because $\sin\theta$ is negative in QIV.

$$\tan\theta = \frac{\sin\theta}{\cos\theta}$$

$$= \frac{-3/\sqrt{13}}{2/\sqrt{13}} = -\frac{3}{2}$$

$$\sec\theta = \frac{1}{\cos\theta}$$

$$= \frac{1}{2/\sqrt{13}}$$

$$= \frac{\sqrt{13}}{2}$$

$$\cot\theta = \frac{1}{\tan\theta}$$

$$= \frac{1}{-3/2} = -\frac{2}{3}$$

$$\csc\theta = \frac{1}{\sin\theta}$$

$$= \frac{1}{-3/\sqrt{13}}$$

$$= -\frac{\sqrt{13}}{3}$$

All six ratios are:

$$\sin\theta = -\frac{3}{\sqrt{13}}$$

$$\cos\theta = \frac{2}{\sqrt{13}}$$

$$\tan\theta = -\frac{3}{2}$$

$$\cot\theta = -\frac{2}{3}$$

$$\sec\theta = \frac{\sqrt{13}}{2}$$

$$\csc\theta = -\frac{\sqrt{13}}{3}$$

51.

$$\cos\theta = \frac{1}{\sec\theta}$$

$$= -\frac{1}{3}$$

$$\sin\theta = \pm\sqrt{1-\cos^2\theta}$$

$$= \pm\sqrt{1-\left(-\frac{1}{3}\right)^2}$$

$$= \pm\sqrt{1-\frac{1}{9}}$$

$$= \pm\sqrt{\frac{8}{9}} = \pm\frac{2\sqrt{2}}{3}$$

Since θ is in QIII, $\sin\theta = -\dfrac{2\sqrt{2}}{3}$.

$$\tan\theta = \frac{\sin\theta}{\cos\theta}$$

$$= \frac{-\dfrac{2\sqrt{2}}{3}}{-\dfrac{1}{3}}$$

$$= 2\sqrt{2}$$

$$\cot\theta = \frac{1}{\tan\theta}$$

$$= \frac{1}{2\sqrt{2}}$$

$$\sec\theta = \frac{1}{\cos\theta}$$

$$= \frac{1}{-\dfrac{1}{3}}$$

$$= -3$$

This problem is continued on the next page

$$\csc\theta = \frac{1}{\sin\theta}$$

$$= \frac{1}{-\frac{2\sqrt{2}}{3}}$$

$$= -\frac{3}{2\sqrt{2}}$$

All six ratios are:

$$\sin\theta = \frac{-2\sqrt{2}}{3}$$

$$\cos\theta = -\frac{1}{3}$$

$$\tan\theta = 2\sqrt{2}$$

$$\cot\theta = \frac{1}{2\sqrt{2}}$$

$$\sec\theta = -3$$

$$\csc\theta = -\frac{3}{2\sqrt{2}}$$

53.

$$\sin\theta = \frac{1}{\csc\theta}$$

$$\sin\theta = \frac{1}{a}, a \neq 0$$

$$\cos\theta = \pm\sqrt{1-\sin^2\theta}$$

$$= \pm\sqrt{1-\left(\frac{1}{a}\right)^2}$$

$$= \pm\sqrt{1-\frac{1}{a^2}}$$

$$= \pm\sqrt{\frac{a^2-1}{a^2}}$$

$$= \pm\frac{\sqrt{a^2-1}}{a} = \frac{\sqrt{a^2-1}}{a} \text{ because } \theta \text{ is in QI.}$$

$$\tan\theta = \frac{\sin\theta}{\cos\theta}$$

$$= \frac{1/a}{\sqrt{a^2-1}/a}$$

$$= \frac{1}{\sqrt{a^2-1}}$$

$$\sec\theta = \frac{1}{\cos\theta} = \frac{1}{\sqrt{a^2-1}/a} = \frac{a}{\sqrt{a^2-1}}$$

$$\cot\theta = \frac{1}{\tan\theta}$$

$$= \frac{1}{1/\sqrt{a^2-1}}$$

$$= \sqrt{a^2-1}$$

All six ratios are:

$$\sin\theta = \frac{1}{a}$$

$$\cos\theta = \frac{\sqrt{a^2-1}}{a}$$

$$\tan\theta = \frac{1}{\sqrt{a^2-1}}$$

$$\cot\theta = \sqrt{a^2-1}$$

$$\sec\theta = \frac{a}{\sqrt{a^2-1}}$$

$$\csc\theta = a$$

55.   All functions are positive in QI.

$$\cos\theta = \sqrt{1-\sin^2\theta}$$

$$= \sqrt{1-(0.23)^2}$$

$$= \sqrt{1-0.0529}$$

$$= \sqrt{0.9471}$$

$$\approx 0.97$$

$$\tan\theta = \frac{\sin\theta}{\cos\theta}$$

$$= \frac{0.23}{0.97}$$

$$\approx 0.24$$

$$\cot\theta = \frac{1}{\tan\theta}$$

$$= \frac{1}{0.24}$$

$$\approx 4.17$$

$$\sec\theta = \frac{1}{\cos\theta}$$

$$= \frac{1}{0.97}$$

$$\approx 1.03$$

$$\csc\theta = \frac{1}{\sin\theta}$$

$$= \frac{1}{0.23}$$

$$\approx 4.35$$

All six ratios are:

$\sin\theta = 0.23$ $\cot\theta = 4.17$

$\cos\theta = 0.97$ $\sec\theta = 1.03$

$\tan\theta = 0.24$ $\csc\theta = 4.35$

57. $$\cos\theta = \frac{1}{\sec\theta}$$

$$= \frac{1}{-1.24}$$

$$= -0.806$$

$$= -0.81 \ \text{(rounded to the nearest hundredth)}$$

$$\sin\theta = \pm\sqrt{1-\cos^2\theta}$$

$$= \pm\sqrt{1-(-0.81)^2}$$

$$= \pm\sqrt{1-0.6561}$$

$$= \pm\sqrt{0.3439}$$

$$= \pm 0.586$$

$$= 0.59 \ \text{because} \ \sin\theta \ \text{is positive in QII}$$

$$\tan\theta = \frac{\sin\theta}{\cos\theta}$$

$$= \frac{0.59}{-0.81}$$

$$= -0.728$$

$$= -0.73$$

$$\cot\theta = \frac{1}{\tan\theta}$$

$$= \frac{1}{-0.73}$$

$$= -1.369$$

$$= -1.37$$

$$\csc\theta = \frac{1}{\sin\theta}$$

$$= \frac{1}{0.59}$$

$$= 1.694$$

$$= 1.69$$

All six ratios are:

$\sin\theta = 0.59$ $\cot\theta = -1.37$

$\cos\theta = -0.81$ $\sec\theta = -1.24$

$\tan\theta = -0.73$ $\csc\theta = 1.69$

59. $m = \dfrac{y_2 - y_1}{x_2 - x_1}$

$= \dfrac{3-0}{1-0}$

$= \dfrac{3}{1}$

$= 3$

61. This line passes through $(0, 0)$ and $(1, m)$

$\text{slope} = \dfrac{y_2 - y_1}{x_2 - x_1}$

$\phantom{\text{slope}}= \dfrac{m-0}{1-0}$

$\phantom{\text{slope}}= \dfrac{m}{1} \text{ or } m$

Problem Set 1.5

1. $\cos\theta = \pm\sqrt{1 - \sin^2\theta}$ Pythagorean identity

3. $\cot\theta = \dfrac{\cos\theta}{\sin\theta}$ Ratio identity

$ = \pm\dfrac{\sqrt{1 - \sin^2\theta}}{\sin\theta}$

5. $\sec\theta = \dfrac{1}{\cos\theta}$ Reciprocal identity

7. $\tan\theta = \dfrac{\sin\theta}{\cos\theta}$ Ratio identity

$ = \pm\dfrac{\sqrt{1 - \cos^2\theta}}{\cos\theta}$

9. $\csc\theta\cot\theta = \dfrac{1}{\sin\theta}\cdot\dfrac{\cos\theta}{\sin\theta}$ Reciprocal and ratio identities

$ = \dfrac{\cos\theta}{\sin^2\theta}$ Multiplication of fractions

11. $\csc\theta\tan\theta = \dfrac{1}{\sin\theta}\cdot\dfrac{\sin\theta}{\cos\theta}$ Reciprocal and ratio identities

$ = \dfrac{\sin\theta}{\sin\theta\cos\theta}$ Multiplication of fractions

$ = \dfrac{1}{\cos\theta}$ Division

13. $\dfrac{\sec\theta}{\csc\theta} = \dfrac{1/\cos\theta}{1/\sin\theta}$ Reciprocal identities

$\phantom{\dfrac{\sec\theta}{\csc\theta}} = \dfrac{\sin\theta}{\cos\theta}$ Division of fractions

15.
$$\dfrac{\sec\theta}{\tan\theta} = \dfrac{\dfrac{1}{\cos\theta}}{\dfrac{\sin\theta}{\cos\theta}} \qquad \text{Reciprocal and ratio identities}$$

$$= \dfrac{\cos\theta}{\sin\theta\cos\theta} \qquad \text{Division of fractions}$$

$$= \dfrac{1}{\sin\theta} \qquad \text{Division}$$

17.
$$\dfrac{\tan\theta}{\cot\theta} = \dfrac{\sin\theta/\cos\theta}{\cos\theta/\sin\theta} \qquad \text{Ratio identities}$$

$$= \dfrac{\sin^2\theta}{\cos^2\theta} \qquad \text{Division of fractions}$$

19.
$$\dfrac{\sin\theta}{\csc\theta} = \dfrac{\sin\theta}{\dfrac{1}{\sin\theta}} \qquad \text{Reciprocal identity}$$

$$= \dfrac{\sin^2\theta}{1} \qquad \text{Division of fractions}$$

$$= \sin^2\theta \qquad \text{Simplify}$$

21.
$$\tan\theta + \sec\theta = \dfrac{\sin\theta}{\cos\theta} + \dfrac{1}{\cos\theta} \qquad \text{Ratio and reciprocal identities}$$

$$= \dfrac{\sin\theta + 1}{\cos\theta} \qquad \text{Addition of fractions}$$

23.
$$\sin\theta\cot\theta + \cos\theta = \sin\theta \cdot \dfrac{\cos\theta}{\sin\theta} + \cos\theta \qquad \text{Ratio identity}$$

$$= \dfrac{\sin\theta\cos\theta}{\sin\theta} + \cos\theta \qquad \text{Multiplication of fractions}$$

$$= \cos\theta + \cos\theta \qquad \text{Division}$$

$$= 2\cos\theta \qquad \text{Addition}$$

25.
$$\sec\theta - \tan\theta\sin\theta = \dfrac{1}{\cos\theta} - \dfrac{\sin\theta}{\cos\theta} \cdot \sin\theta \qquad \text{Reciprocal and ratio identities}$$

$$= \dfrac{1}{\cos\theta} - \dfrac{\sin^2\theta}{\cos\theta} \qquad \text{Multiplication of fractions}$$

$$= \dfrac{1 - \sin^2\theta}{\cos\theta} \qquad \text{Subtraction of fractions}$$

This problem is continued on the next page

$$= \frac{\cos^2 \theta}{\cos \theta} \qquad \text{Pythagorean identity}$$

$$= \cos \theta \qquad \text{Division}$$

27.
$$\frac{\sin \theta}{\cos \theta} + \frac{1}{\sin \theta} = \frac{\sin \theta}{\cos \theta} \cdot \frac{\sin \theta}{\sin \theta} + \frac{1}{\sin \theta} \cdot \frac{\cos \theta}{\cos \theta} \qquad \text{LCD is } \sin \theta \cos \theta$$

$$= \frac{\sin^2 \theta}{\sin \theta \cos \theta} + \frac{\cos \theta}{\sin \theta \cos \theta} \qquad \text{Multiplication of fractions}$$

$$= \frac{\sin^2 \theta + \cos \theta}{\sin \theta \cos \theta} \qquad \text{Addition of fractions}$$

29.
$$\frac{1}{\sin \theta} - \frac{1}{\cos \theta} = \frac{1}{\sin \theta} \cdot \frac{\cos \theta}{\cos \theta} - \frac{1}{\cos \theta} \cdot \frac{\sin \theta}{\sin \theta} \qquad \text{LCD is } \sin \theta \cos \theta$$

$$= \frac{\cos \theta}{\sin \theta \cos \theta} - \frac{\sin \theta}{\sin \theta \cos \theta} \qquad \text{Multiplication}$$

$$= \frac{\cos \theta - \sin \theta}{\sin \theta \cos \theta} \qquad \text{Subtraction of fractions}$$

31.
$$\sin \theta + \frac{1}{\cos \theta} = \sin \theta \cdot \frac{\cos \theta}{\cos \theta} + \frac{1}{\cos \theta} \qquad \text{LCD is } \cos \theta$$

$$= \frac{\sin \theta \cos \theta}{\cos \theta} + \frac{1}{\cos \theta} \qquad \text{Multiplication of fractions}$$

$$= \frac{\sin \theta \cos \theta + 1}{\cos \theta} \qquad \text{Addition of fractions}$$

33.
$$\frac{1}{\sin \theta} - \sin \theta = \frac{1}{\sin \theta} \cdot \sin \theta \cdot \frac{\sin \theta}{\sin \theta} \qquad \text{LCD is } \sin \theta$$

$$= \frac{1}{\sin \theta} - \frac{\sin^2 \theta}{\sin \theta} \qquad \text{Multiplication}$$

$$= \frac{1 - \sin^2 \theta}{\sin \theta} \qquad \text{Subtraction of fractions}$$

$$= \frac{\cos^2 \theta}{\sin \theta} \qquad \text{Pythagorean identity}$$

35.
$$(\sin \theta + 4)(\sin \theta + 3) = \sin^2 \theta + 3\sin \theta + 4\sin \theta + 12$$

$$= \sin^2 \theta + 7\sin \theta + 12$$

37. $(2\cos\theta + 3)(4\cos\theta - 5) = 8\cos^2\theta - 10\cos\theta + 12\cos\theta - 15$
$$= 8\cos^2\theta + 2\cos\theta - 15$$

39. $(1 - \sin\theta)(1 + \sin\theta) = 1 + \sin\theta - \sin\theta - \sin^2\theta$
$$= 1 - \sin^2\theta$$
$$= \cos^2\theta \qquad\qquad \text{Pythagorean identity}$$

41. $(1 - \tan\theta)(1 + \tan\theta) = 1 + \tan\theta - \tan\theta - \tan^2\theta$
$$= 1 - \tan^2\theta$$

43. $(\sin\theta - \cos\theta)^2 = (\sin\theta - \cos\theta)(\sin\theta - \cos\theta)$
$$= \sin^2\theta - \sin\theta\cos\theta - \sin\theta\cos\theta + \cos^2\theta$$
$$= \sin^2\theta - 2\sin\theta\cos\theta + \cos^2\theta$$
$$= (\sin^2\theta + \cos^2\theta) - 2\sin\theta\cos\theta$$
$$= 1 - 2\sin\theta\cos\theta \qquad\qquad \text{Pythagorean identity}$$

45. $(\sin\theta - 4)^2 = (\sin\theta - 4)(\sin\theta - 4)$
$$= \sin^2\theta - 4\sin\theta - 4\sin\theta + 16$$
$$= \sin^2\theta - 8\sin\theta + 16$$

47. $\sqrt{x^2 + 4} = \sqrt{(2\tan\theta)^2 + 4}$ Substitute given value
$$= \sqrt{4\tan^2\theta + 4} \qquad\qquad \text{Multiply}$$
$$= \sqrt{4(\tan^2\theta + 1)} \qquad\qquad \text{Factor}$$
$$= \sqrt{4\sec^2\theta} \qquad\qquad \text{Pythagorean identity}$$
$$= |2\sec\theta| = 2|\sec\theta| \qquad\qquad \text{Simplify}$$

49. $\sqrt{9 - x^2} = \sqrt{9 - (3\sin\theta)^2}$ Substitute given value
$$= \sqrt{9 - 9\sin^2\theta} \qquad\qquad \text{Multiply}$$
$$= \sqrt{9(1 - \sin^2\theta)} \qquad\qquad \text{Factor}$$
$$= \sqrt{9\cos^2\theta} \qquad\qquad \text{Pythagorean identity}$$
$$= |3\cos\theta| = 3|\cos\theta| \qquad\qquad \text{Simplify}$$

51.
$$\sqrt{x^2 - 36} = \sqrt{(6\sec\theta)^2 - 36} \qquad \text{Substitute given value}$$
$$= \sqrt{36\sec^2\theta - 36} \qquad \text{Multiply}$$
$$= \sqrt{36(\sec^2\theta - 1)} \qquad \text{Factor}$$
$$= \sqrt{36\tan^2\theta} \qquad \text{Pythagorean identity}$$
$$= |6\tan\theta| = 6|\tan\theta| \qquad \text{Simplify}$$

53.
$$\sqrt{4x^2 + 16} = \sqrt{4(2\tan\theta)^2 + 16} \qquad \text{Substitute given value}$$
$$= \sqrt{16\tan^2\theta + 16} \qquad \text{Multiply}$$
$$= \sqrt{16(\tan^2\theta + 1)} \qquad \text{Factor}$$
$$= \sqrt{16\sec^2\theta} \qquad \text{Pythagorean identity}$$
$$= |4\sec\theta| = 4|\sec\theta| \qquad \text{Simplify}$$

55.
$$\sqrt{36 - 9x^2} = \sqrt{36 - 9(2\sin\theta)^2} \qquad \text{Substitute given value}$$
$$= \sqrt{36 - 9(4\sin^2\theta)} \qquad \text{Multiply}$$
$$= \sqrt{36 - 36\sin^2\theta} \qquad \text{Multiply}$$
$$= \sqrt{36(1 - \sin^2\theta)} \qquad \text{Factor}$$
$$= \sqrt{36\cos^2\theta} \qquad \text{Pythagorean identity}$$
$$= |6\cos\theta| = 6|\cos\theta| \qquad \text{Simplify}$$

57.
$$\sqrt{9x^2 - 81} = \sqrt{9(3\sec\theta)^2 - 81} \qquad \text{Substitute given value}$$
$$= \sqrt{9(9\sec^2\theta) - 81} \qquad \text{Multiply}$$
$$= \sqrt{81\sec^2\theta - 81} \qquad \text{Multiply}$$
$$= \sqrt{81(\sec^2\theta - 1)} \qquad \text{Factor}$$
$$= \sqrt{81\tan^2\theta} \qquad \text{Pythagorean identity}$$
$$= |9\tan\theta| = 9|\tan\theta| \qquad \text{Simplify}$$

59.
$$\cos\theta\tan\theta = \cos\theta \cdot \frac{\sin\theta}{\cos\theta} \qquad \text{Ratio identity}$$
$$= \frac{\cos\theta\sin\theta}{\cos\theta} \qquad \text{Multiplication of fractions}$$
$$= \sin\theta \qquad \text{Division of common factors}$$

61. $\sin\theta\sec\theta\cot\theta = \sin\theta \cdot \dfrac{1}{\cos\theta} \cdot \dfrac{\cos\theta}{\sin\theta}$ Reciprocal and ratio identities

$\qquad\qquad = \dfrac{\sin\theta\cos\theta}{\cos\theta\sin\theta}$ Multiplication

$\qquad\qquad = 1$ Division of common factors

63. $\dfrac{\sin\theta}{\csc\theta} = \dfrac{\sin\theta}{\dfrac{1}{\sin\theta}}$ Reciprocal identity

$\qquad = \dfrac{\sin^2\theta}{1}$ Division of fractions

$\qquad = \sin^2\theta$

65. $\dfrac{\csc\theta}{\cot\theta} = \dfrac{1/\sin\theta}{\cos\theta/\sin\theta}$ Reciprocal and ratio identities

$\qquad = \dfrac{\sin\theta}{\sin\theta\cos\theta}$ Division of fractions

$\qquad = \dfrac{1}{\cos\theta}$ Division of common factor

$\qquad = \sec\theta$ Reciprocal identity

67. $\dfrac{\sec\theta}{\csc\theta} = \dfrac{\dfrac{1}{\cos\theta}}{\dfrac{1}{\sin\theta}}$ Reciprocal identity

$\qquad = \dfrac{\sin\theta}{\cos\theta}$ Division of fractions

$\qquad = \tan\theta$ Ratio identity

69. $\dfrac{\sec\theta\cot\theta}{\csc\theta} = \dfrac{\dfrac{1}{\cos\theta} \cdot \dfrac{\cos\theta}{\sin\theta}}{\dfrac{1}{\sin\theta}}$ Reciprocal and ratio identities

$\qquad = \dfrac{\cos\theta\sin\theta}{\cos\theta\sin\theta}$ Division of fractions

$\qquad = 1$ Division of common factors

71.

$$\sin\theta\tan\theta + \cos\theta = \sin\theta \cdot \frac{\sin\theta}{\cos\theta} + \cos\theta \qquad \text{Ratio identity}$$

$$= \frac{\sin^2\theta}{\cos\theta} + \cos\theta \qquad \text{Multiplication of fractions}$$

$$= \frac{\sin^2\theta}{\cos\theta} + \frac{\cos\theta}{1} \cdot \frac{\cos\theta}{\cos\theta} \qquad \text{L.C.D. is } \cos\theta$$

$$= \frac{\sin^2\theta}{\cos\theta} + \frac{\cos^2\theta}{\cos\theta} \qquad \text{Multiplication of fractions}$$

$$= \frac{\sin^2\theta + \cos^2\theta}{\cos\theta} \qquad \text{Addition of fractions}$$

$$= \frac{1}{\cos\theta} \qquad \text{Pythagorean identity}$$

$$= \sec\theta \qquad \text{Reciprocal identity}$$

73.

$$\tan\theta + \cot\theta = \frac{\sin\theta}{\cos\theta} + \frac{\cos\theta}{\sin\theta} \qquad \text{Ratio and reciprocal identities}$$

$$= \frac{\sin\theta}{\cos\theta} \cdot \frac{\sin\theta}{\sin\theta} + \frac{\cos\theta}{\sin\theta} \cdot \frac{\cos\theta}{\cos\theta} \qquad \text{L.C.D. is } \sin\theta\cos\theta$$

$$= \frac{\sin^2\theta}{\sin\theta\cos\theta} + \frac{\cos^2\theta}{\sin\theta\cos\theta} \qquad \text{Multiplication}$$

$$= \frac{\sin^2\theta + \cos^2\theta}{\sin\theta\cos\theta} \qquad \text{Addition of fractions}$$

$$= \frac{1}{\sin\theta\cos\theta} \qquad \text{Pythagorean identity}$$

$$= \frac{1}{\sin\theta} \cdot \frac{1}{\cos\theta} \qquad \text{Property of fractions}$$

$$= \csc\theta\sec\theta \qquad \text{Reciprocal identities}$$

75.

$$\csc\theta - \sin\theta = \frac{1}{\sin\theta} - \sin\theta \qquad \text{Reciprocal identity}$$

$$= \frac{1}{\sin\theta} - \frac{\sin\theta}{1} \cdot \frac{\sin\theta}{\sin\theta} \qquad \text{L.C.D. is } \sin\theta$$

$$= \frac{1}{\sin\theta} - \frac{\sin^2\theta}{\sin\theta} \qquad \text{Multiplication of fractions}$$

$$= \frac{1 - \sin^2\theta}{\sin\theta} \qquad \text{Subtraction of fractions}$$

$$= \frac{\cos^2\theta}{\sin\theta} \qquad \text{Pythagorean identity}$$

77.

$$\csc\theta\tan\theta - \cos\theta = \frac{1}{\sin\theta}\cdot\frac{\sin\theta}{\cos\theta} - \cos\theta \qquad \text{Reciprocal and ratio identities}$$

$$= \frac{\sin\theta}{\sin\theta\cos\theta} - \cos\theta \qquad \text{Multiplication}$$

$$= \frac{1}{\cos\theta} - \cos\theta \qquad \text{Division of common factor}$$

$$= \frac{1}{\cos\theta} - \cos\theta\cdot\frac{\cos\theta}{\cos\theta} \qquad \text{L.C.D. is } \cos\theta$$

$$= \frac{1}{\cos\theta} - \frac{\cos^2\theta}{\cos\theta} \qquad \text{Multiplication}$$

$$= \frac{1-\cos^2\theta}{\cos\theta} \qquad \text{Subtraction of fractions}$$

$$= \frac{\sin^2\theta}{\cos\theta} \qquad \text{Pythagorean identity}$$

79.

$$(1-\cos\theta)(1+\cos\theta) = 1-\cos^2\theta \qquad \text{Multiplication}$$

$$= \sin^2\theta \qquad \text{Pythagorean identity}$$

81.

$$(\sin\theta+1)(\sin\theta-1) = \sin^2\theta - 1 \qquad \text{Multiplication}$$

$$= 1-\cos^2\theta - 1 \qquad \text{Pythagorean identity}$$

$$= -\cos^2\theta \qquad \text{Addition}$$

83.

$$\frac{\cos\theta}{\sec\theta} + \frac{\sin\theta}{\csc\theta} = \frac{\cos\theta}{\dfrac{1}{\cos\theta}} + \frac{\sin\theta}{\dfrac{1}{\sin\theta}} \qquad \text{Reciprocal identities}$$

$$= \cos^2\theta + \sin^2\theta \qquad \text{Division of fractions}$$

$$= 1 \qquad \text{Pythagorean identity}$$

85.

$$(\sin\theta-\cos\theta)^2 - 1 = \sin^2\theta - 2\sin\theta\cos\theta + \cos^2\theta - 1 \qquad \text{Multiplication}$$

$$= -2\sin\theta\cos\theta + (\sin^2\theta + \cos^2\theta) - 1 \qquad \text{Commutative property}$$

$$= -2\sin\theta\cos\theta + 1 - 1 \qquad \text{Pythagorean identity}$$

$$= -2\sin\theta\cos\theta \qquad \text{Addition}$$

87.

$$\sin\theta\left(\sec\theta + \csc\theta\right) = \sin\theta \cdot \sec\theta + \sin\theta \cdot \csc\theta$$ Distributive property

$$= \frac{\sin\theta}{1} \cdot \frac{1}{\cos\theta} + \frac{\sin\theta}{1} \cdot \frac{1}{\sin\theta}$$ Reciprocal identities

$$= \frac{\sin\theta}{\cos\theta} + \frac{\sin\theta}{\sin\theta}$$ Multiplication of fractions

$$= \tan\theta + 1$$ Ratio identity and division of common factor

89.

$$\sin\theta\left(\sec\theta + \cot\theta\right) = \sin\theta\sec\theta + \sin\theta\cot\theta$$ Multiplication

$$= \sin\theta \cdot \frac{1}{\cos\theta} + \sin\theta \cdot \frac{\cos\theta}{\sin\theta}$$ Reciprocal and ratio identities

$$= \frac{\sin\theta}{\cos\theta} + \frac{\sin\theta\cos\theta}{\sin\theta}$$ Multiplication

$$= \tan\theta + \cos\theta$$ Ratio identity and division of common factor

91.

$$\sin\theta\left(\csc\theta - \sin\theta\right) = \sin\theta\csc\theta - \sin^2\theta$$ Distributive property

$$= \frac{\sin\theta}{1} \cdot \frac{1}{\sin\theta} - \sin^2\theta$$ Reciprocal identity

$$= \frac{\sin\theta}{\sin\theta} - \sin^2\theta$$ Multiplication of fractions

$$= 1 - \sin^2\theta$$ Division of common factor

$$= \cos^2\theta$$ Pythagorean identity

Chapter 1 Test

1. The complement of $70°$ is $90° - 70° = 20°$.

   The supplement of $70°$ is $180° - 70° = 110°$.

2. | | |
   |---|---|
   | $x^2 + (3)^2 = (6)^2$ | Pythagorean Theorem |
   | $x^2 + 9 = 36$ | Simplify |
   | $x^2 = 27$ | Subtract 9 from both sides |
   | $x = \pm 3\sqrt{3}$ | Take square root of both sides |
   | $x = 3\sqrt{3}$ because x must be positive | |

3. | | |
   |---|---|
   | $x^2 + 4^2 = (x+2)^2$ | Pythagorean Theorem |
   | $x^2 + 16 = x^2 + 4x + 4$ | Simplify |
   | $12 = 4x$ | Subtract 4 and x^2 from both sides |
   | $3 = x$ | Divide both sides by 4 |

4. | | |
   |---|---|
   | $\tan 45° = \dfrac{h}{s}$ | Tangent relationship |
   | $1 = \dfrac{h}{5\sqrt{3}}$ | Substitute known values |
   | $h = 5\sqrt{3}$ | Multiply both sides by $5\sqrt{3}$ |
   | $\cos 45° = \dfrac{s}{r}$ | Cosine relationship |
   | $\dfrac{\sqrt{2}}{2} = \dfrac{5\sqrt{3}}{r}$ | Substitute known values |
   | $\dfrac{\sqrt{2}}{2} r = 5\sqrt{3}$ | Solve for r |
   | $r = \dfrac{10\sqrt{3}}{\sqrt{2}}$ | |
   | $= 5\sqrt{6}$ | Rationalize the denominator |
   | $\tan 60° = \dfrac{h}{y}$ | Tangent relationship |
   | $\sqrt{3} = \dfrac{5\sqrt{3}}{y}$ | Substitute known values |
   | $\sqrt{3}\,y = 5\sqrt{3}$ | Solve for y |
   | $y = 5$ | |

This problem is continued on the next page

$$\sin 30° = \frac{y}{x}$$ Sine relationship

$$\frac{1}{2} = \frac{5}{x}$$ Substitute known values

$$\frac{1}{2}x = 5$$ Solve for x

$$x = 10$$

5. $$\sin 30° = \frac{y}{x}$$ Sine relationship

$$\frac{1}{2} = \frac{3}{x}$$ Substitute known values

$$x = 6$$ Solve for x

$$\cos 30° = \frac{h}{x}$$ Cosine relationship

$$\frac{\sqrt{3}}{2} = \frac{h}{6}$$ Substitute known values

$$h = 3\sqrt{3}$$ Solve for h

$$\tan 45° = \frac{s}{h}$$ Tangent relationship

$$1 = \frac{s}{3\sqrt{3}}$$ Substitute known values

$$s = 3\sqrt{3}$$ Solve for s

$$\sin 45° = \frac{h}{r}$$ Sine relationship

$$\frac{\sqrt{2}}{2} = \frac{3\sqrt{3}}{r}$$ Substitute known values

$$\frac{\sqrt{2}}{2}r = 3\sqrt{3}$$ Solve for r

$$r = \frac{6\sqrt{3}}{\sqrt{2}}$$

$$= 3\sqrt{6}$$ Rationalize the denominator

6. $$(AB)^2 + (BC)^2 = (AC)^2$$ Pythagorean Theorem

$$(AB)^2 = (3)^2 = (5)^2$$ Substitute given values

$$(AB)^2 + 9 = 25$$ Simplify

$$(AB)^2 = 16$$ Subtract 9 from both sides

This problem is continued on the next page

$$AB = \pm 4 \qquad \text{Take square root of both sides}$$
$$AB = 4 \quad \text{because it must be positive}$$

$$(DB)^2 = (DA)^2 + (AB)^2 \qquad \text{Pythagorean Theorem}$$
$$(DB)^2 = (6)^2 + (4)^2 \qquad \text{Substitute known values}$$
$$(DB)^2 = 36 + 16 \qquad \text{Simplify}$$
$$(DB)^2 = 52$$
$$DB = \pm\sqrt{52} \qquad \text{Take square root of both sides}$$
$$DB = 2\sqrt{13} \text{ or } 7.21 \text{ because it must be positive}$$

7. One complete revolution takes 12 hours or $360°$. Let θ represent the number of degrees a clock moves in 3 hours, then:

$$\frac{\theta}{360°} = \frac{3\,hr}{12\,hr}$$
$$\frac{\theta}{360°} = \frac{1}{4}$$
$$\theta = 90°$$

8. The shortest side is $\frac{1}{2}(5)$ or $\frac{5}{2}$.

   The medium side is $\frac{5}{2}\left(\sqrt{3}\right)$ or $\frac{5\sqrt{3}}{2}$.

9. We are looking for the hypotenuse of a $45° - 45° - 90°$ triangle where the side opposite the $45°$ angle is 15 feet. Using the Pythagorean Theorem:

$$c^2 = 15^2 + 15^2$$
$$= 225 + 225$$
$$= 450$$
$$c = \pm\sqrt{450} = 15\sqrt{2}$$

   The length of the escalator is $15\sqrt{2}$ feet.

10. The central angle of each congruent triangle is $\dfrac{360°}{5} = 72°$.

    When we draw in the altitude of this triangle we get a

    $36°$ angle. Therefore, $\dfrac{\theta}{2} = 90° - 36°$

    $$= 54°$$
    $$\theta = 108°$$

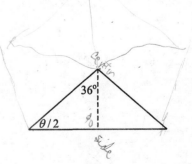

11. The central angle of each congruent triangle is $\dfrac{360^\circ}{6} = 60^\circ$. When we draw in the altitude of this triangle we get a 30° angle.

Therefore, $\dfrac{\theta}{2} = 90^\circ - 30^\circ$

$\qquad\qquad = 60^\circ$

$\qquad \theta = 120^\circ$

12. The x-intercept is $(3, 0)$ and the y-intercept is $(0, 2)$.

13.

$r = \sqrt{\left(x_2 - x_1\right)^2 + \left(y_2 - y_1\right)^2}$	Distance formula
$= \sqrt{\left[4 - (-1)\right]^2 + \left(-2 - 10\right)^2}$	Substitute known values
$= \sqrt{(5)^2 + (-12)^2}$	Simplify
$= \sqrt{25 + 144}$	
$= \sqrt{169}$	
$= 13$	

14.

$d = \sqrt{(a - 0)^2 + (b - 0)^2}$	Distance formula
$= \sqrt{a^2 + b^2}$	Simplify

15.

$\sqrt{\left[x - (-2)\right]^2 + (1 - 3)^2} = \sqrt{13}$	Distance formula
$\sqrt{(x + 2)^2 + (-2)^2} = \sqrt{13}$	Simpify
$(x + 2)^2 + 4 = 13$	Square both sides
$(x + 2)^2 = 9$	Subtract 4 from both sides
$x + 2 = \pm 3$	Take square root of both sides
$x = -2 \pm 3$	Solve for x
$x = -5$ or $x = 1$	

16. The cannonball is at the ground ($y = 0$) when $x = 0$ and when $x = 140$. The x-coordinate of the vertex (the maximum) will be at $\frac{1}{2}(140)$ or 70, and the y-coordinate will be 50. We can now sketch the parabola through the points $(0, 0)$, $(70, 50)$ and $(140, 0)$.

The equation will be in the form $y = a(x - 70)^2 + 50$. We will use the point $(140, 0)$ to find a. Let $x = 140$ and $y = 0$:

$$y = a(x - 70)^2 + 50$$
$$0 = a(140 - 70)^2 + 50$$
$$-50 = a(70)^2$$
$$-50 = 4900a$$
$$a = -\frac{1}{98}$$

Therefore, the equation is $y = -\frac{1}{98}(x - 70)^2 + 50$.

17. A point on the terminal side of $90°$ is $(0, 1)$ and $r = 1$.

$$\sin 90° = \frac{y}{r} = \frac{1}{1} = 1$$

$$\cos 90° = \frac{x}{r} = \frac{0}{1} = 0$$

$$\tan 90° = \frac{y}{x} = \frac{1}{0} \text{ is undefined}$$

18. A point on the terminal side of $-45°$ is $(1, -1)$ and $r = \sqrt{2}$.

$$\sin 45° = \frac{y}{r} = \frac{-1}{\sqrt{2}} = -\frac{1}{\sqrt{2}}$$

$$\cos 45° = \frac{x}{r} = \frac{1}{\sqrt{2}}$$

$$\tan 45° = \frac{y}{x} = \frac{-1}{1} = -1$$

19. $\sin\theta$ is negative in QIII and QIV.
$\cos\theta$ is positive in QI and QIV.
Therefore, θ must lie in QIV.

20. $\csc\theta$ is positive in QI and QII.
$\cos\theta$ is negative in QII and QIII.
Therefore, θ must lie in QII.

21. $(x, y) = (-6, 8)$

$x = -6$ and $y = 8$

$$r = \sqrt{x^2 + y^2}$$
$$= \sqrt{(-6)^2 + 8^2}$$
$$= \sqrt{36 + 64}$$
$$= \sqrt{100}$$
$$= 10$$

$$\sin\theta = \frac{y}{r} = \frac{8}{10} = \frac{4}{5}$$

$$\cos\theta = \frac{x}{r} = -\frac{6}{10} = -\frac{3}{5}$$

$$\tan\theta = \frac{y}{x} = \frac{8}{-6} = -\frac{4}{3}$$

$$\cot\theta = \frac{x}{y} = \frac{-6}{8} = -\frac{3}{4}$$

$$\sec\theta = \frac{r}{x} = \frac{10}{-6} = -\frac{5}{3}$$

$$\csc\theta = \frac{r}{y} = \frac{10}{8} = \frac{5}{4}$$

22. $(x,y) = (-3,-1)$

$\sin\theta = \dfrac{y}{r} = \dfrac{-1}{\sqrt{10}} = -\dfrac{1}{\sqrt{10}}$

$\cot\theta = \dfrac{x}{y} = \dfrac{-3}{-1} = 3$

$x = -3$ and $y = -1$

$\cos\theta = \dfrac{x}{r} = \dfrac{-3}{\sqrt{10}} = -\dfrac{3}{\sqrt{10}}$

$\sec\theta = \dfrac{r}{x} = \dfrac{\sqrt{10}}{-3} = -\dfrac{\sqrt{10}}{3}$

$r = \sqrt{x^2 + y^2}$

$\tan\theta = \dfrac{y}{x} = \dfrac{-1}{-3} = \dfrac{1}{3}$

$\csc\theta = \dfrac{r}{y} = \dfrac{\sqrt{10}}{-1} = -\sqrt{10}$

$= \sqrt{(-3)^2 + (-1)^2}$

$= \sqrt{9+1}$

$= \sqrt{10}$

23. If $\sin\theta = \dfrac{1}{2}$ and θ terminates in QII, then:

$\cos\theta = \sqrt{1 - \left(\dfrac{1}{2}\right)^2}$

$\cot\theta = \dfrac{1}{\tan\theta} = \dfrac{1}{-1/\sqrt{3}} = -\sqrt{3}$

$= -\sqrt{\dfrac{3}{4}}$

$\sec\theta = \dfrac{1}{\cos\theta} = \dfrac{1}{-\sqrt{3}/2} = -\dfrac{2}{\sqrt{3}}$

$= -\dfrac{\sqrt{3}}{2}$

$\csc\theta = \dfrac{1}{\sin\theta} = \dfrac{1}{1/2} = 2$

$\tan\theta = \dfrac{\sin\theta}{\cos\theta} = \dfrac{1/2}{-\sqrt{3}/2} = -\dfrac{1}{\sqrt{3}}$

24. If $\tan\theta = \dfrac{12}{5}$ and θ terminates in QIII, then:

$\sec^2\theta = \tan^2\theta + 1$

$= \dfrac{144}{25} + 1$

$= \dfrac{169}{25}$

$\sec\theta = -\dfrac{13}{5}$ because $\sec\theta$ is negative in QIII.

$\cot\theta = \dfrac{1}{\tan\theta} = \dfrac{1}{12/5} = \dfrac{5}{12}$

$\cos\theta = \dfrac{1}{\sec\theta} = \dfrac{1}{-13/5} = -\dfrac{5}{13}$

$\dfrac{\sin\theta}{\cos\theta} = \tan\theta$

$\csc\theta = \dfrac{1}{\sin\theta}$

$\dfrac{\sin\theta}{-5/13} = \dfrac{12}{5}$

$= \dfrac{1}{-12/13}$

$\sin\theta = -\dfrac{12}{13}$

$= -\dfrac{13}{12}$

25. A point on the terminal side of θ in QIV is $(1,-2)$.

$(x,y)=(1,-2)$

$x=1$ and $y=-2$

$r=\sqrt{x^2+y^2}$

$=\sqrt{1^2+(-2)^2}$

$=\sqrt{1+4}$

$=\sqrt{5}$

$\sin\theta=\dfrac{y}{r}=\dfrac{-2}{\sqrt{5}}$
$\qquad\qquad\qquad$
$\cos\theta=\dfrac{x}{r}=\dfrac{1}{\sqrt{5}}$

26. $\csc\theta=\dfrac{1}{\sin\theta}\qquad$ Reciprocal identity

$=\dfrac{1}{-3/4}\qquad$ Substitute known value

$=-\dfrac{4}{3}\qquad$ Simplify

27. $\cos\theta=\dfrac{1}{\sec\theta}\qquad$ Reciprocal identity

$=\dfrac{1}{-2}$

$=-\dfrac{1}{2}$

28. $\sin^3\theta=\left(\dfrac{1}{3}\right)^3=\dfrac{1}{27}$

29. $\cos\theta=\dfrac{1}{\sec\theta}\qquad$ Reciprocal identity

$=\dfrac{1}{3}$

$\sin\theta=-\sqrt{1-\cos^2\theta}$ because θ is in QIV $\qquad$ $\tan\theta=\dfrac{\sin\theta}{\cos\theta}\qquad$ Ratio identity

$=-\sqrt{1-\dfrac{1}{9}}\qquad\qquad\qquad\qquad\qquad=-\dfrac{2\sqrt{2}/3}{1/3}$

$=-\sqrt{\dfrac{8}{9}}\qquad\qquad\qquad\qquad\qquad\qquad=-2\sqrt{2}$

$=-\dfrac{2\sqrt{2}}{3}$

30.

$$\cos\theta = \pm\sqrt{1-\sin^2\theta} \qquad \text{Pythagorean identity}$$

$$= \pm\sqrt{1-\left(\frac{1}{a}\right)^2} \qquad \text{Substitute known value}$$

$$= \pm\sqrt{1-\frac{1}{a^2}} \qquad \text{Simplify}$$

$$= \pm\sqrt{\frac{a^2-1}{a^2}}$$

$$= \pm\frac{\sqrt{a^2-1}}{a}$$

$$\cos\theta = \frac{\sqrt{a^2-1}}{a} \quad \text{because } \cos\theta \text{ is positive in QI}$$

$$\csc\theta = \frac{1}{\sin\theta} \qquad \text{Reciprocal identity}$$

$$= \frac{1}{1/a} \qquad \text{Substitute known value}$$

$$= a \qquad \text{Simplify}$$

$$\cot\theta = \frac{\cos\theta}{\sin\theta} \qquad \text{Ratio identity}$$

$$= \frac{\sqrt{a^2-1}/a}{1/a} \qquad \text{Substitute known values}$$

$$= \sqrt{a^2-1} \qquad \text{Simplify}$$

31.

$$(\sin\theta + 3)(\sin\theta - 7) = \sin^2\theta - 7\sin\theta + 3\sin\theta - 21$$

$$= \sin^2\theta - 4\sin\theta - 21$$

32.

$$(\cos\theta - \sin\theta)^2 = \cos^2\theta - 2\sin\theta\cos\theta + \sin^2\theta \qquad \text{Multiplication}$$

$$= (\cos^2\theta + \sin^2\theta) - 2\sin\theta\cos\theta \qquad \text{Commutative Property}$$

$$= 1 - 2\sin\theta\cos\theta \qquad \text{Pythagorean identity}$$

33.

$$\frac{1}{\sin\theta} - \sin\theta = \frac{1}{\sin\theta} - \frac{\sin^2\theta}{\sin\theta} \qquad \text{L.C.D. is } \sin\theta$$

$$= \frac{1-\sin^2\theta}{\sin\theta} \qquad \text{Subtraction of fractions}$$

$$= \frac{\cos^2\theta}{\sin\theta} \qquad \text{Pythagorean identity}$$

34.
$$\sqrt{4-x^2} = \sqrt{4-(2\sin\theta)^2} \qquad \text{Substitute known value}$$
$$= \sqrt{4-4\sin^2\theta} \qquad \text{Multiplication}$$
$$= \sqrt{4(1-\sin^2\theta)} \qquad \text{Factor}$$
$$= \sqrt{4\cos^2\theta} \qquad \text{Pythagorean identity}$$
$$= |2\cos\theta| = 2|\cos\theta| \qquad \text{Simplify}$$

35.
$$\frac{\cot\theta}{\csc\theta} = \frac{\cos\theta/\sin\theta}{1/\sin\theta} \qquad \text{Ratio and reciprocal identities}$$
$$= \frac{\cos\theta\sin\theta}{\sin\theta} \qquad \text{Division of fractions}$$
$$= \cos\theta \qquad \text{Division of common factor}$$

36.
$$\cot\theta + \tan\theta = \frac{\cos\theta}{\sin\theta} + \frac{\sin\theta}{\cos\theta} \qquad \text{Ratio identities}$$
$$= \frac{\cos^2\theta + \sin^2\theta}{\sin\theta\cos\theta} \qquad \text{L.C.D. is } \sin\theta\cos\theta$$
$$= \frac{1}{\sin\theta\cos\theta} \qquad \text{Pythagorean identity}$$
$$= \frac{1}{\sin\theta}\cdot\frac{1}{\cos\theta} \qquad \text{Separate fractions}$$
$$= \csc\theta\sec\theta \qquad \text{Reciprocal identities}$$

37.
$$(1-\sin\theta)(1+\sin\theta) = 1-\sin^2\theta \qquad \text{Multitiplication}$$
$$= \cos^2\theta \qquad \text{Pythagorean identity}$$

38.
$$\sin\theta(\csc\theta+\cot\theta) = \sin\theta\left(\frac{1}{\sin\theta} + \frac{\cos\theta}{\sin\theta}\right) \qquad \text{Reciprocal and ratio identities}$$
$$= 1+\cos\theta \qquad \text{Multiplication}$$

CHAPTER 2   Right Triangle Trigonometry

Problem Set 2.1

1.
$$a = \sqrt{c^2 - b^2} \quad \text{Pythagorean Theorem}$$
$$= \sqrt{(5)^2 - (3)^2} \quad \text{Substitute known values}$$
$$= \sqrt{25 - 9} \quad \text{Simplify}$$
$$= \sqrt{16} = 4$$

$$\sin A = \frac{a}{c} = \frac{4}{5} \qquad \cot A = \frac{b}{a} = \frac{3}{4}$$

$$\cos A = \frac{b}{c} = \frac{3}{5} \qquad \sec A - \frac{c}{b} = \frac{5}{3}$$

$$\tan A = \frac{a}{b} = \frac{4}{3} \qquad \csc A = \frac{c}{a} = \frac{5}{4}$$

3.
$$c = \sqrt{a^2 + b^2} \quad \text{Pythagorean Theorem}$$
$$= \sqrt{(2)^2 + (1)^2} \quad \text{Substitute known values}$$
$$= \sqrt{4 + 1} \quad \text{Simplify}$$
$$= \sqrt{5}$$

$$\sin A = \frac{a}{c} = \frac{2}{\sqrt{5}} \qquad \cot A = \frac{b}{a} = \frac{1}{2}$$

$$\cos A = \frac{b}{c} = \frac{1}{\sqrt{5}} \qquad \sec A = \frac{c}{b} = \frac{\sqrt{5}}{1} = \sqrt{5}$$

$$\tan A = \frac{a}{b} = \frac{2}{1} = 2 \qquad \csc A = \frac{c}{a} = \frac{\sqrt{5}}{2}$$

5.
$$c = \sqrt{a^2 + b^2} \quad \text{Pythagorean Theorem}$$
$$= \sqrt{(2)^2 + \left(\sqrt{5}\right)^2} \quad \text{Substitute known values}$$
$$= \sqrt{4 + 5} \quad \text{Simplify}$$
$$= \sqrt{9} = 3$$

$$\sin A = \frac{a}{c} = \frac{2}{3} \qquad \cot A = \frac{b}{a} = \frac{\sqrt{5}}{2}$$

$$\cos A = \frac{b}{c} = \frac{\sqrt{5}}{3} \qquad \sec A = \frac{c}{b} = \frac{3}{\sqrt{5}}$$

$$\tan A = \frac{a}{b} = \frac{2}{\sqrt{5}} \qquad \csc A = \frac{c}{a} = \frac{3}{2}$$

7.
$$b = \sqrt{c^2 - a^2} \quad \text{Pythagorean Theorem}$$
$$= \sqrt{(6)^2 - (5)^2} \quad \text{Substitute known values}$$
$$= \sqrt{36 - 25} \quad \text{Simplify}$$
$$= \sqrt{11}$$

$$\sin A = \frac{a}{c} = \frac{5}{6} \qquad \sin B = \frac{b}{c} = \frac{\sqrt{11}}{6}$$

$$\cos A = \frac{b}{c} = \frac{\sqrt{11}}{6} \qquad \cos B = \frac{a}{c} = \frac{5}{6}$$

$$\tan A = \frac{a}{b} = \frac{5}{\sqrt{11}} \qquad \tan B = \frac{b}{a} = \frac{\sqrt{11}}{5}$$

9.
$$c = \sqrt{a^2 + b^2} \quad \text{Pythagorean Theorem}$$
$$= \sqrt{(1)^2 + (1)^2} \quad \text{Substitute known values}$$
$$= \sqrt{1 + 1} \quad \text{Simplify}$$
$$= \sqrt{2}$$

$$\sin A = \frac{a}{c} = \frac{1}{\sqrt{2}} \qquad \sin B = \frac{b}{c} = \frac{1}{\sqrt{2}}$$

$$\cos A = \frac{b}{c} = \frac{1}{\sqrt{2}} \qquad \cos B = \frac{a}{c} = \frac{1}{\sqrt{2}}$$

$$\tan A = \frac{a}{b} = \frac{1}{1} = 1 \qquad \tan B = \frac{b}{a} = \frac{1}{1} = 1$$

11. $b = \sqrt{c^2 - a^2}$ Pythagorean Theorem

$\quad\quad = \sqrt{10^2 - 6^2}$ Substitute known values

$\quad\quad = \sqrt{100 - 36}$ Simplify

$\quad\quad = \sqrt{64} = 8$

$\sin A = \dfrac{a}{c} = \dfrac{6}{10} = \dfrac{3}{5}$ $\quad\sin B = \dfrac{b}{c} = \dfrac{8}{10} = \dfrac{4}{5}$

$\cos A = \dfrac{b}{c} = \dfrac{8}{10} = \dfrac{4}{5}$ $\quad\cos B = \dfrac{a}{c} = \dfrac{6}{10} = \dfrac{3}{5}$

$\tan A = \dfrac{a}{b} = \dfrac{6}{8} = \dfrac{3}{4}$ $\quad\tan B = \dfrac{b}{a} = \dfrac{8}{6} = \dfrac{4}{3}$

13. $a = \sqrt{c^2 - b^2}$ Pythagorean Theorem

$\quad\quad = \sqrt{(2x)^2 - (x)^2}$ Substitute known values

$\quad\quad = \sqrt{4x^2 - x^2}$ Simplify

$\quad\quad = \sqrt{3x^2}$

$\quad\quad = x\sqrt{3}$

$\sin A = \dfrac{a}{c} = \dfrac{x\sqrt{3}}{2x} = \dfrac{\sqrt{3}}{2}$ $\quad\sin B = \dfrac{b}{c} = \dfrac{x}{2x} = \dfrac{1}{2}$

$\cos A = \dfrac{b}{c} = \dfrac{x}{2x} = \dfrac{1}{2}$ $\quad\cos B = \dfrac{a}{c} = \dfrac{x\sqrt{3}}{2x} = \dfrac{\sqrt{3}}{2}$

$\tan A = \dfrac{a}{b} = \dfrac{x\sqrt{3}}{x} = \sqrt{3}$ $\quad\tan B = \dfrac{b}{a} = \dfrac{x}{x\sqrt{3}} = \dfrac{1}{\sqrt{3}}$

15. The coordinates of B are (4, 3).

$a = 3, \quad b = 4, \quad c = 5$

$\sin A = \dfrac{a}{c} = \dfrac{3}{5}$

$\cos A = \dfrac{b}{c} = \dfrac{4}{5}$

$\tan A = \dfrac{a}{b} = \dfrac{3}{4}$

17. $\sin 10^\circ = \cos(90^\circ - 10^\circ) = \cos 80^\circ$

19. $\tan 8^\circ = \cot(90^\circ - 8^\circ) = \cot 82^\circ$

21. $\sin x^\circ = \cos(90^\circ - x^\circ)$

23. $\tan(90^\circ - x^\circ) = \cot x^\circ$

25. $\csc x = \dfrac{1}{\sin x}$

$\csc 0^\circ = \dfrac{1}{0}$ undefined

$\csc 30^\circ = \dfrac{1}{1/2} = 2$

$\csc 45^\circ = \dfrac{1}{1/\sqrt{2}} = \sqrt{2}$

$\csc 60^\circ = \dfrac{1}{\sqrt{3}/2} = \dfrac{2}{\sqrt{3}}$

$\csc 90^\circ = \dfrac{1}{1} = 1$

27. $4\sin 30^\circ = 4\left(\dfrac{1}{2}\right) = 2$

29.
$$\left(2\cos 30°\right)^2 = \left[2\left(\frac{\sqrt{3}}{2}\right)\right]^2 = \left(\sqrt{3}\right)^2 = 3$$

31.
$$\left(\sin 60° + \cos 60°\right)^2 = \left(\frac{\sqrt{3}}{2} + \frac{1}{2}\right)^2$$
$$= \left(\frac{\sqrt{3}+1}{2}\right)^2$$
$$= \frac{\left(\sqrt{3}+1\right)\left(\sqrt{3}+1\right)}{4}$$
$$= \frac{3+2\sqrt{3}+1}{4}$$
$$= \frac{4+2\sqrt{3}}{4}$$
$$= \frac{2\left(2+\sqrt{3}\right)}{4} = \frac{2+\sqrt{3}}{2}$$

33.
$$\sin^2 45° - 2\sin 45° \cos 45° + \cos^2 45° = \left(\frac{\sqrt{2}}{2}\right)^2 - 2\left(\frac{\sqrt{2}}{2}\right)\left(\frac{\sqrt{2}}{2}\right) + \left(\frac{\sqrt{2}}{2}\right)^2$$
$$= \frac{2}{4} - 2\left(\frac{2}{4}\right) + \frac{2}{4}$$
$$= \frac{1}{2} - 1 + \frac{1}{2} = 0$$

35.
$$\left(\tan 45° + \tan 60°\right)^2 = \left(1+\sqrt{3}\right)^2$$
$$= \left(1+\sqrt{3}\right)\left(1+\sqrt{3}\right)$$
$$= 1 + 2\sqrt{3} + 3$$
$$= 4 + 2\sqrt{3}$$

37.
$$2\sin 30° = 2\left(\frac{1}{2}\right)$$
$$= 1$$

39.
$$4\cos\left(z - 30°\right) = 4\cos\left(60° - 30°\right)$$
$$= 4\cos 30°$$
$$= 4\left(\frac{\sqrt{3}}{2}\right) = 2\sqrt{3}$$

41.
$$-3\sin 2(30°) = -3\sin 60°$$
$$= -3\left(\frac{\sqrt{3}}{2}\right)$$
$$= -\frac{3\sqrt{3}}{2}$$

43.
$$2\cos(3x - 45°) = 2\cos(3 \cdot 30° - 45°)$$
$$= 2\cos(90° - 45°)$$
$$= 2\cos 45°$$
$$= 2 \cdot \frac{\sqrt{2}}{2} = \sqrt{2}$$

45.
$$\sec 30° = \frac{1}{\cos 30°} \qquad \text{Reciprocal identity}$$
$$= \frac{1}{\sqrt{3}/2} \qquad \text{Substitute exact value from Table 1}$$
$$= \frac{2}{\sqrt{3}} \qquad \text{Division of fractions}$$

47.
$$\csc 60° = \frac{1}{\sin 60°}$$
$$= \frac{1}{\sqrt{3}/2}$$
$$= \frac{2}{\sqrt{3}}$$

49.
$$\cot 45° = \frac{\cos 45°}{\sin 45°} \qquad \text{Ratio identity}$$
$$= \frac{\sqrt{2}/2}{\sqrt{2}/2} \qquad \text{Substitute values from Table 1}$$
$$= 1 \qquad \text{Simplify}$$

51.
$$\sec 45° = \frac{1}{\cos 45°}$$
$$= \frac{1}{1/\sqrt{2}}$$
$$= \sqrt{2}$$

53.
$$b = \sqrt{c^2 - a^2}$$
$$= \sqrt{(5.70)^2 - (3.42)^2}$$
$$= \sqrt{20.7936}$$
$$= 4.56$$

$$\sin A = \frac{a}{c} = \frac{3.42}{5.70} = 0.60$$
$$\cos A = \frac{b}{c} = \frac{4.56}{5.70} = 0.80$$
$$\sin B = \frac{b}{c} = \frac{4.56}{5.70} = 0.80$$
$$\cos B = \frac{a}{c} = \frac{3.42}{5.70} = 0.60$$

55.

$$c = \sqrt{a^2 + b^2}$$

$$= \sqrt{(19.44)^2 + (5.67)^2}$$

$$= \sqrt{410.0625}$$

$$= 20.25$$

$$\sin A = \frac{a}{c} = \frac{19.44}{20.25} = 0.96$$

$$\cos A = \frac{b}{c} = \frac{5.67}{20.25} = 0.28$$

$$\sin B = \frac{b}{c} = \frac{5.67}{20.25} = 0.28$$

$$\cos B = \frac{a}{c} = \frac{19.44}{20.25} = 0.96$$

57.

$$CH = \sqrt{(CD^2) + (DH)^2}$$

$$= \sqrt{5^2 + 5^2}$$

$$= \sqrt{25 + 25}$$

$$= \sqrt{50} = 5\sqrt{2}$$

$$CF = \sqrt{(CH)^2 + (FH)^2}$$

$$= \sqrt{(5\sqrt{2})^2 + (5)^2}$$

$$= \sqrt{50 + 25}$$

$$= \sqrt{75} = 5\sqrt{3}$$

$$\sin \theta = \frac{FH}{CF}$$

$$= \frac{5}{5\sqrt{3}}$$

$$= \frac{1}{\sqrt{3}}$$

$$\cos \theta = \frac{CH}{CF}$$

$$= \frac{5\sqrt{2}}{5\sqrt{3}}$$

$$= \frac{\sqrt{2}}{\sqrt{3}} \text{ or } \frac{\sqrt{6}}{3}$$

59.

$$CH = \sqrt{(CD^2) + (DH)^2}$$

$$= \sqrt{x^2 + x^2}$$

$$= \sqrt{2x^2}$$

$$= x\sqrt{2}$$

$$CF = \sqrt{(CH)^2 + (FH)^2}$$

$$= \sqrt{(x\sqrt{2})^2 + x^2}$$

$$= \sqrt{2x^2 + x^2}$$

$$= \sqrt{3x^2} = x\sqrt{3}$$

$$\sin \theta = \frac{FH}{CF}$$

$$= \frac{x}{x\sqrt{3}}$$

$$= \frac{1}{\sqrt{3}}$$

$$\cos \theta = \frac{CH}{CF}$$

$$= \frac{x\sqrt{2}}{x\sqrt{3}}$$

$$= \frac{\sqrt{2}}{\sqrt{3}} \text{ or } \frac{\sqrt{6}}{3}$$

61.

$$r = \sqrt{(x_2 - x_1)^2 + (y_2 - y_1)^2}$$ Distance formula

$$= \sqrt{(5-2)^2 + (1-5)^2}$$ Substitute known values

$$= \sqrt{(3)^2 + (-4)^2}$$ Simplify

$$= \sqrt{9 + 16}$$

$$= \sqrt{25} = 5$$

63.

$$r = \sqrt{(x_2 - x_1)^2 + (y_2 - y_1)^2}$$ Distance formula

$$\sqrt{13} = \sqrt{(x-1)^2 + (2-5)^2}$$ Substitute known values

$$\sqrt{13} = \sqrt{(x-1)^2 + 9}$$ Simplify

$$13 = (x-1)^2 + 9$$ Square both sides

$$4 = (x-1)^2$$ Subtract 9 from both sides

$$\pm 2 = x - 1$$ Solve using the Square Root Method

$$x - 1 = 2 \text{ or } x - 1 = -2$$

$$x = 3 \text{ or } \quad x = -1$$

65.   If $x = 0$, then $y = 2(0) - 1$

$$y = -1$$

Therefore, the point $(0, -1)$ satisfies the equation.

If $x = 2$, then $y = 2(2) - 1$

$$y = 4 - 1$$

$$y = 3$$

Therefore, the point $(2, 3)$ satisfies the equation.

Plot the points $(0, -1)$ and $(2, 3)$ and draw the line through these points.

67.   The terminal side is the line $y = x$. Some points in quadrant I on the line $y = x$ are $(1,1)$, $(2,2)$, and $(3,3)$.

69. $-135° + 360° = 225°$

71. $-300° + 360° = 60°$

Problem Set 2.2

1.

$$37° \, 45'$$
$$+26° \, 24'$$
$$\overline{63° \, 69'} = 64° \, 9' \text{ since } 60' = 1°$$

3.

$$51° \, 55'$$
$$+37° \, 45'$$
$$\overline{88° \, 100'} = 89° \, 40'$$

5.

$$61° \, 33'$$
$$+45° \, 16'$$
$$\overline{106° \, 49'}$$

7.

$$90° \qquad = \qquad 89° \, 60'$$
$$-34° \, 12' \qquad \quad -34° \, 12'$$
$$\overline{\qquad\qquad\qquad 55° \, 48'}$$

9.

$$180° \qquad = \qquad 179° \, 60' \qquad \text{Change } 1° \text{ to } 60'$$
$$-120° \, 17' \qquad -120° \, 17'$$
$$\overline{\qquad\qquad\qquad 59° \, 43'}$$

11.

$$76° \, 24' \qquad = \qquad 75° \, 84'$$
$$-22° \, 34' \qquad \quad -22° \, 34'$$
$$\overline{\qquad\qquad\qquad 53° \, 50'}$$

13.

$$70° \, 40' \qquad = \qquad 69° \, 100' \qquad \text{Change } 1° \text{ to } 60'$$
$$-30° \, 50' \qquad \quad -30° \, 50'$$
$$\overline{\qquad\qquad\qquad 39° \, 50'}$$

15.

$$35.4° = 35° + 0.4(60)'$$
$$= 35° + 24'$$
$$= 35° \, 24'$$

17.

$$16.25° = 16° + 0.25(60)'$$
$$= 16° + 15'$$
$$= 16° \, 15'$$

19.

$$92.55° = 92° + 0.55(60)'$$
$$= 92° + 33'$$
$$= 92° \, 33'$$

21.

$$19.9° = 19° + 0.9(60)'$$
$$= 19° + 54'$$
$$= 19° 54'$$

23.

$$45° 12' = 45 + \frac{12}{60}$$
$$= 45.2°$$

25.

$$62° \, 36' = 62 + \frac{36}{60}$$
$$= 62.6°$$

27.

$$17° \, 20' = 17 + \frac{20}{60}$$
$$= 17.33°$$

29.
$$48° \, 27' = 48 + \frac{27}{60}$$
$$= 48.45°$$

31. Scientific Calculator: 27.2 $\boxed{\sin}$

Graphing Calculator: $\boxed{\sin}$ $\boxed{(}$ 27.2 $\boxed{)}$ $\boxed{\text{ENTER}}$

Answer to 4 places: 0.4571

33. Scientific Calculator: 18 $\boxed{\cos}$

Graphing Calculator: $\boxed{\cos}$ $\boxed{(}$ 18 $\boxed{)}$ $\boxed{\text{ENTER}}$

Answer to 4 places: 0.9511

35. Scientific Calculator: 87.32 $\boxed{\tan}$

Graphing Calculator: $\boxed{\tan}$ $\boxed{(}$ 87.32 $\boxed{)}$ $\boxed{\text{ENTER}}$

Answer to 4 places: 21.3634

37.
$$\cot 31° = \frac{1}{\tan 31°}$$

Scientific Calculator: 31 $\boxed{\tan}$ $\boxed{1/x}$

Graphing Calculator: $\boxed{\tan}$ $\boxed{(}$ 31 $\boxed{)}$ $\boxed{x^{-1}}$ $\boxed{\text{ENTER}}$

Answer: 1.6643

39.
$$\sec 48.2° = \frac{1}{\cos 48.2°}$$

Scientific Calculator: 48.2 $\boxed{\cos}$ $\boxed{1/x}$

Graphing Calculator: $\boxed{\cos}$ $\boxed{(}$ 48.2 $\boxed{)}$ $\boxed{x^{-1}}$

Answer: 1.5003

41.
$$\csc 14.15° = \frac{1}{\sin 14.15°}$$

Scientific Calculator: 14.15 $\boxed{\sin}$ $\boxed{1/x}$

Graphing Calculator: $\boxed{\sin}$ $\boxed{(}$ 14.15 $\boxed{)}$ $\boxed{x^{-1}}$ $\boxed{\text{ENTER}}$

Answer: 4.0906

43.
$$24° \, 30' = 24 + \frac{30}{60} = 24.5°$$

Scientific Calculator: 24.5 $\boxed{\cos}$

Graphing Calculator: $\boxed{\cos}$ $\boxed{(}$ 24.5 $\boxed{)}$ $\boxed{\text{ENTER}}$

Answer: 0.9100

45. $42° 15' = 42 + \dfrac{15}{60}$

$\qquad\quad = 42.25°$

Scientific Calculator: 42.25 $\boxed{\text{tan}}$

Graphing Calculator: $\boxed{\text{tan}}$ $\boxed{(}$ 42.25 $\boxed{)}$ $\boxed{\text{ENTER}}$

Answer: 0.9083

47. $56° 40' = 56 + \dfrac{40}{60} = 56.67°$

Scientific Calculator: 56.67 $\boxed{\text{sin}}$

Graphing Calculator: $\boxed{\text{sin}}$ $\boxed{(}$ 56.67 $\boxed{)}$ $\boxed{\text{ENTER}}$

Answer: 0.8355

49. $45° 54' = 45 + \dfrac{54}{60}$

$\qquad\quad = 45.9°$

$\sec 45.9° = \dfrac{1}{\cos 45.9°}$

Scientific Calculator: 45.9 $\boxed{\text{cos}}$ $\boxed{1/x}$

Graphing Calculator: $\boxed{\text{cos}}$ $\boxed{(}$ 45.9 $\boxed{)}$ $\boxed{x^{-1}}$ $\boxed{\text{ENTER}}$

Answer: 1.4370

51.

$\sin 0° = 0 \qquad\qquad\qquad \csc 0° = $ undefined

$\sin 30° = 0.5 \qquad\qquad\quad \csc 30° = 2$

$\sin 45° = 0.7071 \qquad\qquad \csc 45° = 1.4142$

$\sin 60° = 0.8660 \qquad\qquad \csc 60° = 1.1547$

$\sin 90° = 1 \qquad\qquad\qquad \csc 90° = 1$

53.

$\sin 0° = 0$	$\cos 0° = 1$	$\tan 0° = 0$
$\sin 15° = 0.2588$	$\cos 15° = 0.9659$	$\tan 15° = 0.2679$
$\sin 30° = 0.5$	$\cos 30° = 0.8660$	$\tan 30° = 0.5774$
$\sin 45° = 0.7071$	$\cos 45° = 0.7071$	$\tan 45° = 1$
$\sin 60° = 0.8660$	$\cos 60° = 0.5$	$\tan 60° = 1.7321$
$\sin 75° = 0.9659$	$\cos 75° = 0.2588$	$\tan 75° = 3.7321$
$\sin 90° = 1$	$\cos 90° = 0$	$\tan 90°$ is undefined (error on calculator)

55. Scientific Calculator: 0.9770 [inv] [cos]

Graphing Calculator: [2nd] [cos] [(] 0.9770 [)] [ENTER]

Answer: 12.3°

57. Scientific Calculator: 0.6873 [inv] [tan]

Graphing Calculator: [2nd] [tan] [(] 0.6873 [)] [ENTER]

Answer: 34.5°

59. Scientific Calculator: 0.9813 [inv] [sin]

Graphing Calculator: [2nd] [sin] [(] 0.9813 [)] [ENTER]

Answer: 78.9°

61. $\sec\theta = 1.0191$

$\dfrac{1}{\cos\theta} = 1.0191$

$\cos\theta = \dfrac{1}{1.0191}$

Scientific Calculator: 1 [÷] 1.0191 [=] [inv] [cos]

Graphing Calculator: [2nd] [cos] [(] 1 [÷] 1.0191 [)] [ENTER]

Answer: 11.1°

63. $\csc\theta = 1.8214$

$\dfrac{1}{\sin\theta} = 1.8214$

$\sin\theta = \dfrac{1}{1.8214}$

Scientific Calculator: 1 [÷] 1.8214 [=] [inv] [sin]

Graphing Calculator: [2nd] [sin] [(] 1 [÷] 1.8214 [)] [ENTER]

Answer: 33.3°

65. $\cot\theta = 0.6873$

$\dfrac{1}{\tan\theta} = 0.6873$

$\tan\theta = \dfrac{1}{0.6873}$

Scientific Calculator: 1 [÷] 0.6873 [=] [inv] [tan]

Graphing Calculator: [2nd] [tan] [(] 1 [÷] 0.6873 [)] [ENTER]

Answer: 55.5°

67. Scientific Calculator: 0.7038 [inv] [sin]

Answer in decimal degrees is 44.733°

Convert the decimal part to minutes: 0.733 [x] 60 [=]

Graphing Calculator: [2nd] [sin] [(] 0.7038 [)] [2nd] [APPS] [▷ DMS] [ENTER]

Answer: $\theta = 44°44'$

69.   Scientific Calculator: 0.4112 $\boxed{\text{inv}}$ $\boxed{\cos}$

Answer in decimal degrees is 65.719°

Convert the decimal part to minutes: 0.719 $\boxed{\times}$ 60 $\boxed{=}$

Graphing Calculator: $\boxed{\text{2nd}}$ $\boxed{\cos}$ $\boxed{(}$ 0.4112 $\boxed{)}$ $\boxed{\text{2nd}}$ $\boxed{\text{APPS}}$ $\boxed{\triangleright \text{DMS}}$ $\boxed{\text{ENTER}}$

Answer: $\theta = 65°43'$

71. $\cot\theta = 5.5764$

$\dfrac{1}{\tan\theta} = 5.5764$

$\tan\theta = \dfrac{1}{5.5764}$

Scientific Calculator: 1 $\boxed{\div}$ 5.5764 $\boxed{=}$ $\boxed{\text{inv}}$ $\boxed{\tan}$

Answer in decimal degrees is 10.1666°

Convert the decimal part to minutes: 0.1666 $\boxed{\times}$ 60 $\boxed{=}$

Graphing Calculator: $\boxed{\text{2nd}}$ $\boxed{\tan}$ $\boxed{(}$ 5.5764 $\boxed{)}$ $\boxed{\text{2nd}}$ $\boxed{\text{APPS}}$ $\boxed{\triangleright \text{DMS}}$ $\boxed{\text{ENTER}}$

Answer: $\theta = 10°10'$

73. $\sec\theta = 1.0129$

$\dfrac{1}{\cos\theta} = 1.0129$

$\cos\theta = \dfrac{1}{1.0129}$

Scientific Calculator: 1 $\boxed{\div}$ 1.0129 $\boxed{=}$ $\boxed{\text{inv}}$ $\boxed{\cos}$

Answer in decimal degrees is 9.154°

Convert the decimal part to minutes: 0.154 $\boxed{\times}$ 60 $\boxed{=}$

Graphing Calculator: $\boxed{\text{2nd}}$ $\boxed{\cos}$ $\boxed{(}$ 1 $\boxed{\div}$ 1.0129 $\boxed{)}$ $\boxed{\text{2nd}}$ $\boxed{\text{APPS}}$ $\boxed{\triangleright \text{DMS}}$ $\boxed{\text{ENTER}}$

Answer: $\theta = 9°9'$

75.   Scientific Calculator: 23 $\boxed{\sin}$ and   67 $\boxed{\cos}$

Graphing Calculator: $\boxed{\sin}$ $\boxed{(}$ 23 $\boxed{)}$ $\boxed{\text{ENTER}}$ $\boxed{\cos}$ $\boxed{(}$ 67 $\boxed{)}$ $\boxed{\text{ENTER}}$

Both answers should be 0.3907.

77.   To calculate $\sec 34.5° = \dfrac{1}{\cos 34.5°}$:

Scientific Calculator: 34.5 $\boxed{\cos}$ $\boxed{1/x}$

Graphing Calculaltor: $\boxed{\cos}$ $\boxed{(}$ 34.5 $\boxed{)}$ $\boxed{x^{-1}}$ $\boxed{\text{ENTER}}$

This problem is continued on the next page

To calculate $\csc 55.5° = \dfrac{1}{\sin 55.5°}$:

Scientific Calculator: 55.5 $\boxed{\sin}$ $\boxed{1/\text{x}}$

Graphing Calculaltor: $\boxed{\sin}$ $\boxed{(}$ 55.5 $\boxed{)}$ $\boxed{\text{x}^{-1}}$ $\boxed{\text{ENTER}}$

Both answers should be 1.2134.

79.   Scientific Calculator: 4.5 $\boxed{\tan}$

        Graphing Calculaltor: $\boxed{\tan}$ $\boxed{(}$ 4.5 $\boxed{)}$ $\boxed{\text{ENTER}}$

To calculate $\cot 85.5° = \dfrac{1}{\tan 85.5°}$:

Scientific Calculator: 85.5 $\boxed{\tan}$ $\boxed{1/\text{x}}$

Graphing Calculaltor: $\boxed{\tan}$ $\boxed{(}$ 85.5 $\boxed{)}$ $\boxed{\text{x}^{-1}}$ $\boxed{\text{ENTER}}$

Both answers should be 0.0787.

81.   Scientific Calculator: 37 $\boxed{\cos}$ $\boxed{\text{x}^2}$ $\boxed{+}$ 37 $\boxed{\sin}$ $\boxed{\text{x}^2}$ $\boxed{=}$

        Graphing Calculator: $\boxed{\cos}$ $\boxed{(}$ 37 $\boxed{)}$ $\boxed{\text{x}^2}$ $\boxed{+}$ $\boxed{\sin}$ $\boxed{(}$ 37 $\boxed{)}$ $\boxed{\text{x}^2}$ $\boxed{\text{ENTER}}$

83.   Scientific Calculator: 10 $\boxed{\sin}$ $\boxed{\text{x}^2}$ $\boxed{+}$ 10 $\boxed{\cos}$ $\boxed{\text{x}^2}$ $\boxed{=}$

        Graphing Calculator: $\boxed{\sin}$ $\boxed{(}$ 10 $\boxed{)}$ $\boxed{\text{x}^2}$ $\boxed{+}$ $\boxed{\cos}$ $\boxed{(}$ 10 $\boxed{)}$ $\boxed{\text{x}^2}$ $\boxed{\text{ENTER}}$

85.   Scientific Calculator: 1.234 $\boxed{\text{inv}}$ $\boxed{\sin}$

        Graphing Calculator: $\boxed{\text{2nd}}$ $\boxed{\sin}$ 1.234 $\boxed{\text{ENTER}}$

        You should get an error message. The sine of an angle can never be greater than 1.

91. $(x, y) = (3, -2)$ $\sin\theta = \dfrac{y}{r} = -\dfrac{2}{\sqrt{13}}$

        $x = 3$ and $y = -2$ $\cos\theta = \dfrac{x}{r} = \dfrac{3}{\sqrt{13}}$

        $r = \sqrt{3^2 + (-2)^2}$ $\tan\theta = \dfrac{y}{x} = -\dfrac{2}{3}$

            $= \sqrt{9 + 4}$

            $= \sqrt{13}$

93. A point on the terminal side of an angle of 90° in standard position is (0, 1), where $x = 0, y = 1$, and $r = 1$.

$$\sin 90° = \frac{y}{r} = \frac{1}{1} = 1$$

$$\cos 90° = \frac{x}{r} = \frac{0}{1} = 0$$

$$\tan 90° = \frac{y}{x} = \frac{1}{0} \text{ (undefined)}$$

95. $\cos \theta = -\frac{5}{13}$ and θ is in QIII. In QIII, both x and y are negative.

$$\cos \theta = \frac{x}{r} = \frac{-5}{13}$$

$x = -5$ and $r = 13$

$\qquad\qquad\qquad\qquad\qquad\qquad\qquad\qquad$ $\sin \theta = \frac{y}{r} = -\frac{12}{13}$

$$x^2 + y^2 = r^2$$

$\qquad\qquad\qquad\qquad\qquad\qquad\qquad\qquad$ $\tan \theta = \frac{y}{x} = \frac{-12}{-5} = \frac{12}{5}$

$$(-5)^2 + y^2 = 13^2$$

$\qquad\qquad\qquad\qquad\qquad\qquad\qquad\qquad$ $\cot \theta = \frac{x}{y} = \frac{-5}{-12} = \frac{5}{12}$

$$25 + y^2 = 169$$

$\qquad\qquad\qquad\qquad\qquad\qquad\qquad\qquad$ $\sec \theta = \frac{r}{x} = -\frac{13}{5}$

$$y^2 = 144$$

$\qquad\qquad\qquad\qquad\qquad\qquad\qquad\qquad$ $\csc \theta = \frac{r}{y} = -\frac{13}{12}$

$$y = \pm 12$$

$y = -12$ because θ is in QIII

97. The $\sin \theta$ is positive in QI and QII.
The $\cos \theta$ is negative in QII and QIII.
Therefore, θ must lie in QII.

Problem Set 2.3

1.

$\sin 42° = \dfrac{a}{15}$ $\qquad\qquad$ Sine relationship

$\qquad a = 15 \sin 42°$ $\qquad\qquad$ Multiply both sides by 15

$\qquad\quad = 15(0.6691)$ $\qquad\quad$ Substitute value for $\sin 42°$

$\qquad\quad = 10 \text{ ft}$ $\qquad\qquad\quad$ Answer rounded to 2 significant digits

3.
$$\sin 34° = \frac{22}{c}$$ Sine relationship

$$c \sin 34° = 22$$ Multiply both sides by c

$$c = \frac{22}{\sin 34°}$$ Divide both sides by $\sin 34°$

$$c = \frac{22}{0.5592}$$ Substitute value for $\sin 34°$

$$c = 39\,\text{m}$$ Answer rounded to 2 significant digits

5.
$$\cos 24.5° = \frac{a}{2.34}$$ Cosine relationship

$$a = 2.34 \sin 24.5°$$ Multiply both sides by 2.34

$$= 2.34(0.9099)$$ Substitute value for $\cos 24.5°$

$$= 2.13\,\text{ft}$$ Answer rounded to 3 significant digits

7.
$$\tan 55.33° = \frac{12.34}{a}$$ Tangent relationship

$$a \tan 55.33° = 12.34$$ Multiply both sides by a

$$a = \frac{12.34}{\tan 55.33°}$$ Divide both sides by $\tan 55.33°$

$$a = \frac{12.34}{1.4458}$$ Substitute value for $\tan 55.33°$

$$a = 8.535\ \text{yd}$$ Answer rounded to 4 significant digits

9.
$$\tan A = \frac{16}{26}$$ Tangent relationship

$$= 0.6153$$ Divide 16 by 26

$$A = \tan^{-1}(0.6153)$$ Use calculator to find angle

$$A = 32°$$ Answer rounded to the nearest degree

11.
$$\cos A = \frac{6.7}{7.7}$$ Cosine relationship

$$= 0.8701$$ Divide 6.7 by 7.7

$$A = \cos^{-1}(0.8701)$$ Use calculator to find angle

$$= 30°$$ Answer rounded to the nearest degree

13.

$$\cos B = \frac{23.32}{45.54}$$ Cosine relationship

$$= 0.5120$$ Divide 23.32 by 45.54

$$B = \cos^{-1}(0.5120)$$ Use calculator to find angle

$$= 59.20°$$ Answer rounded to the nearest hundredth of a degree

15. First, we find $\angle B$:

$$\angle B = 90° - \angle A$$

$$= 90° - 25°$$

$$= 65°$$

Next, we find side a:

$$\sin 25° = \frac{a}{24}$$ Sine relationship

$$a = 24 \sin 25°$$ Multiply both sides by 24

$$a = 10\,\text{m}$$ Answer rounded to 2 significant digits

Last, we find side b:

$$\cos 25° = \frac{b}{24}$$ Cosine relationship

$$b = 24 \sin 25°$$ Multiply both sides by 24

$$b = 22\,\text{m}$$ Answer rounded to 2 significant digits

17. First, we find $\angle B$:

$$\angle B = 90° - \angle A$$

$$= 90° - 32.6°$$

$$= 57.4°$$

Next, we find side c:

$$\sin 32.6° = \frac{43.4}{c}$$ Sine relationship

$$c = \frac{43.4}{\sin 32.6°}$$ Multiply both sides by c then divide by $\sin 32.6°$

$$= 80.6\,\text{in}$$ Answer rounded to 3 significant digits

Last, we find side b:

$$\tan 57.4° = \frac{b}{43.4}$$ Tangent relationship

$$b = 43.4 \tan 57.4°$$ Multiply both sides by 43.4

$$= 67.9\,\text{in}$$ Answer rounded to 2 significant digits

19. First, we find $\angle B$:

$$\angle B = 90° - \angle A$$

$$= 90° - 10°42'$$

$$= 79°18'$$

This problem is continued on the next page

Next, we find side a:

$$\tan 10°42' = \frac{a}{5.932}$$
Tangent relationship

$$\tan 10.7° = \frac{a}{5.932}$$
Change angle to decimal degrees

$$a = 5.932 \tan 10.7°$$
Multiply both sides by 5.932

$$a = 1.121 \, \text{cm}$$
Answer rounded to 4 significant digits

Last, we find side c:

$$\cos 10.7° = \frac{5.932}{c}$$
Cosine relationship

$$c = \frac{5.932}{\cos 10.7°}$$
Multiply both sides by c then divide by $\cos 10.7°$

$$c = 6.037 \, \text{cm}$$
Answer rounded to 4 significant digits

21. First, we find $\angle A$:

$$\angle A = 90° - 76°$$
$$= 14°$$

Next, we find side a:

$$\cos 76° = \frac{a}{5.8}$$
Cosine relationship

$$a = 5.8 \cos 76°$$
Multiply both sides by 5.8

$$= 1.4 \, \text{ft}$$
Answer rounded to 2 significant digits

Last, we find side b:

$$\sin 76° = \frac{b}{5.8}$$
Sine relationship

$$b = 5.8 \sin 76°$$
Multiply both sides by 5.8

$$= 5.6 \, \text{ft}$$
Answer rounded to 2 significant digits

23. First, we find $\angle A$:

$$\angle A = 90° - \angle B$$
$$= 90° - 26°30'$$
$$= 63°30'$$

Next, we find side a:

$$\tan 26°30' = \frac{324}{a}$$
Tangent relationship

$$\tan 26.5° = \frac{324}{a}$$
Change angle to decimal degrees

$$a = \frac{324}{\tan 26.5°}$$
Multiply both sides by a then divide by $\tan 26.5°$

$$a = 650 \, \text{mm}$$
Answer rounded to 3 significant digits

This problem is continued on the next page

Last, we find side c:

$$\sin 26.5° = \frac{324}{c}$$

Sine relationship

$$c = \frac{324}{\sin 26.5°}$$

Multiply both sides by c then divide by $\sin 26.5°$

$$= 726 \text{ mm}$$

Answer rounded to 3 significant digits

25. First, we find $\angle A$:

$$\angle A = 90° - 23.45°$$
$$= 66.55°$$

Next, we find side b:

$$\tan 23.45° = \frac{b}{5.432}$$

Tangent relationship

$$b = 5.432 \tan 23.45°$$

Multiply both sides by 5.432

$$= 2.356 \text{ mi}$$

Answer rounded to 4 significant digits

Last, we find side c:

$$\cos 23.45° = \frac{5.432}{c}$$

Cosine relationship

$$c = \frac{5.432}{\cos 23.45°}$$

Multiply both sides by c and then divide by $\cos 23.45°$

$$= 5.921 \text{ mi}$$

Answer rounded to 4 significant digits

27. First, we find $\angle A$:

$$\tan A = \frac{37}{87}$$

Tangent relationship

$$= 0.4253$$

Divide 37 by 87

$$A = \tan^{-1}(0.4253)$$

Use calculator to find angle

$$= 23°$$

Answer rounded to nearest degree

Next, we find $\angle B$:

$$\angle B = 90° - \angle A$$
$$= 90° - 23° = 67°$$

Last, we find c:

$$c^2 = 37^2 + 87^2$$

Pythagorean Theorem

$$= 1369 + 7569$$

Simplify

$$= 8938$$

Simplify

$$c = \pm 95$$

Take square root of both sides

$$= 95 \text{ ft}$$

c must be positive

29. First, we find $\angle A$:

$\sin A = \dfrac{2.75}{4.05}$ Sine relationship

$\quad\quad = 0.6790$ Divide 2.75 by 4.05

$\quad A = \sin^{-1}(0.6790)$ Use calculator to find angle

$\quad A = 42.8°$ Answer rounded to nearest degree

Next, we find $\angle B$:

$\angle B = 90° - 42.8°$

$\quad\quad = 47.2°$

Last, we find side b:

$b^2 + (2.75)^2 = (4.05)^2$ Pythagorean Theorem

$b^2 + 7.5625 = 16.4025$ Simplify

$b^2 = 8.84$ Subtract 7.5625 from both sides

$b = \pm 2.97$ Take square root of both sides

$\quad = 2.97 \text{ cm}$ b must be positive

31. First, we find $\angle A$:

$\cos A = \dfrac{12.21}{25.52}$ Cosine relationship

$\quad\quad = 0.4784$ Divide 12.21 by 25.52

$\quad A = \cos^{-1}(0.4784)$ Use calculator to find angle

$\quad\quad = 61.42°$ Answer rounded to nearest hundredth of a degree

Next, we find $\angle B$:

$\angle B = 90° - \angle A$

$\quad\quad = 90° - 61.42°$

$\quad\quad = 28.58°$

Last, we find side a:

$a^2 + 12.21^2 = 25.52^2$ Pythagorean Theorem

$a^2 + 149.0841 = 651.2704$ Simplify

$a^2 = 502.1863$ Subtract 149.0841 from both sides

$a = \pm 22.41$ Take square root of both sides

$\quad = 22.41 \text{ in}$ a must be positive

33. Using $\triangle BCD$, we find BD:

$\sin 30° = \dfrac{BD}{6}$ Sine relationship

$BD = 6\sin 30°$ Multiply both sides by 6

$\quad = 3$ Exact answer

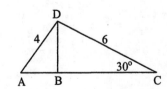

This problem is continued on the next page

Next, we find $\angle A$

$\sin A = \dfrac{3}{4}$	Sine relationship
$= 0.75$	Divide 3 by 4
$A = \sin^{-1}(0.75)$	Use calculator to find angle
$A = 49°$	Answer rounded to the nearest degree

35.

$\sin 31° = \dfrac{12}{x+12}$	Sine relationship
$(x+12)\sin 31° = 12$	Multiply both sides by $x + 12$
$x+12 = \dfrac{12}{\sin 31°}$	Divide both sides by $\sin 31°$
$x = \dfrac{12}{\sin 31°} - 12$	Subtract 12 from both sides
$x = 11$	Answer rounded to 2 significant digits

37.

$\sin 45° = \dfrac{r}{r+15}$	Sine relationship
$r = (r+15)\sin 45°$	Multiply both side by $r +15$
$r = r\sin 45° + 15\sin 45°$	Use distributive property
$r - r\sin 45° = 15\sin 45°$	Subtract $r\sin 45°$ from both sides
$r(1-\sin 45°) = 15\sin 45°$	Factor left side
$r = \dfrac{15\sin 45°}{1-\sin 45°}$	Divide both sides by $1-\sin 45°$
$= 36$	Answer rounded to 2 significant digits

39. Using $\triangle ABC$, we find side x:

$\tan 62° = \dfrac{x}{42}$	Tangent relationship
$x = 42\tan 62°$	Multiply both sides by 42
$= 79$	Answer rounded to 2 significant digits

Next, using $\triangle ABD$, we find side h:

$\tan 27° = \dfrac{h}{x}$	Tangent relationship
$= \dfrac{h}{79}$	Substitute value for x
$h = 79\tan 27°$	Multiply both sides by 79
$h = 40$	Answer rounded to 2 significant digits

41. Using $\triangle ABC$, we find side x:

$$\sin 41° = \frac{x}{32}$$ Sine relationship

$$x = 32 \sin 41°$$ Multiply both sides by 32

$$= 21$$ Answer rounded to 2 significant digits

Next, using $\triangle ABD$, we find $\angle ABD$:

$$\tan \angle ABD = \frac{h}{x}$$ Tangent relationship

$$= \frac{19}{21}$$ Substitute known values

$$= 0.9047$$ Divide 19 by 21

$$\angle ABD = \tan^{-1}(0.9047)$$ Use calculator to find angle

$$\angle ABD = 42°$$ Answer rounded to the nearest degree

43. Using $\triangle BCD$, we find side x:

$$\cos 58° = \frac{x}{14}$$ Cosine relationship

$$x = 14 \cos 58°$$ Multiply both sides by 14

$$x = 7.4$$ Answer rounded to 2 significant digits

Next, using $\triangle ABC$, we find y:

$$\cos 41° = \frac{x+y}{18}$$ Cosine relationship

$$x + y = 18 \cos 41°$$ Multiply both sides by 18

$$x + y = 13.58$$ Evaluate right side

$$7.4 + y = 13.58$$ Substitute value for x

$$y = 6.18$$ Subtract 7.4 from both sides

$$y = 6.2$$ Answer rounded to 2 significant digits

45. Using $\triangle ABC$, we find side h:

$$\sin 41° = \frac{h}{28}$$ Sine relationship

$$h = 28 \sin 41°$$ Multiply both sides by 28

$$= 18$$ Answer rounded to 2 significant digits

Next, using $\triangle BCD$, we find side x:

$$\tan 58° = \frac{h}{x}$$ Tangent relationship

$$\tan 58° = \frac{18}{x}$$ Substitute value found for h

$$x = \frac{18}{\tan 58°} = 11$$ Solve for x and round to 2 significant digits

47. Since h is in both $\triangle ABC$ and $\triangle BCD$, we will solve for h in the two triangles:

In $\triangle BCD$, $\tan 57^\circ = \dfrac{h}{x}$ Tangent relationship

$h = x \tan 57^\circ$ Multiply both sides by x

In $\triangle ABC$, $\tan 43^\circ = \dfrac{h}{x+y}$ Tangent relationship

$h = (x+y)\tan 43^\circ$ Multiply both sides by $x + y$

$h = (x+10)\tan 43^\circ$ Substitute value for y

Therefore, $x \tan 57^\circ = (x+10)\tan 43^\circ$ $h = h$

$x \tan 57^\circ = x \tan 43 + 10 \tan 43^\circ$ Distribution Property

$x \tan 57^\circ - x \tan 43^\circ = 10 \tan 43^\circ$ Subtract $x \tan 43^\circ$ from both sides

$x\left(\tan 57^\circ - \tan 43^\circ\right) = 10 \tan 43^\circ$ Factor left side

$x = \dfrac{10 \tan 43^\circ}{\tan 57^\circ - \tan 43^\circ}$ Divide both sides by $\tan 57^\circ - \tan 43^\circ$

$= 15$ Answer rounded to 2 significant digits

49. From Problem 57 in Problem Set 2.1, we found that

$\sin \theta = \dfrac{1}{\sqrt{3}}$

$= 0.5774$

$\theta = \sin^{-1}(0.5774)$

$= 35.3^\circ$

51. From Problem 57 in Problem Set 2.1, we found that

$\cos \theta = \dfrac{\sqrt{2}}{\sqrt{3}}$

$= 0.8165$

$\theta = \cos^{-1}(0.8165)$

$= 35.3^\circ$

53. $\cos 30^\circ = \dfrac{x}{125} = \dfrac{139 - h}{125}$

$125 \cos 30^\circ = 139 - h$

$h = 139 - 125 \cos 30^\circ$ Solve for h

$= 139 - 108.25$

$= 30.7 \text{ ft}$ Round to 3 significant digits

55.

$$\cos 120° = \frac{x}{125} = \frac{139 - h}{125}$$

$$125 \cos 120° = 139 - h$$

$$h = 139 - 125 \cos 120° \qquad \text{Solve for } h$$

$$= 139 - (-62.5)$$

$$= 201.5 \text{ ft}$$

57. $r = 98.5$

a. $h = 12 + 98.5 + x$

$$\cos 60° = \frac{x}{98.5}$$

$$x = 98.5 \cos 60°$$

$$= 49.25$$

$$h = 12 + 98.5 + 49.25$$

$$= 159.8 \text{ ft}$$

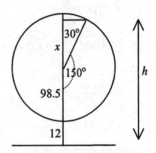

b. $h = 12 + 98.5 + x$

$$\cos 30° = \frac{x}{98.5}$$

$$x = 98.5 \cos 30°$$

$$= 85.3$$

$$h = 12 + 98.5 + 85.3$$

$$= 195.8 \text{ ft}$$

c. $r + 12 = 98.5 + 12 = 110.5$

$$h = 110.5 - x$$

$$\cos 45° = \frac{x}{98.5}$$

$$x = 98.5 \cos 45°$$

$$= 69.7$$

$$h = 110.5 - 69.7$$

$$= 40.8 \text{ ft}$$

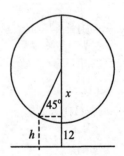

59. $\sec \theta = 2$

$$\cos \theta = \frac{1}{\sec \theta} \qquad \text{Reciprocal identity}$$

$$= \frac{1}{2} \qquad \text{Substitute known value}$$

$$\cos^2 \theta = \left(\frac{1}{2}\right)^2 = \frac{1}{4} \qquad \text{Square both sides}$$

61. $\cos\theta = \sqrt{1-\sin^2\theta}$ $\qquad\qquad$ θ is in QI

$$= \sqrt{1-\left(\frac{1}{3}\right)^2}$$

$$= \sqrt{1-\frac{1}{9}}$$

$$= \sqrt{\frac{8}{9}}$$

$$= \frac{2\sqrt{2}}{3}$$

63. $\sin A = \pm\sqrt{1-\cos^2 A}$ $\qquad$ Pythagorean identity

$$= \pm\sqrt{1-\left(\frac{2}{5}\right)^2} \qquad \text{Substitute known value}$$

$$= \pm\sqrt{1-\frac{4}{25}} \qquad \text{Simplify}$$

$$= \pm\sqrt{\frac{21}{25}} = \pm\frac{\sqrt{21}}{5} \qquad \text{Simplify}$$

$$= -\frac{\sqrt{21}}{5} \qquad \text{The sine is negative in QIV}$$

65. $\cos\theta = -\sqrt{1-\sin^2\theta}$ $\qquad$ Pythagorean identity, θ in QII

$$= -\sqrt{1-\left(\frac{\sqrt{3}}{2}\right)^2} \qquad \text{Substitute known value}$$

$$= -\sqrt{1-\frac{3}{4}} \qquad \text{Simplify}$$

$$= -\sqrt{\frac{1}{4}} = -\frac{1}{2}$$

$\tan\theta = \dfrac{\sin\theta}{\cos\theta}$ $\quad$ Ratio identity $\qquad\qquad$ $\csc\theta = \dfrac{1}{\sin\theta}$ $\quad$ Reciprocal identity

$\quad= \dfrac{\sqrt{3}/2}{-1/2}$ $\quad$ Substitute known values $\qquad\quad = \dfrac{1}{\sqrt{3}/2} = \dfrac{2}{\sqrt{3}}$

$\quad= -\sqrt{3}$ $\quad$ Simplify

$\sec\theta = \dfrac{1}{\cos\theta}$ $\quad$ Reciprocal identity $\qquad\qquad$ $\cot\theta = \dfrac{1}{\tan\theta}$ $\quad$ Reciprocal identity

$\quad= \dfrac{1}{-1/2} = -2$ $\qquad\qquad\qquad\qquad\qquad\quad = \dfrac{1}{-\sqrt{3}} = -\dfrac{1}{\sqrt{3}}$

67.

$$\cos\theta = \frac{1}{\sec\theta} \qquad \text{Reciprocal identity}$$

$$= -\frac{1}{2} \qquad \text{Substitute known values}$$

$$\sin\theta = -\sqrt{1-\cos^2\theta} \qquad \text{Pythagorean identity, } \theta \text{ in QIII}$$

$$= -\sqrt{1-\left(-\frac{1}{2}\right)^2} \qquad \text{Substitute value for } \cos\theta$$

$$= -\sqrt{1-\frac{1}{4}} = -\sqrt{\frac{3}{4}} \qquad \text{Simplify}$$

$$= -\frac{\sqrt{3}}{2}$$

$$\tan\theta = \frac{\sin\theta}{\cos\theta} \qquad \text{Ratio identity}$$

$$= \frac{-\sqrt{3}/2}{-1/2} \qquad \text{Substitute values}$$

$$= \sqrt{3}$$

$$\cot\theta = \frac{1}{\tan\theta} \qquad \text{Reciprocal identity}$$

$$= \frac{1}{\sqrt{3}}$$

$$\csc\theta = \frac{1}{\sin\theta} \qquad \text{Reciprocal identity}$$

$$= \frac{1}{-\sqrt{3}/2} = -\frac{2}{\sqrt{3}}$$

Problem Set 2.4

1. To find the height, *h,* we can use the Pythagorean Theorem:

$$h^2 + (15)^2 = (42)^2$$
$$h^2 + 225 = 1,764$$
$$h^2 = 1,539$$
$$h = \pm\sqrt{1,539}$$
$$= 39 \text{ cm}$$

To find angle θ, we can use the cosine ratio:

$$\cos\theta = \frac{15}{42}$$
$$= 0.3571$$
$$\theta = \cos^{-1}(0.3571)$$
$$\theta = 69°$$

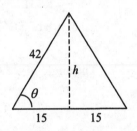

The height is 39 cm and the two equal angles are 69°.

3. Consider the right triangle with sides of 25.3 cm and 5.2 cm (one-half of the diameter):

$$\tan\theta = \frac{25.3}{5.2}$$
$$= 4.8654$$
$$\theta = \tan^{-1}(4.8654)$$
$$= 78.4°$$

The angle the side makes with the base is 78.4°.

5. To find the length of the escalator, x, we use the sine ratio:

$$\sin 33° = \frac{21}{x}$$

$$x = \frac{21}{\sin 33°}$$

$$= \frac{21}{0.5446}$$

$$= 39 \text{ ft}$$

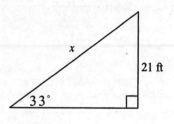

The length of the escalator is 39 feet.

7.
$$\sin \theta = \frac{43.2}{72.5}$$

$$= 0.5959$$

$$\theta = \sin^{-1}(0.5959)$$

$$= 36.6°$$

The angle the rope makes with the pole is $36.6°$

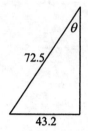

9. We use the tangent ratio to find the angle of elevation to the sun, θ:

$$\tan \theta = \frac{73.0}{51.0}$$

$$= 1.4313$$

$$\theta = \tan^{-1}(1.4313)$$

$$= 55.1°$$

The angle of elevation to the sun is $55.1°$.

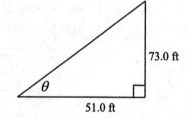

11.
$$\tan 11° = \frac{x}{150}$$

$$x = 150 \tan 11°$$

$$= 29 \text{ cm}$$

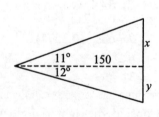

$$\tan 12° = \frac{y}{150}$$

$$y = 150 \tan 12°$$

$$= 32 \text{ cm}$$

The vertical dimension of the mirror is $x + y$ or 61 cm.

13. a.) horizontal distance $= 0.5(1,600) = 800$ ft

b.) vertical distance $=$ (number of contour intervals)(40)

$$= 5(40)$$
$$= 200 \text{ ft}$$

c.)
$$\tan\theta = \frac{\text{vertical distance}}{\text{horizontal distance}}$$
$$= \frac{200}{800}$$
$$= 0.25$$
$$\theta = \tan^{-1}(0.25)$$
$$= 14.0°$$

15.
$$\tan 59° = \frac{10}{y}$$

$$y = \frac{10}{\tan 59°}$$
$$= \frac{10}{1.6643}$$
$$= 6.0$$

$$\tan 47° = \frac{10}{\tan 59°}$$

$$x = y \tan 47°$$
$$= 6.0(1.0724)$$
$$= 6.4 \text{ ft}$$

The vertical dimension of the door is 6.4 feet.

17. We use the Pythagorean Theorem to find the distance x:

$$x^2 = 25^2 + 18^2$$
$$= 625 + 324$$
$$= 949$$
$$x = 31 \text{ mi}$$

We use the tangent relationship to find angle θ:

$$\tan\theta = \frac{18}{25}$$
$$= 0.72$$
$$\theta = \tan^{-1}(0.72)$$
$$= 36°$$

To find the bearing we add $42° + 36° = 78°$. The boat is 31 miles from the harbor entrance and its bearing is N 78° E.

19.

$$\tan 65° = \frac{x}{18}$$

$$x = 18 \tan 65°$$

$$= 18(2.1445)$$

$$= 39 \text{ mi}$$

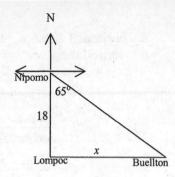

The distance from Lompoc to Buellton is 39 miles.

21. We will call the east distance, x and the north distance, y:

$$\sin 37° \, 10' = \frac{x}{79.5} \qquad \cos 37°10' = \frac{y}{79.5}$$

$$x = 79.5 \sin 37°10' \qquad y = 79.5 \cos 37°10'$$

$$= 48.0 \text{ mi} \qquad = 63.4 \text{ mi}$$

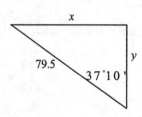

The boat has traveled 48.0 miles west and 63.4 miles north.

23. In $\triangle ABC$, $\tan 42.17° = \dfrac{h}{x+33}$

$$h = (x+33)\tan 42.17°$$

In $\triangle BCD$, $\tan 47.5° = \dfrac{h}{x}$

$$h = x \tan 47.5°$$

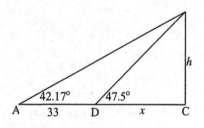

Therefore, $x \tan 47.5° = (x+33)\tan 42.17°$

$$x \tan 47.5° = x \tan 42.17° + 33 \tan 42.17°$$

$$x \tan 47.5° - x \tan 42.17° = 33 \tan 42.17°$$

$$x\left(\tan 47.5° - \tan 42.17°\right) = 33 \tan 42.17°$$

$$x = \frac{33 \tan 42.17°}{\tan 47.5° - \tan 42.17°}$$

$$= 161 \text{ ft}$$

The person at point A is 161 feet from the base of the antenna.

25.

$$\tan 86.6° = \frac{x}{24.8}$$

$$x = 24.8 \tan 86.6°$$

$$= 24.8(16.8319)$$

$$= 417.431$$

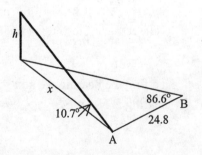

This problem is continued on the next page

$$\tan 10.7° = \frac{h}{x}$$
$$h = x \tan 10.7°$$
$$= (417.431)(0.18895)$$
$$= 78.9 \text{ ft}$$

The tree is 78.9 feet high.

27. First, we will find each person's distance from the pole, x, using the Pythagorean Theorem:
$$x^2 + x^2 = 25^2$$
$$2x^2 = 625$$
$$x^2 = 312.5$$
$$x = 17.678 \text{ ft}$$

Next, we will find the height of the pole, h, using the tangent relationship:
$$\tan 56° = \frac{h}{17.678}$$
$$h = 17.678 \tan 56°$$
$$= 26 \text{ ft}$$

The height of the pole is 26 feet.

29.
$$\sin 76.6° = \frac{r}{r + 112}$$
$$r = (r + 112)\sin 76.6°$$
$$r = r \sin 76.6° + 112 \sin 76.6°$$
$$r - r \sin 76.6° = 112 \sin 76.6°$$
$$r(1 - \sin 76.6°) = 112 \sin 76.6°$$
$$r = \frac{112 \sin 76.6°}{1 - \sin 76.6°}$$
$$= \frac{112(0.9728)}{1 - 0.9728}$$
$$= \frac{108.9509}{0.02722}$$
$$= 4,000 \text{ mi}$$
The radius of the earth is 4,000 miles.

31. We want to find x and y in terms of h

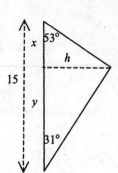

$$\tan 53° = \frac{h}{x} \qquad\qquad \tan 31° = \frac{h}{y}$$

$$x \tan 53° = h \qquad\qquad y \tan 31° = h$$

$$x = \frac{h}{\tan 53°} \qquad\qquad y = \frac{h}{\tan 31°}$$

We know that $x + y = 15$. Therefore,

$$\frac{h}{\tan 53°} + \frac{h}{\tan 31°} = 15$$

$$h\left(\frac{1}{\tan 53°} + \frac{1}{\tan 31°}\right) = 15$$

$$h(0.7536 + 1.6643) = 15$$

$$2.4179h = 15$$

$$h = \frac{15}{2.4179} = 6.2 \text{ mi}$$

The ship is 6.2 miles from the shore.

33.
$$\tan \theta_1 = \frac{1}{1} \qquad\qquad \tan \theta_2 = \frac{1}{\sqrt{2}} \qquad\qquad \tan \theta_3 = \frac{1}{\sqrt{3}}$$
$$= 1 \qquad\qquad\quad\; = 0.7071 \qquad\qquad\; = 0.5774$$
$$\theta_1 = \tan^{-1}(1) \qquad\quad \theta_2 = \tan^{-1}(0.7071) \qquad \theta_3 = \tan^{-1}(0.5774)$$
$$\theta_1 = 45.00° \qquad\qquad \theta_2 = 35.26° \qquad\qquad\quad \theta_3 = 30.00°$$

35.
$$(\sin \theta - \cos \theta)^2 = (\sin \theta - \cos \theta)(\sin \theta - \cos \theta)$$
$$= \sin^2 \theta - 2\sin \theta \cos \theta + \cos^2 \theta$$
$$= \sin^2 \theta + \cos^2 \theta - 2\sin \theta \cos \theta$$
$$= 1 - 2\sin \theta \cos \theta$$

37.
$$\sin \theta \cot \theta = \sin \theta \cdot \frac{\cos \theta}{\sin \theta} \qquad\qquad \text{Ratio identity}$$

$$= \frac{\sin \theta \cos \theta}{\sin \theta} \qquad\qquad\quad \text{Multiplication of fractions}$$

$$= \cos \theta \qquad\qquad\qquad\quad \text{Division of common factor}$$

39.
$$\frac{\sec\theta}{\tan\theta} = \frac{\dfrac{1}{\cos\theta}}{\dfrac{\sin\theta}{\cos\theta}}$$
Reciprocal and ratio identity

$$= \frac{1}{\cos\theta} \cdot \frac{\cos\theta}{\sin\theta}$$
Division of fractions

$$= \frac{1}{\sin\theta}$$
Multiplication of fractions and divide common factor

$$= \csc\theta$$

41.
$$\sec\theta - \cos\theta = \frac{1}{\cos\theta} - \cos\theta$$
Reciprocal identity

$$= \frac{1}{\cos\theta} - \cos\theta \cdot \frac{\cos\theta}{\cos\theta}$$
L.C.D. is $\cos\theta$

$$= \frac{1 - \cos^2\theta}{\cos\theta}$$
Subtraction of fractions

$$= \frac{\sin^2\theta}{\cos\theta}$$
Pythagorean identity

Problem Set 2.5

9. The first hour and a half, the distance traveled is
(22.0 mph)(1.5 hr) = 33 miles
The next two hours, the distance traveled is
(18.5 mph)(2 hr) = 37 miles

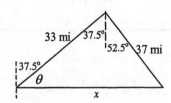

We will use the Pythagorean Theorem to find x:
$$x^2 = 33^2 + 37^2$$
$$x^2 = 1089 + 1369$$
$$x^2 = 2458$$
$$x = 49.6 \text{ mi}$$

We will use the tangent ratio to find θ and then add 37.5°:
$$\tan\theta = \frac{37}{33}$$
$$= 1.1212$$
$$\theta = \tan^{-1}(1.1212)$$
$$= 48.3° \qquad\qquad 48.3° + 37.5° = 85.8°$$

The balloon is 49.6 miles from its starting point. The bearing is N85.8° E.

11. To find the distance, x, the plane has flown off its course, we can use the sine ratio:

$$\sin 3° = \frac{x}{130}$$

$$x = 130 \sin 3°$$

$$= 6.80 \text{ miles}$$

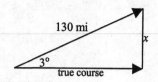

130 mi
x
3°
true course

13.

$$|V_x| = |V| \cos \theta$$
$$= 13.8 \cos 24.2°$$
$$= 12.6$$

$$|V_y| = |V| \sin \theta$$
$$= 13.8 \cos 24.2°$$
$$= 5.66$$

15.

$$|V_x| = |V| \cos \theta$$
$$= 420 \cos 36°10'$$
$$= 420 \cos 36.17°$$
$$= 420(0.8073)$$
$$= 339$$

$$|V_y| = |V| \sin \theta$$
$$= 420 \sin 36°10'$$
$$= 420 \cos 36.17°$$
$$= 420(0.5901)$$
$$= 248$$

17.

$$|V_x| = |V| \cos \theta$$
$$= 64 \cos 0°$$
$$= 64(1) = 64$$

$$|V_y| = |V| \sin \theta$$
$$= 64 \sin 0°$$
$$= 64(0) = 0$$

19.

$$|V| = \sqrt{|V_x|^2 + |V_y|^2}$$
$$= \sqrt{(35.0)^2 + (26.0)^2}$$
$$= \sqrt{1,225 + 676}$$
$$= \sqrt{1,901}$$
$$= 43.6$$

21.

$$|V| = \sqrt{|V_x|^2 + |V_y|^2}$$
$$= \sqrt{(4.5)^2 + (3.8)^2}$$
$$= \sqrt{20.25 + 14.44}$$
$$= \sqrt{34.69}$$
$$= 5.9$$

23.

$$|V_x| = |V| \cos \theta$$
$$= 1,200 \cos 45°$$
$$= 1200(0.7071)$$
$$= 850 \text{ feet per second}$$

$$|V_y| = |V| \sin \theta$$
$$= 1,200 \sin 45°$$
$$= 1200(0.7071)$$
$$= 850 \text{ feet per second}$$

25. In 3 seconds, the bullet travels 3(850 ft/sec)=2,550 ft.

27. We are given that $|V_x| = 35.0$ and $|V_y| = 15.0$

$$|V| = \sqrt{|V_x|^2 + |V_y|^2}$$
$$= \sqrt{(35.0)^2 + (15.0)^2}$$
$$= \sqrt{1,225 + 225}$$
$$= \sqrt{1,450}$$
$$= 38.1 \text{ feet per second}$$

$$\tan\theta = \frac{|V_y|}{|V_x|}$$
$$= \frac{15.0}{35.0}$$
$$= 0.4285$$
$$\theta = 23.2°$$

Therefore, the velocity of the arrow is 38.1 feet per second at an elevation of $23.2°$.

29. $|V_x| = 135\cos 48°$ $|V_y| = 135\sin 48°$

$\quad\ = 90$ $\qquad\qquad = 100$

The ship has traveled 90 km east and 100 km south.

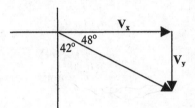

31. To find the total distance traveled north, we must find the sum of $|V_y|$ and $|W_y|$ and to find the total distance traveled east, we must find the sum of $|V_x|$ and $|W_x|$.

We are given that $|V|$ is 175 mi. at an angle of inclination of $90° - 18°$ or $72°$ and also that $|W|$ is 120 mi. at an angle of inclination of $90° - 49°$ or $41°$

$$|V_x| = |V|\cos\theta_1$$
$$= 175\cos 72°$$
$$= 175(0.3090)$$
$$= 50 \text{ mi}$$

$$|V_y| = |V|\sin\theta_1$$
$$= 175\sin 72°$$
$$= 175(0.9510)$$
$$= 170 \text{ mi}$$

$$|W_x| = |W|\cos\theta_2$$
$$= 120\cos 41°$$
$$= 120(0.7547)$$
$$= 90 \text{ mi}$$

$$|W_y| = |W|\sin\theta_2$$
$$= 120\sin 41°$$
$$= 120(0.6560)$$
$$= 80 \text{ mi}$$

Therefore, the total distance north is $|V_y| + |W_y| = 170 + 80 = 250$ miles and the total distance east is $|V_x| + |W_x| = 50 + 90 = 140$ miles.

33. $|W| = 42.0$

$$\cos 45.0° = \frac{|W|}{|T|}$$

$$|T| = \frac{|W|}{\cos 45.0°}$$

$$= \frac{42.0}{\cos 45.0°}$$

$$= 59.4 \text{ lb.}$$

$$\tan 45.0° = \frac{|H|}{|W|}$$

$$|H| = |W| \tan 45.0°$$

$$= 42.0 \tan 45.0°$$

$$= 42.0 \text{ lb.}$$

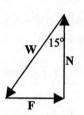

35. We are given that $|W| = 10$

$$\cos 15° = \frac{|N|}{|W|}$$

$$|N| = |W| \cos 15°$$

$$= 10(0.9659)$$

$$= 9.7 \text{ pounds}$$

$$\sin 15° = \frac{|F|}{|W|}$$

$$|F| = |W| \sin 15°$$

$$= 10(0.2588)$$

$$= 2.6 \text{ pounds}$$

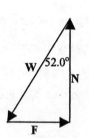

37. $|W| = 42.0$

$$\sin 52.0° = \frac{|F|}{|W|}$$

$$|F| = |W| \sin 52.0°$$

$$= 42.0 \sin 52.0°$$

$$= 33.1 \text{ lb}$$

39. $\theta = 20°, |F| = 40$ lb, and $d = 75$ ft

$$|F_x| = |F| \cos \theta$$

$$= 40 \cos 20°$$

$$\text{Work} = |F_x| \cdot d$$

$$= (40 \cos 20°)(75)$$

$$= 2800 \text{ ft - lb.}$$

41. $\theta = 15°, |F| = 85$ lb, and $d = 100$ ft

$$|F_x| = |F| \cos \theta$$

$$= 85 \cos 15°$$

$$\text{Work} = |F_x| \cdot d$$

$$= (85 \cos 15°)(100)$$

$$= 8200 \text{ ft - lb.}$$

43. $(x, y) = (-1, 1)$

$x = -1, y = 1$ and $r = \sqrt{2}$

$\sin 135° = \dfrac{y}{r} = \dfrac{1}{\sqrt{2}}$

$\cos 135° = \dfrac{x}{r} = -\dfrac{1}{\sqrt{2}}$

$\tan 135° = \dfrac{y}{x} = \dfrac{1}{-1} = -1$

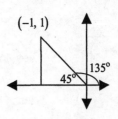

45.   A point on the line $y = 2x$ in quadrant I is $(1, 2)$. $x = 1, y = 2$, and $r = \sqrt{1^2 + 2^2} = \sqrt{5}$

$\sin \theta = \dfrac{y}{r} = \dfrac{2}{\sqrt{5}}$ $\qquad\qquad\qquad$ $\cos \theta = \dfrac{x}{r} = \dfrac{1}{\sqrt{5}}$

47. $\sin \theta = \dfrac{y}{r} = \dfrac{-4}{5} = \dfrac{-8}{10}$ $\qquad\qquad\qquad$ $x^2 + 64 = 100$

$y = -8$ and $r = 10$ $\qquad\qquad\qquad\qquad\qquad$ $x^2 = 36$

$x^2 + y^2 = r^2$ $\qquad\qquad\qquad\qquad\qquad\qquad$ $x = \pm 6$
$x^2 + (-8)^2 = 10^2$

Chapter 2 Test

1. $c = \sqrt{a^2 + b^2}$ $\qquad$ $\sin A = \dfrac{a}{c} = \dfrac{1}{\sqrt{5}}$ $\qquad$ $\sin B = \dfrac{b}{c} = \dfrac{2}{\sqrt{5}}$

$\quad\quad = \sqrt{(1)^2 + (2)^2}$ $\qquad$ $\cos A = \dfrac{b}{c} = \dfrac{2}{\sqrt{5}}$ $\qquad$ $\cos B = \dfrac{a}{c} = \dfrac{1}{\sqrt{5}}$

$\quad\quad = \sqrt{1 + 4}$

$\quad\quad = \sqrt{5}$ $\qquad\qquad$ $\tan A = \dfrac{a}{b} = \dfrac{1}{2}$ $\qquad$ $\tan B = \dfrac{b}{a} = \dfrac{2}{1} = 2$

2. $a = \sqrt{c^2 - b^2}$ $\qquad$ $\sin A = \dfrac{a}{c} = \dfrac{3\sqrt{3}}{6} = \dfrac{\sqrt{3}}{2}$ $\qquad$ $\sin B = \dfrac{b}{c} = \dfrac{3}{6} = \dfrac{1}{2}$

$\quad\quad = \sqrt{6^2 - 3^2}$

$\quad\quad = \sqrt{36 - 9}$ $\qquad$ $\cos A = \dfrac{b}{c} = \dfrac{3}{6} = \dfrac{1}{2}$ $\qquad$ $\cos B = \dfrac{a}{c} = \dfrac{3\sqrt{3}}{6} = \dfrac{\sqrt{3}}{2}$

$\quad\quad = \sqrt{27}$

$\quad\quad = 3\sqrt{3}$ $\qquad$ $\tan A = \dfrac{a}{b} = \dfrac{3\sqrt{3}}{3} = \sqrt{3}$ $\qquad$ $\tan B = \dfrac{b}{a} = \dfrac{3}{3\sqrt{3}} = \dfrac{1}{\sqrt{3}}$

3. $b = \sqrt{c^2 - a^2}$ $\qquad$ $\sin A = \dfrac{a}{c} = \dfrac{3}{5}$ $\qquad$ $\sin B = \dfrac{b}{c} = \dfrac{4}{5}$

$\quad\quad = \sqrt{5^2 - 3^2}$

$\quad\quad = \sqrt{25 - 9}$ $\qquad$ $\cos A = \dfrac{b}{c} = \dfrac{4}{5}$ $\qquad$ $\cos B = \dfrac{a}{c} = \dfrac{3}{5}$

$\quad\quad = \sqrt{16} = 4$ $\qquad$ $\tan A = \dfrac{a}{b} = \dfrac{3}{4}$ $\qquad$ $\tan B = \dfrac{b}{a} = \dfrac{4}{3}$

4. $c = \sqrt{a^2 + b^2}$ $\qquad$ $\sin A = \dfrac{a}{c} = \dfrac{5}{13}$ $\qquad$ $\sin B = \dfrac{b}{c} = \dfrac{12}{13}$

$\quad\quad = \sqrt{5^2 + 12^2}$

$\quad\quad = \sqrt{25 + 144}$ $\qquad$ $\cos A = \dfrac{b}{c} = \dfrac{12}{13}$ $\qquad$ $\cos B = \dfrac{a}{c} = \dfrac{5}{13}$

$\quad\quad = \sqrt{169} = 13$ $\qquad$ $\tan A = \dfrac{a}{b} = \dfrac{5}{12}$ $\qquad$ $\tan B = \dfrac{b}{a} = \dfrac{12}{5}$

5. $\sin 14° = \cos(90° - 14°)$ $\qquad\qquad$ **6.** $\csc 73° = \sec(90° - 73°)$

$\qquad\quad = \cos 76°$ $\qquad\qquad\qquad\qquad\qquad = \sec 17°$

7. $\sin^2 45° + \cos^2 30° = \left(\dfrac{1}{\sqrt{2}}\right)^2 + \left(\dfrac{\sqrt{3}}{2}\right)^2$ $\qquad$ **8.** $\tan 45° + \cot 45° = 1 + 1 = 2$

$\qquad\qquad\qquad\qquad = \dfrac{1}{2} + \dfrac{3}{4} = \dfrac{5}{4}$

9. $\sin^2 60° - \cos^2 30° = \left(\dfrac{\sqrt{3}}{2}\right)^2 - \left(\dfrac{\sqrt{3}}{2}\right)^2$

$\qquad\qquad\qquad\quad = 0$

10. $\dfrac{1}{\sec 30°} = \cos 30°$

$\qquad\qquad = \dfrac{\sqrt{3}}{2}$

11. $\begin{array}{r} 48°18' \\ +24°52' \\ \hline 72°70' = 73°10' \quad (60'=1°) \end{array}$

12. $\begin{array}{rl} 25°15' & = 24°75' \quad (1°=60') \\ -15°32' & \underline{-15°32'} \\ & 9°43' \end{array}$

13. $73.2° = 73° + 0.2(60)'$

$\qquad = 73°12'$

14. $16.45° = 16° + 0.45(60)'$

$\qquad = 16°27'$

15. $2°48' = 2° + \left(\dfrac{48}{60}\right)°$

$\qquad = 2° + 0.8°$

$\qquad = 2.8°$

16. $79°30' = 79° + \left(\dfrac{30}{60}\right)°$

$\qquad = 79° + 0.5°$

$\qquad = 79.5°$

17. $\sin 24°20' = \sin 24.33°$

$\qquad\qquad = 0.4120$

18. $\cos 37.8° = 0.7902$

19. $\tan 63°50' = \tan 63.833°$

$\qquad\qquad = 2.0353$

20. $\cot 71°20' = \cot 71.333°$

$\qquad\qquad = 0.3378$

27. $c = \sqrt{a^2 + b^2}$

$\quad = \sqrt{(68.0)^2 + (104)^2}$

$\quad = \sqrt{4,624 + 10,816}$

$\quad = \sqrt{15,440}$

$\quad = 124$

$\tan A = \dfrac{68.0}{104}$

$\qquad = 0.6538$

$A = 33.2°$

$B = 90° - 33.2°$

$\quad = 56.8°$

28. $b = \sqrt{c^2 - a^2}$

$\quad = \sqrt{(48.1)^2 + (24.3)^2}$

$\quad = \sqrt{2,313.61 + 590.49}$

$\quad = \sqrt{1,723.12}$

$\quad = 41.5$

$\sin A = \dfrac{24.3}{48.1}$

$\qquad = 0.5052$

$A = 30.3°$

$B = 90° - 30.3°$

$\quad = 59.7°$

29.

$$A = 90° - 24.9°$$
$$= 65.1°$$

$$\sin 24.9° = \frac{305}{c}$$

$$c = \frac{305}{\sin 24.9°}$$
$$= \frac{305}{0.4210}$$
$$= 724$$

$$\tan 65.1° = \frac{a}{305}$$

$$a = 305 \tan 65.1°$$
$$= 305(2.1543)$$
$$= 657$$

30.

$$b = 90° - 35°30'$$
$$= 54°30'$$

$$\sin 35°30' = \frac{a}{0.462}$$
$$a = 0.462 \sin 35.5°$$
$$= 0.462(0.5807)$$
$$= 0.268$$

$$\sin 54°30' = \frac{a}{0.462}$$
$$b = 0.462 \sin 54.5°$$
$$= 0.462(0.8141)$$
$$= 0.376$$

31.

$$\sin 17° = \frac{25}{x}$$

$$x = \frac{25}{\sin 17°}$$
$$= \frac{25}{0.2924}$$
$$= 86 \text{ cm}$$

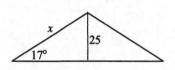

32.

$$\tan 75°30' = \frac{x}{1.5}$$

$$x = 1.5 \tan 75.5°$$
$$= 1.5(3.8667)$$
$$= 5.8 \text{ ft}$$

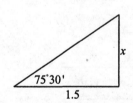

33.

$$\tan 43° = \frac{35}{x}$$

$$x = \frac{35}{\tan 43°}$$
$$= \frac{35}{0.9325}$$
$$= 37.5 \text{ ft}$$

$$\tan 47° = \frac{35}{y}$$

$$y = \frac{35}{\tan 47°}$$
$$= \frac{35}{1.0724}$$
$$= 32.6 \text{ ft}$$

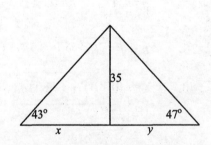

$$x + y = 70 \text{ feet (rounded to 2 significant digits)}$$

34. $\quad |V_x| = 5.0 \cos 30°$ $\qquad |V_y| = 5.0 \sin 30°$

$\qquad\qquad = 5.0(0.8660)$ $\qquad\qquad = 5.0(0.5)$

$\qquad\qquad = 4.3$ $\qquad\qquad\qquad = 2.5$

35. $\quad \tan \theta = \dfrac{|V_y|}{|V_x|}$

$\qquad\qquad = \dfrac{31}{11}$

$\qquad\qquad = 2.8182$

$\qquad\theta = \tan^{-1}(2.8182)$

$\qquad\qquad = 70°$

36. $\quad |V_x| = 800 \cos 62°$ $\qquad\qquad\qquad\qquad |V_y| = 800 \sin 62°$

$\qquad\qquad = 800(0.4695)$ $\qquad\qquad\qquad\qquad\qquad = 800(0.8829)$

$\qquad\qquad = 380 \text{ ft}$ $\qquad\qquad\qquad\qquad\qquad\quad = 710 \text{ ft}$

37. $\quad \theta = -30° + 360° = 330°$

$\qquad |V_x| = 120 \cos 330°$ $\qquad\qquad\qquad\qquad |V_y| = 120 \sin 330°$

$\qquad\qquad = 120(0.8660)$ $\qquad\qquad\qquad\qquad\qquad = 120(-0.5)$

$\qquad\qquad = 100$ $\qquad\qquad\qquad\qquad\qquad\qquad = -60$

The ship travels 100 miles east and 60 miles south.

38. $\quad \tan 25.5° = \dfrac{|H|}{|W|}$

$\qquad\qquad\quad = \dfrac{|H|}{95.5}$

$\qquad |H| = 95.5 \tan 25.5°$

$\qquad\qquad = 95.5(0.4770) = 45.6 \text{ pounds}$

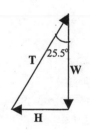

39. $\quad \sin 8.5° = \dfrac{|F|}{|W|}$

$\qquad\qquad\quad = \dfrac{|F|}{58.0}$

$\qquad |F| = 58.0 \sin 8.5°$

$\qquad\qquad = 58.0(0.1478) = 8.57 \text{ pounds}$

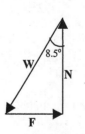

40. $\quad \theta = 40°, |F| = 35 \text{ lb, and } d = 80 \text{ ft}$

$\qquad |F_x| = |F| \cos \theta$ $\qquad\qquad$ Work $= |F_x| \cdot d$

$\qquad\qquad = 35 \cos 40°$ $\qquad\qquad\qquad\qquad = (35 \cos 40°)(80)$

$\qquad\qquad\qquad\qquad\qquad\qquad\qquad\qquad = 2100 \text{ ft - lb}$

CHAPTER 3   Radian Measure

Problem Set 3.1

1. $210° - 180° = 30°$

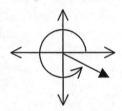

3. $180° - 143.4° = 36.6°$

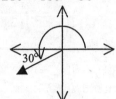

5. $360° - 311.7° = 48.3°$

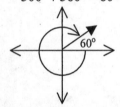

7. $195°10' - 180° = 15°10'$

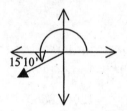

9. $-300° + 360° = 60°$

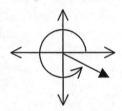

11. $-120° + 180° = 60°$

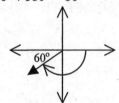

13. $\hat{\theta} = 225° - 180°$
    $= 45°$

Since θ terminates in QIII, $\cos\theta$ is negative.

$\cos 225° = -\cos 45°$

$\qquad = -\dfrac{1}{\sqrt{2}}$

15. $\hat{\theta} = 180° - 120°$
    $= 60°$

Since θ terminates in QII, $\sin\theta$ is positive.

$\sin 120° = \sin 60°$

$\qquad = \dfrac{\sqrt{3}}{2}$

17. $\hat{\theta} = 180° - 135°$
    $= 45°$

Since θ terminates in QII, $\tan\theta$ is negative.

$\tan 135° = -\tan 45°$

$\qquad = -1$

19. $\hat{\theta} = 240° - 180°$
    $= 60°$

Since θ terminates in QIII, $\cos\theta$ is negative.

$\cos 240° = -\cos 60°$

$\qquad = -\dfrac{1}{2}$

21. $\hat{\theta} = 360° - 330°$
$\quad = 30°$

Since θ terminates in QIV, $\csc\theta$ is negative.

$\csc 330° = -\csc 30°$
$\quad\quad\quad = -2$

23. $\hat{\theta} = 360° - 300°$
$\quad = 60°$

Since θ terminates in QIV, $\sec\theta$ is positive.

$\sec 300° = \sec 60°$
$\quad\quad\quad = 2$

25. $\hat{\theta} = 390° - 360°$
$\quad = 30°$

Since θ terminates in QI, $\sin\theta$ is positive.

$\sin 390° = \sin 30°$
$\quad\quad\quad = \dfrac{1}{2}$

27. $\theta = 480° - 360° = 120°$
$\hat{\theta} = 180° - 120° = 60°$

Since θ terminates in QII, $\cot\theta$ is negative.

$\cot 480° = -\cot 60°$
$\quad\quad\quad = -\dfrac{1}{\sqrt{3}}$

49. First, we find $\hat{\theta}$:
$$\sin\hat{\theta} = 0.3090$$
$$\hat{\theta} = \sin^{-1}(0.3090)$$
$$\hat{\theta} = 18.0°$$

Scientific Calculator: 0.3090 $\boxed{\text{INV}}$ $\boxed{\text{SIN}}$

Graphing Calculator: $\boxed{\text{2nd}}$ $\boxed{\text{SIN}}$ $\boxed{(}$ 0.3090 $\boxed{)}$ $\boxed{\text{ENTER}}$

Since θ is in QIII, $\theta = 180° + 18.0°$
$\quad\quad\quad\quad\quad = 198.0°$

51. First, we find $\hat{\theta}$:
$$\cos\hat{\theta} = 0.7660$$
$$\hat{\theta} = \cos^{-1}(0.7660)$$
$$= 40.0°$$

Scientific Calculator: 0.7660 $\boxed{\text{INV}}$ $\boxed{\text{COS}}$

Graphing Calculator: $\boxed{\text{2nd}}$ $\boxed{\text{COS}}$ $\boxed{(}$ 0.7660 $\boxed{)}$ $\boxed{\text{ENTER}}$

Since θ is in QII, $\theta = 180° - 40.0°$
$\quad\quad\quad\quad\quad = 140.0°$

53. First we find $\hat{\theta}$:
$$\tan\hat{\theta} = 0.5890$$
$$\hat{\theta} = \tan^{-1}(0.5890)$$
$$\hat{\theta} = 30.5°$$

Scientific Calculator: 0.5890 $\boxed{\text{INV}}$ $\boxed{\text{TAN}}$

Graphing Calculator: $\boxed{\text{2nd}}$ $\boxed{\text{TAN}}$ $\boxed{(}$ 0.5890 $\boxed{)}$ $\boxed{\text{ENTER}}$

Since θ is in QIII, $\theta = 180° + 30.5°$
$\quad\quad\quad\quad\quad = 210.5°$

55. Since θ is in QI, $\hat{\theta} = \theta$
$$\cos\theta = 0.2644$$
$$\theta = \cos^{-1}(0.2664)$$
$$= 74.7°$$

Scientific Calculator: 0.2644 [INV] [COS]

Graphing Calculator: [2nd] [COS] [(] 0.2644 [)] [ENTER]

57. First , we find $\hat{\theta}$:
$$\sin\hat{\theta} = 0.9652$$
$$\hat{\theta} = 74.8°$$

Scientific Calculator: 0.9652 [INV] [SIN]

Graphing Calculator: [2nd] [SIN] [(] 0.9652 [)] [ENTER]

Since θ is in QII, $\theta = 180° - 74.8°$
$$= 105.2°$$

59. First, we find $\hat{\theta}$:
$$\sec\hat{\theta} = 1.4325$$
$$\cos\hat{\theta} = \frac{1}{1.4325}$$
$$\hat{\theta} = \cos^{-1}\left(\frac{1}{1.4325}\right)$$
$$= 45.7°$$

Scientific Calculator: 1.4325 [1/x] [INV] [COS]

Graphing Calculator: [2nd] [COS] [(] 1.4325 [x⁻¹] [)] [ENTER]

Since θ is in QIV, $\theta = 360° - 45.7°$
$$= 314.3°$$

61. First, we find $\hat{\theta}$:
$$\csc\hat{\theta} = 2.4957$$
$$\sin\hat{\theta} = \frac{1}{2.4957}$$
$$\hat{\theta} = \sin^{-1}\left(\frac{1}{2.4957}\right)$$
$$\hat{\theta} = 23.6°$$

Scientific Calculator: 2.4957 [1/x] [INV] [SIN]

Graphing Calculator: [2nd] [SIN] [(] 2.4957 [x⁻¹] [)] [ENTER]

Since θ is in QII, $\theta = 180° - 23.6°$
$$= 156.4°$$

63.   First, we find $\hat{\theta}$:

$$\cot\hat{\theta} = 0.7366$$

$$\tan\hat{\theta} = \frac{1}{0.7366}$$

$$\hat{\theta} = \tan^{-1}\left(\frac{1}{0.7366}\right)$$

$$\hat{\theta} = 53.6°$$

Scientific Calculator: 0.7366 $\boxed{1/\text{x}}$ $\boxed{\text{INV}}$ $\boxed{\text{TAN}}$

Graphing Calculator: $\boxed{\text{2nd}}$ $\boxed{\text{TAN}}$ $\boxed{(}$ 0.7366 $\boxed{\text{x}^{-1}}$ $\boxed{)}$ $\boxed{\text{ENTER}}$

Since θ is in QII, $\theta = 180° - 53.6°$
$ = 126.4°$

65.   First, we find $\hat{\theta}$:

$$\sec\hat{\theta} = 1.7876$$

$$\cos\hat{\theta} = \frac{1}{1.7876}$$

$$\hat{\theta} = \cos^{-1}\left(\frac{1}{1.7876}\right)$$

$$\hat{\theta} = 56.0°$$

Scientific Calculator: $0.1.7876$ $\boxed{1/\text{x}}$ $\boxed{\text{INV}}$ $\boxed{\text{COS}}$

Graphing Calculator: $\boxed{\text{2nd}}$ $\boxed{\text{COS}}$ $\boxed{(}$ 1.7876 $\boxed{\text{x}^{-1}}$ $\boxed{)}$ $\boxed{\text{ENTER}}$

Since θ is in QIII, $\theta = 180° - 56.0°$
$ = 236.0°$

67.   First, we find $\hat{\theta}$: $\sin\hat{\theta} = \dfrac{\sqrt{3}}{2}$

$$\hat{\theta} = 60°$$

Since θ is in QIII, $\theta = 180° + 60°$
$ = 240°$

69.   First, we find $\hat{\theta}$: $\cos\hat{\theta} = \dfrac{1}{\sqrt{2}}$

$$\hat{\theta} = 45°$$

Since θ is in QII, $\theta = 180° - 45°$
$ = 135°$

71.   First, we find $\hat{\theta}$: $\sin\hat{\theta} = \dfrac{\sqrt{3}}{2}$

$$\hat{\theta} = 60°$$

Since θ is in QIV, $\theta = 360° - 60°$
$ = 300°$

73.   First, we find $\hat{\theta}$: $\tan\hat{\theta} = \sqrt{3}$

$$\hat{\theta} = 60°$$

Since θ is in QIII, $\theta = 180° + 60°$
$ = 240°$

75. First, we find $\hat{\theta}$: $\sec\hat{\theta} = 2$

$$\cos\hat{\theta} = \frac{1}{2}$$

$$\hat{\theta} = 60°$$

Since θ is in QII, $\theta = 180° - 60°$

$$= 120°$$

77. First, we find $\hat{\theta}$: $\csc\hat{\theta} = \sqrt{2}$

$$\sin\hat{\theta} = \frac{1}{\sqrt{2}}$$

$$\hat{\theta} = 45°$$

Since θ is in QII, $\theta = 180° - 45°$

$$= 135°$$

79. First, we find $\hat{\theta}$: $\cot\hat{\theta} = 1$

$$\tan\hat{\theta} = 1$$

$$\hat{\theta} = 45°$$

Since θ is in QIV, $\theta = 360° - 45°$

$$= 315°$$

81. The complement of $70°$ is $20°$ because $70° + 20° = 90°$.

The supplement of $70°$ is $110°$ because $70° + 110° = 180°$.

83. The complement of x is $90° - x$.

The supplement of x is $180° - x$.

85. The side opposite the $30°$ angle is one-half of the longest side, or $\frac{1}{2} \cdot 10 = 5$.

The side opposite the $60°$ angle is $\sqrt{3}$ times the shortest side, or $5\sqrt{3}$.

87. $\sin 30° \cos 60° = \frac{1}{2} \cdot \frac{1}{2}$

$$= \frac{1}{4}$$

89. $\sin^2 45° + \cos^2 45° = \left(\frac{1}{\sqrt{2}}\right)^2 + \left(\frac{1}{\sqrt{2}}\right)^2$

$$= \frac{1}{2} + \frac{1}{2}$$

$$= 1$$

Problem Set 3.2

1. $\theta = \dfrac{s}{r}$ Definition of radian measure

     $= \dfrac{9\,\text{cm}}{3\,\text{cm}}$ Substitute given values

     $= 3$ Divide

3. $\theta = \dfrac{s}{r}$ Definition of radian measure

     $= \dfrac{5\,\text{in}}{10\,\text{in}}$ Substitute given values

     $= \dfrac{1}{2}$ Divide

5. $\theta = \dfrac{s}{r}$ Definition of radian measure

     $= \dfrac{12\pi \ \text{inches}}{4\ \text{inches}}$ Substitute given values

     $= 3\pi$ Divide

7. $\theta = \dfrac{s}{r}$ Definition of radian measure

     $= \dfrac{\frac{1}{2}\ \text{cm}}{\frac{1}{4}\ \text{cm}}$ Substitute given values

     $= 2$ Divide

9. $\theta = \dfrac{s}{r}$ Definition of radian measure

     $= \dfrac{450}{4000}$ Substitute given values

     $= 0.1125$ Divide

11.   (b) $30° = 30\left(\dfrac{\pi}{180}\right)$

          $= \dfrac{\pi}{6}$ radians

     (c)   Reference angle is itself :

          $30° = \dfrac{\pi}{6}$ radians

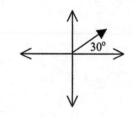

13.   (b) $90° = 90\left(\dfrac{\pi}{180}\right)$

          $= \dfrac{\pi}{2}$ radians

     (c)   Reference angle is undefined.

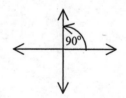

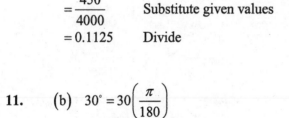

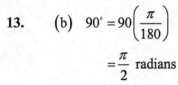

15. (b) $260° = 260\left(\dfrac{\pi}{180}\right)$

$\qquad = \dfrac{13\pi}{9}$ radians

(c) $\hat{\theta} = 260° - 180°$

$\qquad = 80°$

$80° = 80\left(\dfrac{\pi}{180}\right)$

$\qquad = \dfrac{4\pi}{9}$ radians

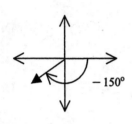

17. (b) $-150° = -150\left(\dfrac{\pi}{180}\right)$

$\qquad = -\dfrac{5\pi}{6}$

(c) $-150° = -150 + 360°$

$\qquad = 210°$

$\hat{\theta} = 210° - 180° = 30°$

$30° = 30\left(\dfrac{\pi}{180}\right)$

$\qquad = \dfrac{\pi}{6}$ radians

19. (b) $420° = 420\left(\dfrac{\pi}{180}\right)$

$\qquad = \dfrac{7\pi}{3}$ radians

(c) $\hat{\theta} = 420° - 360° = 60°$

$60° = 60\left(\dfrac{\pi}{180}\right)$

$\qquad = \dfrac{\pi}{3}$ radians

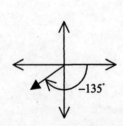

21. (b) $-135° = -135\left(\dfrac{\pi}{180}\right)$

$\qquad = -\dfrac{3\pi}{4}$ radians

This problem is continued on the next page

(c) $-135° = -135° + 360°$

$\qquad = 225°$

$\qquad \hat{\theta} = 225° - 180° = 45°$

$\qquad 45° = 45\left(\dfrac{\pi}{180}\right)$

$\qquad = \dfrac{\pi}{4}$ radians

23. $120°40' = \left(120 + \dfrac{40}{60}\right)° = 120.67°$

$\qquad 120.67° = 120.67\left(\dfrac{\pi}{180}\right)$

$\qquad\qquad = 2.11$

25.   1 minute $= \dfrac{1}{60}$ degree

$\qquad = \dfrac{1}{60}\left(\dfrac{\pi}{180}\right)$

$\qquad = \dfrac{\pi}{10,800}$

$\qquad = 0.000291$

27.   From problem 25, we know that $1' = \dfrac{\pi}{10,800}$ radians.

$\theta = \dfrac{s}{r}$

$s = r\theta$

$\qquad = 4,000\left(\dfrac{\pi}{10,800}\right)$

$\qquad = 1.16$ miles

29. $\theta = \dfrac{5 \text{ min}}{60 \text{ min}} \cdot 2\pi$

$\qquad = \dfrac{1}{12} \cdot 2\pi$

$\qquad = \dfrac{\pi}{6}$

31.   (a) $\dfrac{\pi}{3} = \dfrac{\pi}{3}\left(\dfrac{180}{\pi}\right)°$

$\qquad\qquad = 60°$

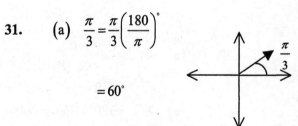

(c)   Reference angle is itself:

$\qquad \dfrac{\pi}{3} = 60°$

33. (a) $\dfrac{2\pi}{3} = \dfrac{2\pi}{3}\left(\dfrac{180}{\pi}\right)^{\circ}$

$\qquad = 120^{\circ}$

(c) $\hat{\theta} = 180^{\circ} - 120^{\circ} = 60^{\circ}$

$\qquad 60^{\circ} = \dfrac{\pi}{3}$

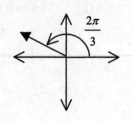

35. (a) $-\dfrac{7\pi}{6} = -\dfrac{7\pi}{6}\left(\dfrac{180}{\pi}\right)^{\circ}$

$\qquad = -210^{\circ}$

(c) $-210^{\circ} = -210^{\circ} + 360^{\circ}$

$\qquad = 150^{\circ}$

$\qquad \hat{\theta} = 180^{\circ} - 150^{\circ} = 30^{\circ}$

$\qquad 30^{\circ} = 30\left(\dfrac{\pi}{180}\right)$

$\qquad = \dfrac{\pi}{6}$

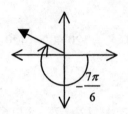

37. (a) $\dfrac{5\pi}{3} = \dfrac{5\pi}{3}\left(\dfrac{180}{\pi}\right)^{\circ}$

$\qquad = 300^{\circ}$

(c) $\hat{\theta} = 360^{\circ} - 300^{\circ} = 60^{\circ}$

$\qquad 60^{\circ} = \dfrac{\pi}{3}$

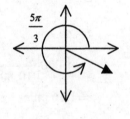

39. (a) $4\pi = 4\pi\left(\dfrac{180}{\pi}\right)^{\circ}$

$\qquad = 720^{\circ}$

(c) $720^{\circ} = 720^{\circ} - 360^{\circ} = 360^{\circ}$

$\qquad 360^{\circ} = 360^{\circ} - 360^{\circ}$

$\qquad = 0^{\circ}$

Reference angle is undefined.

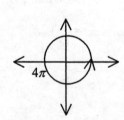

41. (a) $\dfrac{\pi}{12} = \dfrac{\pi}{12}\left(\dfrac{180}{\pi}\right)^{\circ} = 15^{\circ}$

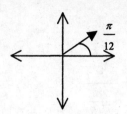

(c) $\hat{\theta} = 15^{\circ} = \dfrac{\pi}{12}$

43. $1 = 1\left(\dfrac{180}{\pi}\right)^{\circ}$

$\quad = \left(\dfrac{180}{\pi}\right)^{\circ}$

$\quad = 57.3^{\circ}$

45. $1.3 = 1.3\left(\dfrac{180}{\pi}\right)^{\circ}$

$\quad = 74.5^{\circ}$

47. $0.75 = 0.75\left(\dfrac{180}{\pi}\right)^{\circ}$

$\quad = \left(\dfrac{135}{\pi}\right)^{\circ}$

$\quad = 43.0^{\circ}$

49. $5 = 5\left(\dfrac{180}{\pi}\right)^{\circ}$

$\quad = 286.5^{\circ}$

51. Since $\dfrac{4\pi}{3}$ terminates in QIII, its sine will be negative.

$\hat{\theta} = \dfrac{4\pi}{3} - \pi$

$\quad = \dfrac{\pi}{3}$

$\sin\dfrac{4\pi}{3} = -\sin\dfrac{\pi}{3} \qquad \left(\dfrac{\pi}{3} = 60^{\circ}\right)$

$\qquad\quad = -\dfrac{\sqrt{3}}{2}$

53. $\tan\dfrac{\pi}{6} = \dfrac{1}{\sqrt{3}}$

55. Since $\dfrac{2\pi}{3}$ terminates in QII, its secant will be negative.

$\hat{\theta} = \pi - \dfrac{2\pi}{3} = \dfrac{\pi}{3}$

$\sec\dfrac{2\pi}{3} = -\sec\dfrac{\pi}{3}$

$\qquad\quad = -2$

57. Since $\dfrac{5\pi}{6}$ terminates in QII, its cosecant is positive.

$\hat{\theta} = \pi - \dfrac{5\pi}{6} = \dfrac{\pi}{6}$

This problem is continued on the next page

$$\csc\frac{5\pi}{6} = \csc\frac{\pi}{6}$$

$$= \frac{1}{\sin\frac{\pi}{6}}$$

$$= \frac{1}{\frac{1}{2}} = 2$$

59. Since $-\frac{\pi}{4}$ terminates in QIV, its sine will be negative.

$$\theta = -\frac{\pi}{4} + 2\pi = \frac{7\pi}{4}$$

$$\hat{\theta} = 2\pi - \frac{7\pi}{4} = \frac{\pi}{4}$$

$$4\sin\left(-\frac{\pi}{4}\right) = -4\sin\frac{\pi}{4}$$

$$= -4\left(\frac{\sqrt{2}}{2}\right)$$

$$= -2\sqrt{2}$$

61. $\quad -\sin\frac{\pi}{4} = -\frac{1}{\sqrt{2}}$

63. $\quad 2\cos\frac{\pi}{6} = 2\left(\frac{\sqrt{3}}{2}\right)$

$$= \sqrt{3}$$

65. $\quad \sin 2x = \sin 2\left(\frac{\pi}{6}\right)$

$$= \sin\frac{\pi}{3}$$

$$= \frac{\sqrt{3}}{2}$$

67. $\quad 6\cos 3\left(\frac{\pi}{6}\right) = 6\cos\frac{\pi}{2}$

$$= 6(0)$$

$$= 0$$

69. $\quad \sin\left(x + \frac{\pi}{2}\right) = \sin\left(\frac{\pi}{6} + \frac{\pi}{2}\right)$

$$= \sin\frac{2\pi}{3}$$

This problem is continued on the next page

Since $\dfrac{2\pi}{3}$ terminates in QII, its sine is positive.

$$\hat{\theta} = \pi - \frac{2\pi}{3} = \frac{\pi}{3}$$

$$\sin\frac{2\pi}{3} = \sin\frac{\pi}{3} = \frac{\sqrt{3}}{2}$$

71. $\quad 4\cos\left[2\left(\dfrac{\pi}{6}\right) + \dfrac{\pi}{3}\right] = 4\cos\left(\dfrac{\pi}{3} + \dfrac{\pi}{3}\right) = 4\cos\dfrac{2\pi}{3}$

Since $\dfrac{2\pi}{3}$ terminates in QII, its cosine will be negative.

$$\hat{\theta} = \pi - \frac{2\pi}{3} = \frac{\pi}{3} \qquad\qquad\qquad 4\cos\frac{2\pi}{3} = -4\cos\frac{\pi}{3}$$

$$= -4\left(\frac{1}{2}\right) = -2$$

73. For $x = 0$, $y = \sin 0$ $\qquad\qquad\qquad$ For $x = \dfrac{3\pi}{4}$, $y = \sin\dfrac{3\pi}{4}$

$\qquad\qquad\qquad = 0 \qquad (x, y) = (0, 0) \qquad\qquad = \sin\dfrac{\pi}{4} = \dfrac{1}{\sqrt{2}} \qquad (x, y) = \left(\dfrac{3\pi}{4}, \dfrac{1}{\sqrt{2}}\right)$

For $x = \dfrac{\pi}{4}$, $y = \sin\dfrac{\pi}{4}$ $\qquad\qquad\qquad$ For $x = \pi$, $y = \sin\pi$

$\qquad\qquad\qquad = \dfrac{1}{\sqrt{2}} \quad (x, y) = \left(\dfrac{\pi}{4}, \dfrac{1}{\sqrt{2}}\right) \qquad\qquad = 0 \qquad\qquad (x, y) = (\pi, 0)$

For $x = \dfrac{\pi}{2}$, $y = \sin\dfrac{\pi}{2}$

$\qquad\qquad\qquad = 1 \qquad (x, y) = \left(\dfrac{\pi}{2}, 1\right)$

75. For $x = 0$ $\qquad y = 2\sin 0$

$\qquad\qquad\qquad\qquad = 2(0)$

$\qquad\qquad\qquad\qquad = 0 \qquad\qquad (x, y) = (0, 0)$

For $x = \dfrac{\pi}{2}$, $\qquad y = 2\sin\dfrac{\pi}{2}$

$\qquad\qquad\qquad\qquad = 2(1) = 2 \quad (x, y) = \left(\dfrac{\pi}{2}, 2\right)$

For $x = \pi$, $\qquad y = 2\sin\pi$

$\qquad\qquad\qquad\qquad = 2(0) = 0 \quad (x, y) = (\pi, 0)$

This problem is continued on the next page

For $x = \dfrac{3\pi}{2},$ $y = 2\sin\dfrac{3\pi}{2}$

$$= 2(-1) = -2 \qquad (x,y) = \left(\dfrac{3\pi}{2}, -2\right)$$

For $x = 2\pi,$ $y = 2\sin 2\pi$

$$= 2(0) = 0 \qquad (x,y) = (2\pi, 0)$$

77.   For $x = 0,$ $y = \sin 2(0) = \sin 0 = 0$ $(x,y) = (0,0)$

For $x = \dfrac{\pi}{4},$ $y = \sin 2\left(\dfrac{\pi}{4}\right)$ $(x,y) = \left(\dfrac{\pi}{4}, 1\right)$

$$= \sin\dfrac{\pi}{2} = 1$$

For $x = \dfrac{\pi}{2},$ $y = \sin 2\left(\dfrac{\pi}{2}\right)$ $(x,y) = \left(\dfrac{\pi}{2}, 0\right)$

$$= \sin\pi = 0$$

For $x = \dfrac{3\pi}{4},$ $y = \sin 2\left(\dfrac{3\pi}{4}\right)$ $(x,y) = \left(\dfrac{3\pi}{4}, -1\right)$

$$= \sin\dfrac{3\pi}{2} = -1$$

For $x = \pi,$ $y = \sin 2(\pi) = 0$ $(x,y) = (\pi, 0)$

79.   For $x = \dfrac{\pi}{2},$ $y = \sin\left(\dfrac{\pi}{2} - \dfrac{\pi}{2}\right)$ $(x, y) = \left(\dfrac{\pi}{2}, 0\right)$

$$= \sin 0 = 0$$

For $x = \pi,$ $y = \sin\left(\pi - \dfrac{\pi}{2}\right)$ $(x, y) = (\pi, 1)$

$$= \sin\dfrac{\pi}{2} = 1$$

For $x = \dfrac{3\pi}{2},$ $y = \sin\left(\dfrac{3\pi}{2} - \dfrac{\pi}{2}\right)$ $(x, y) = \left(\dfrac{3\pi}{2}, 0\right)$

$$= \sin\pi = 0$$

For $x = 2\pi,$ $y = \sin\left(2\pi - \dfrac{\pi}{2}\right)$ $(x, y) = (2\pi, -1)$

$$= \sin\dfrac{3\pi}{2} = -1$$

For $x = \dfrac{5\pi}{2},$ $y = \sin\left(\dfrac{5\pi}{2} - \dfrac{\pi}{2}\right)$ $(x, y) = \left(\dfrac{5\pi}{2}, 0\right)$

$$= \sin 2\pi = 0$$

81. For $x = -\dfrac{\pi}{4}$, $y = 3\sin\left[2\left(-\dfrac{\pi}{4}\right) + \dfrac{\pi}{2}\right]$ $\qquad (x, y) = \left(-\dfrac{\pi}{4}, 0\right)$

$\qquad\qquad = 3\sin\left(-\dfrac{\pi}{2} + \dfrac{\pi}{2}\right)$

$\qquad\qquad = 3\sin 0 = 3 \cdot 0 = 0$

For $x = 0$, $y = 3\sin\left[2(0) + \dfrac{\pi}{2}\right]$ $\qquad (x, y) = (0, 3)$

$\qquad\qquad = 3\sin\dfrac{\pi}{2} = 3(1) = 3$

For $x = \dfrac{\pi}{4}$, $y = 3\sin\left[2\left(\dfrac{\pi}{4}\right) + \dfrac{\pi}{2}\right]$ $\qquad (x, y) = \left(\dfrac{\pi}{4}, 0\right)$

$\qquad\qquad = 3\sin\left(\dfrac{\pi}{2} + \dfrac{\pi}{2}\right)$

$\qquad\qquad = 3\sin \pi = 3(0) = 0$

For $x = \dfrac{\pi}{2}$, $y = 3\sin\left[2\left(\dfrac{\pi}{2}\right) + \dfrac{\pi}{2}\right]$ $\qquad (x, y) = \left(\dfrac{\pi}{2}, -3\right)$

$\qquad\qquad = 3\sin\left(\pi + \dfrac{\pi}{2}\right)$

$\qquad\qquad = 3\sin\dfrac{3\pi}{2} = 3(-1) = -3$

For $x = \dfrac{3\pi}{4}$, $y = 3\sin\left[2\left(\dfrac{3\pi}{4}\right) + \dfrac{\pi}{2}\right]$ $\qquad (x, y) = \left(\dfrac{3\pi}{4}, 0\right)$

$\qquad\qquad = 3\sin\left(\dfrac{3\pi}{2} + \dfrac{\pi}{2}\right)$

$\qquad\qquad = 3\sin 2\pi = 3(0) = 0$

83. $\theta = \dfrac{2\pi}{8} = \dfrac{\pi}{4}$ radians

85. $(x, y) = (1, -3)$

$r = \sqrt{x^2 + y^2} = \sqrt{(1)^2 + (-3)^2} = \sqrt{1 + 9} = \sqrt{10}$

$\sin\theta = \dfrac{y}{r} = \dfrac{-3}{\sqrt{10}}$ $\qquad\qquad \cot\theta = \dfrac{x}{y} = \dfrac{1}{-3}$

$\cos\theta = \dfrac{x}{r} = \dfrac{1}{\sqrt{10}}$ $\qquad\qquad \sec\theta = \dfrac{r}{x} = \dfrac{\sqrt{10}}{1} = \sqrt{10}$

$\tan\theta = \dfrac{y}{x} = \dfrac{-3}{1} = -3$ $\qquad\qquad \csc\theta = \dfrac{r}{y} = \dfrac{\sqrt{10}}{-3}$

87. $(x, y) = (m, n)$

$$r = \sqrt{x^2 + y^2} = \sqrt{m^2 + n^2}$$

$$\sin\theta = \frac{y}{r} = \frac{n}{\sqrt{m^2 + n^2}} \qquad\qquad \cot\theta = \frac{x}{y} = \frac{m}{n}$$

$$\cos\theta = \frac{x}{r} = \frac{m}{\sqrt{m^2 + n^2}} \qquad\qquad \sec\theta = \frac{r}{x} = \frac{\sqrt{m^2 + n^2}}{m}$$

$$\tan\theta = \frac{y}{x} = \frac{n}{m} \qquad\qquad\qquad \csc\theta = \frac{r}{y} = \frac{\sqrt{m^2 + n^2}}{n}$$

89. $\sin\theta = \dfrac{1}{2}$ and θ terminates in QII.

$$\cos\theta = -\sqrt{1 - \sin^2\theta} \qquad \text{because } \cos\theta \text{ is negative in QII}$$

$$= -\sqrt{1 - \left(\frac{1}{2}\right)^2}$$

$$= -\sqrt{1 - \frac{1}{4}}$$

$$= -\sqrt{\frac{3}{4}} = -\frac{\sqrt{3}}{2}$$

$$\tan\theta = \frac{\sin\theta}{\cos\theta} \qquad \cot\theta = \frac{1}{\tan\theta} \qquad \sec\theta = \frac{1}{\cos\theta} \qquad \csc\theta = \frac{1}{\sin\theta}$$

$$= \frac{1/2}{-\sqrt{3}/2} \qquad\quad = \frac{1}{-1/\sqrt{3}} \qquad\quad = \frac{1}{-\sqrt{3}/2} \qquad\quad = \frac{1}{1/2}$$

$$= -\frac{1}{\sqrt{3}} \qquad\qquad = -\sqrt{3} \qquad\qquad = -\frac{2}{\sqrt{3}} \qquad\qquad = 2$$

91. A point on the line $y = 2x$ in QI is (1, 2).

$$(x, y) = (1, 2)$$

$$r = \sqrt{1^2 + 2^2}$$

$$= \sqrt{1 + 4} = \sqrt{5}$$

$$\sin\theta = \frac{y}{r} = \frac{2}{\sqrt{5}} \qquad\qquad \cot\theta = \frac{x}{y} = \frac{1}{2}$$

$$\cos\theta = \frac{x}{r} = \frac{1}{\sqrt{5}} \qquad\qquad \sec\theta = \frac{r}{x} = \frac{\sqrt{5}}{1} = \sqrt{5}$$

$$\tan\theta = \frac{y}{x} = \frac{2}{1} = 2 \qquad\qquad \csc\theta = \frac{r}{y} = \frac{\sqrt{5}}{2}$$

Problem Set 3.3

1. The point on the unit circle is $\left(-\dfrac{\sqrt{3}}{2}, \dfrac{1}{2}\right)$

$$\sin 150° = \dfrac{1}{2}$$

$$\cos 150° = \dfrac{-\sqrt{3}}{2}$$

$$\tan 150° = \dfrac{\sin 150°}{\cos 150°} = \dfrac{1/2}{-\sqrt{3}/2} = -\dfrac{1}{\sqrt{3}}$$

$$\cot 150° = \dfrac{1}{\tan 150°} = \dfrac{1}{-1/\sqrt{3}} = -\sqrt{3}$$

$$\sec 150° = \dfrac{1}{\cos 150°} = \dfrac{1}{-\sqrt{3}/2} = -\dfrac{2}{\sqrt{3}}$$

$$\csc 150° = \dfrac{1}{\sin 150°} = \dfrac{1}{1/2} = 2$$

3. The point on the unit circle is $\left(\dfrac{\sqrt{3}}{2}, -\dfrac{1}{2}\right)$

$$\sin \dfrac{11\pi}{6} = -\dfrac{1}{2}$$

$$\cos \dfrac{11\pi}{6} = \dfrac{\sqrt{3}}{2}$$

$$\tan \dfrac{11\pi}{6} = \dfrac{-1/2}{\sqrt{3}/2} = -\dfrac{1}{\sqrt{3}}$$

$$\cot \dfrac{11\pi}{6} = \dfrac{1}{-1/\sqrt{3}} = -\sqrt{3}$$

$$\sec \dfrac{11\pi}{6} = \dfrac{1}{\sqrt{3}/2} = \dfrac{2}{\sqrt{3}}$$

$$\csc \dfrac{11\pi}{6} = \dfrac{1}{-1/2} = -2$$

5. The point on the unit circle is $(-1, 0)$.

$$\sin 180° = 0$$

$$\cos 180° = -1$$

$$\tan 180° = \dfrac{0}{-1} = 0$$

$$\cot 180° = \dfrac{-1}{0} \text{ (undefined)}$$

$$\sec 180° = \dfrac{1}{-1} = -1$$

$$\csc 180° = \dfrac{1}{0} \text{ (undefined)}$$

7. The point on the unit circle is $\left(-\dfrac{1}{\sqrt{2}}, \dfrac{1}{\sqrt{2}}\right)$

$$\sin \dfrac{3\pi}{4} = \dfrac{1}{\sqrt{2}}$$

$$\cos \dfrac{3\pi}{4} = -\dfrac{1}{\sqrt{2}}$$

$$\tan \dfrac{3\pi}{4} = \dfrac{1/\sqrt{2}}{-1/\sqrt{2}} = -1$$

$$\cot \dfrac{3\pi}{4} = \dfrac{1}{-1} = -1$$

$$\sec \dfrac{3\pi}{4} = \dfrac{1}{-1/\sqrt{2}} = -\sqrt{2}$$

$$\csc \dfrac{3\pi}{4} = \dfrac{1}{1/\sqrt{2}} = \sqrt{2}$$

9. $\cos(-60°) = \cos 60°$

$$= \dfrac{1}{2}$$

11. $\cos\left(-\dfrac{5\pi}{6}\right) = \cos\left(\dfrac{5\pi}{6}\right)$

$$= -\dfrac{\sqrt{3}}{2}$$

13. $\sin(-30°) = -\sin 30°$

$\qquad = -\dfrac{1}{2}$

15. $\sin\left(-\dfrac{3\pi}{4}\right) = -\sin\left(\dfrac{3\pi}{4}\right)$

$\qquad = -\dfrac{1}{\sqrt{2}}$

17. On the unit circle, we locate all points with a y-coordinate of $\dfrac{1}{2}$. The angles associated with these points are $\dfrac{\pi}{6}$ and $\dfrac{5\pi}{6}$.

19. On the unit circle, we locate all points with an x-coordinate of $-\dfrac{\sqrt{3}}{2}$. The angles associated with these points are $\dfrac{5\pi}{6}$ and $\dfrac{7\pi}{6}$.

21. We look for points on the unit circle where the ratio, $\dfrac{y}{x}$, equals $-\sqrt{3}$. The angles associated with these points are $\dfrac{2\pi}{3}$ and $\dfrac{5\pi}{3}$.

39. $(x,y) = \left(\dfrac{1}{\sqrt{5}}, -\dfrac{2}{\sqrt{5}}\right)$

$\sin\theta = y \qquad\qquad \cos\theta = x \qquad\qquad \tan\theta = \dfrac{y}{x}$

$\quad = -\dfrac{2}{\sqrt{5}} \qquad\qquad = \dfrac{1}{\sqrt{5}} \qquad\qquad = \dfrac{-2/\sqrt{5}}{1/\sqrt{5}} = -2$

41. $\sin(-\theta) = -\sin\theta$

$\qquad = -\left(-\dfrac{1}{3}\right) = \dfrac{1}{3}$

43. $\sin\left(2\pi + \dfrac{\pi}{2}\right) = \sin\dfrac{\pi}{2}$

$\qquad\qquad = 1$

45. $\sin\left(2\pi + \dfrac{\pi}{6}\right) = \sin\dfrac{\pi}{6}$

$\qquad\qquad = \dfrac{1}{2}$

47. $\sin\left(\dfrac{5\pi}{2}\right) = \sin\left(2\pi + \dfrac{\pi}{2}\right)$

$\qquad\qquad = \sin\left(\dfrac{\pi}{2}\right) = 1$

49. $\sin\left(\dfrac{13\pi}{6}\right) = \sin\left(2\pi + \dfrac{\pi}{6}\right)$

$\qquad\qquad = \sin\dfrac{\pi}{6}$

$\qquad\qquad = \dfrac{1}{2}$

53. $\tan(-\theta) = \dfrac{\sin(-\theta)}{\cos(-\theta)}$

$\qquad\qquad = \dfrac{-\sin(\theta)}{\cos(\theta)}$

$\qquad\qquad = -\tan(\theta)$

Therefore, the tangent is an odd function.

55. $\sin(-\theta)\cot(-\theta) = (-\sin\theta)(-\cot\theta)$

$\qquad\qquad = \sin\theta\cot\theta$

$\qquad\qquad = \sin\theta\,\dfrac{\cos\theta}{\sin\theta}$

$\qquad\qquad = \cos\theta$

57. $\sin(-\theta)\sec(-\theta)\cot(-\theta) = \dfrac{\sin(-\theta)}{1}\cdot\dfrac{1}{\cos(-\theta)}\cdot\dfrac{\cos(-\theta)}{\sin(-\theta)} = 1$

59. $\csc\theta + \sin(-\theta) = \dfrac{1}{\sin\theta} + (-\sin\theta)$

$\qquad\qquad = \dfrac{1}{\sin\theta} - \dfrac{\sin\theta}{1}\cdot\dfrac{\sin\theta}{\sin\theta}$

$\qquad\qquad = \dfrac{1}{\sin\theta} - \dfrac{\sin^2\theta}{\sin\theta}$

$\qquad\qquad = \dfrac{1 - \sin^2\theta}{\sin\theta}$

$\qquad\qquad = \dfrac{\cos^2\theta}{\sin\theta}$

63. $B = 90° - 42° = 48°$

$\sin 42° = \dfrac{a}{36}$

$\qquad a = 36\sin 42°$

$\qquad\quad = 24$

$\cos 42° = \dfrac{b}{36}$

$\qquad b = 36\cos 42°$

$\qquad\quad = 27$

65. $A = 90° - 22° = 68°$

$\tan 22° = \dfrac{320}{a}$

$\qquad a = \dfrac{320}{\tan 22°} = 790$

$\sin 22° = \dfrac{320}{c}$

$\qquad c = \dfrac{320}{\sin 22°} = 850$

67.

$$\tan A = \frac{20.5}{31.4}$$

$$= 0.6529$$

$$A = \tan^{-1}(0.6529)$$

$$= 33.1°$$

$$B = 90° - 33.1°$$

$$= 56.9°$$

$$c = \sqrt{a^2 + b^2}$$

$$= \sqrt{(20.5)^2 + (31.4)^2}$$

$$= \sqrt{1406.21}$$

$$= 37.5$$

69.

$$b = \sqrt{c^2 - a^2}$$

$$= \sqrt{(6.21)^2 - (4.37)^2}$$

$$= \sqrt{19.4672}$$

$$= 4.41$$

$$\sin A = \frac{4.37}{6.21}$$

$$= 0.7037$$

$$A = 44.7°$$

$$B = 90° - 44.7°$$

$$= 45.3°$$

Problem Set 3.4

1.

$s = r\theta$	Formula for arc length
$= 3(2)$	Substitute given values
$= 6 \text{ in}$	Simplify

3.

$s = r\theta$	Formula for arc length
$= 1.5(1.5)$	Substitute given values
$= 2.25 \text{ ft}$	Simplify

5.

$s = r\theta$	Formula for arc length
$= 12\left(\dfrac{\pi}{6}\right)$	Substitute given values
$= 2\pi \text{ cm}$	Simplify
$= 6.28 \text{ cm}$	Rounded to 3 significant digits

7. Remember to change θ to radians by multiplying by $\dfrac{\pi}{180}$:

$s = r\theta$ Formula for arc length

$= 4\left[(60)\left(\dfrac{\pi}{180}\right)\right]$ Substitute given values (θ in radians)

$= \dfrac{4\pi}{3}$ mm   Simplify

$= 4.19$ mm   Rounded to 3 significant digits

9. Remember to change θ to radians by multiplying by $\dfrac{\pi}{180}$:

$s = r\theta$ Formula for arc length

$= 10\left[(240)\left(\dfrac{\pi}{180}\right)\right]$ Substitute given values (θ in radians)

$= \dfrac{40\pi}{3}$ in   Simplify

$= 41.9$ in   Rounded to 3 significant digits

11. First, we find θ: $\dfrac{\theta}{2\pi} = \dfrac{20}{60}$ One complete rotation is 60 minutes or 2π radians

$\theta = \dfrac{20}{60} \cdot 2\pi$ Multiply both sides by 2π

$= \dfrac{2\pi}{3}$ Simplify

The radius is 2.4 cm. Therefore, $s = r\theta$

$= 2.4\left(\dfrac{2\pi}{3}\right)$

$= 5.03$ cm

13. First, we find θ: $\dfrac{\theta}{2\pi} = \dfrac{1}{6}$ One complete rotation is 6 hours or 2π radians

$\theta = \dfrac{2\pi}{6}$ Multiply both sides by 2π

$= \dfrac{\pi}{3}$ Simplify

Also, the radius is 200 + 4,000 or 4,200 miles.

Therefore, $s = r\theta = 4,200\left(\dfrac{\pi}{3}\right) = 1,400\pi$ miles $= 4,400$ miles

15. Remember to change θ to radians by multiplying by $\dfrac{\pi}{180}$:

$$s = r\theta$$
$$= 4\left[20\left(\dfrac{\pi}{180}\right)\right]$$
$$= \dfrac{4\pi}{9} \text{ ft}$$
$$= 1.40 \text{ ft}$$

17. We are given that the diameter is 14 ft. Therefore, $r = \dfrac{1}{2}(14) = 7$ ft.

$$s = r\theta$$
$$= 7\left[270\left(\dfrac{\pi}{180}\right)\right]$$
$$= \dfrac{21\pi}{2} \text{ ft}$$
$$= 33.0 \text{ ft}$$

19. We are given that the diameter is 320 mm. Therefore, $r = \dfrac{1}{2}(320) = 160$ mm.

$$\theta = \dfrac{s}{r} \qquad\qquad \theta = 1.92\left(\dfrac{180}{\pi}\right)$$
$$= \dfrac{307}{160} \qquad\qquad = 110°$$
$$= 1.92 \text{ radians}$$

21. We convert $0.5°$ to radians by multiplying by $\dfrac{\pi}{180}$.

$$s = r\theta$$
$$= 240{,}000\left[(0.5)\left(\dfrac{\pi}{180}\right)\right]$$
$$= \dfrac{2000\pi}{3} \text{ mi}$$
$$= 2{,}100 \text{ mi}$$

23.

$$s = r\theta$$

$$= 125\left[30\left(\frac{\pi}{180}\right)\right]$$

$$= \frac{125\pi}{6}\text{ ft}$$

$$= 65.4\text{ ft}$$

25.

$$s = r\theta$$

$$= 125\left[(220)\left(\frac{\pi}{180}\right)\right]$$

$$= \frac{1{,}375\pi}{9}\text{ ft}$$

$$= 480\text{ ft}$$

27. $r = \dfrac{1}{2}(197) = 98.5\text{ ft}$

(a) $s = r\theta$

$$= 98.5\left[(60)\left(\frac{\pi}{180}\right)\right]$$

$$= 103\text{ ft}$$

(b) $s = r\theta$

$$= 98.5\left[(210)\left(\frac{\pi}{180}\right)\right]$$

$$= 361\text{ ft}$$

(c) $s = r\theta$

$$= 98.5\left[(285)\left(\frac{\pi}{180}\right)\right]$$

$$= 490\text{ ft}$$

29.

$$r = \frac{s}{\theta}\qquad\text{Formula for arc length}$$

$$= \frac{3}{6}\qquad\text{Substitute known values}$$

$$= 0.5\text{ ft}\qquad\text{Simplify}$$

31.

$$r = \frac{s}{\theta}\qquad\text{Formula for arc length}$$

$$= \frac{4.2}{1.4}\qquad\text{Substitute known values}$$

$$= 3\text{ in}\qquad\text{Simplify}$$

33.

$$r = \frac{s}{\theta}\qquad\text{Formula for arc length}$$

$$= \frac{\pi}{\pi/4}\qquad\text{Substitute known values}$$

$$= 4\text{ cm}\qquad\text{Simplify}$$

35.   Remember to convert θ to radians:

$$r = \frac{s}{\theta}\qquad\text{Formula for arc length}$$

$$= \frac{\dfrac{\pi}{2}}{90\left(\dfrac{\pi}{180}\right)}\qquad\text{Substitute known values}$$

$$= 1\text{ m}\qquad\text{Simplify}$$

37.   Remember to convert θ to radians:

$$r = \frac{s}{\theta}\qquad\text{Formula for arc length}$$

$$= \frac{4}{225\left(\dfrac{\pi}{180}\right)}\qquad\text{Substitute known values}$$

$$= \frac{16}{5\pi}\text{ or }1.02\text{ km}\qquad\text{Simplify}$$

39.

$$A = \frac{1}{2}r^2\theta \qquad \text{Formula for area of a sector}$$

$$= \frac{1}{2}(3)^2(2) \qquad \text{Substitute known values}$$

$$= 9\,\text{cm}^2 \qquad \text{Simplify}$$

41.

$$A = \frac{1}{2}r^2\theta \qquad \text{Formula for area of a sector}$$

$$= \frac{1}{2}(4)^2(2.4) \qquad \text{Substitute known values}$$

$$= 19.2\,\text{in}^2 \qquad \text{Simplify}$$

43.

$$A = \frac{1}{2}r^2\theta \qquad \text{Formula for area of a sector}$$

$$= \frac{1}{2}(3)^2\left(\frac{\pi}{5}\right) \qquad \text{Substitute known values}$$

$$= \frac{9\pi}{10} = 2.83\,\text{m}^2 \qquad \text{Simplify}$$

45.

$$A = \frac{1}{2}r^2\theta \qquad \text{Formula for area of a sector}$$

$$= \frac{1}{2}(5)^2\left(15 \cdot \frac{\pi}{180}\right) \qquad \text{Substitute known values}$$

$$= \frac{25\pi}{24} \text{ or } 3.27\,\text{m}^2 \qquad \text{Simplify}$$

47.

$$r = \frac{s}{\theta} = \frac{4\,\text{in}}{2\,\text{rad}} = 2\,\text{in}$$

$$A = \frac{1}{2}r^2\theta \qquad \text{Formula for area of a sector}$$

$$= \frac{1}{2}(2)^2(2) \qquad \text{Substitute known values}$$

$$= 4\,\text{in}^2 \qquad \text{Simplify}$$

49.

$$A = \frac{1}{2}r^2\theta \qquad \text{Formula for area of a sector}$$

$$r^2 = \frac{2A}{\theta} \qquad \text{Solve for } r^2$$

$$= \frac{2\left(\dfrac{\pi}{3}\right)}{30\left(\dfrac{\pi}{180}\right)} \qquad \text{Substitute known values}$$

$$= 4 \qquad \text{Simplify right side}$$

$$r = 2 \text{ cm} \qquad \text{Take square root of both sides } (r \text{ must be positive.})$$

51. $\theta = 45° = \dfrac{\pi}{4}$ radians

$A = \dfrac{1}{2}r^2\theta$ Formula for area of a sector

$r^2 = \dfrac{2A}{\theta}$ Solve for r^2

$= \dfrac{2\left(\dfrac{2\pi}{3}\right)}{\dfrac{\pi}{4}}$ Substitute known values

$= \dfrac{16}{3}$ Simplify right side

$r = \dfrac{4}{\sqrt{3}} = 2.31$ in   Take square root of both sides. (r must be positive)

53. $A = \dfrac{1}{2}r^2\theta$ Formula for area of a sector

$= \dfrac{1}{2}(60)^2\left(90\cdot\dfrac{\pi}{180}\right)$ Substitute known values

$= 900\pi$ or $2{,}830$ ft^2 Simplify

55. $r = \dfrac{700}{2} = 350$ mm

$\theta = \dfrac{2\pi}{8} = \dfrac{\pi}{4}$

$s = r\theta$

$= 350\left(\dfrac{\pi}{4}\right) = 275$ mm

57. $\tan\theta = \dfrac{75}{43}$

$= 1.7442$

$\theta = \tan^{-1}(1.7442)$

$= 60.2°$

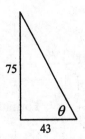

59.
$$\tan 12° = \frac{x}{5.2}$$
$$x = 5.2 \tan 12°$$
$$= 5.2(0.2126)$$
$$= 1.11$$
$$\tan 13° = \frac{y}{5.2}$$
$$y = 5.2 \tan 13°$$
$$= 5.2(0.2309)$$
$$= 1.20$$

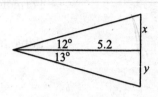

The vertical dimension of the mirror is
$$x + y = 1.11 + 1.20 = 2.31 \, \text{ft}$$

61.
$$\tan \theta = \frac{35.8}{10.25}$$
$$= 3.4927$$
$$\theta = \tan^{-1}(3.4927)$$
$$= 74.0°$$

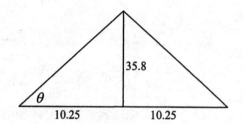

63.
$$\theta = 90° - 43.2° \qquad \tan 14.5° = \frac{y}{x}$$
$$= 46.8° \qquad\qquad\quad = \frac{y}{91.6}$$
$$\tan \theta = \frac{x}{86} \qquad\quad y = 91.6 \tan 14.5°$$
$$x = 86 \tan 46.8° \qquad = 23.7$$
$$= 91.6$$

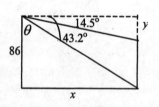

The height of the building is $86 - y = 86 - 23.7$
$$= 62.3 \, \text{ft}$$

Problem Set 3.5

1.
$$v = \frac{s}{t} \qquad \text{Formula for linear velocity}$$
$$= \frac{3}{2} \qquad \text{Substitute known values}$$
$$= 1.5 \, \text{ft/min} \quad \text{Simplify}$$

3.
$$v = \frac{s}{t} \qquad \text{Formula for linear velocity}$$
$$= \frac{12}{4} \qquad \text{Substitute known values}$$
$$= 3 \, \text{cm/sec} \quad \text{Simplify}$$

5.

$v = \dfrac{s}{t}$ Formula for linear velocity

$= \dfrac{30}{2}$ Substitute known values

$= 15 \, \text{mi/hr}$ Simplify

7.

$s = vt$ Formula for linear velocity

$= 20(4)$ Substitute known values

$= 80 \, \text{ft}$ Simplify

9.

$s = vt$ Formula for linear velocity

$= 45\left(\dfrac{1}{2}\right)$ Substitute known values

$= 22.5 \, \text{mi}$ Simplify

11. The time must be in hours

$t = \dfrac{20}{60} = \dfrac{1}{3} \, \text{hr}$

$s = vt$

$= 21\left(\dfrac{1}{3}\right)$

$= 7 \, \text{mi}$

13.

$\omega = \dfrac{\theta}{t}$ Formula for angular velocity

$= \dfrac{2\pi/3}{5}$ Substitute known values

$= \dfrac{2\pi}{15} \, \text{rad/sec}$ Simplify

$= 0.419 \, \text{rad/sec}$ Rounded to 3 significant digits

15.

$\omega = \dfrac{\theta}{t}$ Formula for angular velocity

$= \dfrac{12}{3}$ Substitute known values

$= 4 \, \text{rad/min}$

17.

$\omega = \dfrac{\theta}{t}$ Formula for angular velocity

$= \dfrac{8\pi}{3\pi}$ Substitute known values

$= \dfrac{8}{3} \, \text{rad/sec}$ Simplify

$= 2.67 \, \text{rad/sec}$ Rounded to 3 significant digits

19.

$\omega = \dfrac{\theta}{t}$ Formula for angular velocity

$= \dfrac{45\pi}{1.2}$ Substitute known values

$= 37.5\pi = 118 \, \text{rad/hr}$

21. We know that $\tan\theta = \dfrac{d}{100}$.

Therefore, $d = 100\tan\theta$.

Next, we must find θ. We know that $\omega = \dfrac{\theta}{t}$. Therefore, $\theta = \omega t$. We also know that

$$\omega = \frac{2\pi \text{ radians}}{4 \text{ seconds}} = \frac{1}{2}\pi \text{ radians per second}$$

We have $\theta = \dfrac{1}{2}\pi t$.

Substituting θ, we get: $d = 100\tan\theta$

$$= 100\tan\frac{1}{2}\pi t$$

When $t = \dfrac{1}{2}$, then $d = 100\tan\left[\dfrac{1}{2}(\pi)\left(\dfrac{1}{2}\right)\right]$

$$= 100\tan\frac{\pi}{4}$$

$$= 100(1)$$

$$= 100 \text{ feet}$$

When $t = \dfrac{3}{2}$, then $d = 100\tan\left[\dfrac{1}{2}(\pi)\left(\dfrac{3}{2}\right)\right]$

$$= 100\tan\frac{3\pi}{4}$$

$$= 100(-1)$$

$$= -100 \text{ feet}$$

When $t = 1$, then $d = 100\tan\left[\dfrac{1}{2}(\pi)(1)\right]$

$$= 100\tan\frac{\pi}{2}$$

d is undefined because $\tan\dfrac{\pi}{2}$ is undefined.

When $t = 1$, $\theta = \dfrac{\pi}{2}$ and the light rays are parallel to the wall.

23. First, we find θ using the formula for angular velocity. Then we apply the formula for arc length.

$\theta = \omega t$	Formula for angular velocity
$= 4(5)$	Substitute known values
$= 20 \text{ rad}$	Simplify
$s = r\theta$	Formula for arc length
$= 2(20)$	Substitute values
$= 40 \text{ in}$	Simplify

25. First, we find θ, using the formula for angular velocity. Then we apply the formula for arc length.

$\theta = \omega t$	Formula for angular velocity	$s = r\theta$	Formula for arc length
$= \dfrac{3\pi}{2}(30)$	Substitute known values	$= 4(45\pi)$	Substitute known values
$= 45\pi$	Simplify	$= 180\pi$ meters	Simplify
		$= 565$ meters	Rounded to 3 significant digits

27. Time must be in seconds: $t = 1(60) = 60$ seconds.

$\theta = \omega t$	Formula for angular velocity	$s = r\theta$	Formula for arc length
$= 15(60)$	Substitute known values	$= 5(900)$	Substitute known values
$= 900$ rad	Simplify	$= 4,500$ ft	Simplify

In problems 29 through 33, we convert revolutions per minute to radians per minute by multiplying by 2π.

29. $\omega = 10(2\pi)$

$= 20\pi$ rad / min

$= 62.8$ rad / min

31. $\omega = 33\dfrac{1}{3}(2\pi)$

$= \dfrac{200\pi}{3}$ rad / min

$= 209$ rad / min

33. $\omega = 5.8(2\pi)$

$= 11.6\pi$ rad / min

$= 36.4$ rad / min

Use the relationship between angular velocity and linear velocity, $V = r\omega$, for problems 35 – 39.

35. $v = r\omega$

$= 2(5)$

$= 10$ in / sec

37. $\omega = \dfrac{v}{r}$

$= \dfrac{3}{6}$

$= 0.5$ rad / sec

39. $v = r\omega$

$= 4\big[10(2\pi)\big]$

$= 80\pi$ ft / min

$= 251$ ft / min

41. Angular velocity, ω, is $\dfrac{1}{24}$ revolutions per hour. To convert this to radians per hour, we multiply by 2π.

$\omega = \dfrac{1}{24}(2\pi)$

$= \dfrac{\pi}{12}$ radians per hour

$= 0.262$ radians per hour

43. We are given that the radius is 3 inches and the angular velocity is 600 rev/min.
First, we will convert the angular velocity to rad/min:
$$\omega = 600 \cdot 2\pi = 1200\pi \frac{\text{rad}}{\text{min}}$$
Then we find the linear velocity:
$$v = r\omega$$
$$= 3(1200\pi)$$
$$= 3600\pi \text{ in / min}$$
We want the answer in feet, so we have to divide by 12:
$$v = \frac{3600\pi}{12}$$
$$= 300\pi \text{ ft / min}$$
$$= 942 \text{ ft / min}$$

45. Angular velocity, ω, is 19 revolutions per minute. We want to convert ω to radians per hour:
$$\omega = 19 \frac{\text{rev}}{\text{min}} \cdot 2\pi \frac{\text{rad}}{\text{rev}} \cdot \frac{60 \text{ min}}{1 \text{ hr}}$$
$$= 2280\pi \frac{\text{rad}}{\text{hr}}$$
Next we want to find the radius and convert it to miles:
$$r = \frac{1}{2}(14 \text{ ft}) \cdot \frac{1 \text{ mi}}{5,280 \text{ ft}}$$
$$= \frac{7}{5,280} \text{ mi}$$
Now we are ready to compute the linear velocity:
$$v = r\omega$$
$$= \frac{7}{5,280} \text{ mi} \cdot 2,280\pi \frac{\text{rad}}{\text{hr}}$$
$$= \frac{133\pi}{44} \cdot \frac{\text{mi}}{\text{hr}} = 9.50 \text{ mph}$$

47. Linear velocity is 10 miles per hour. We want to find angular velocity in revolutions per minute:
First we want to find the radius and convert it to miles:
$$r = \frac{1}{2}(12 \text{ ft}) \cdot \frac{1 \text{ mi}}{5,280 \text{ ft}}$$
$$= \frac{1}{880} \text{ mi}$$

This problem is continued on the next page.

Next, we find the angular velocity in radians per hour:

$$\omega = \frac{v}{r}$$

$$= \frac{10 \text{ mi/hr}}{\dfrac{1}{880} \text{ mi}}$$

$$= 8800 \frac{\text{rad}}{\text{hr}}$$

Now we convert ω to revolutions per minute:

$$\omega = 8800 \frac{\text{rad}}{\text{hr}} \cdot \frac{1 \text{ hr}}{60 \text{ min}} \cdot \frac{1 \text{ rev}}{2\pi \text{ rad}}$$

$$= \frac{220}{3\pi} \frac{\text{rev}}{\text{min}} = 23.3 \text{ rpm}$$

49.   Angular velocity, ω, is 9 revolutions per minute. We want to convert ω to radians per second:

$$\omega = 9 \frac{\text{rev}}{\text{min}} \cdot \frac{2\pi \text{ rad}}{1 \text{ rev}} \cdot \frac{1 \text{ min}}{60 \text{ sec}}$$

$$= \frac{3\pi}{10} \frac{\text{rad}}{\text{sec}}$$

Next, we want to find the radius in feet:

$$r = \frac{1}{2}(12 \text{ ft})$$

$$= 6 \text{ ft}$$

Now we can compute the linear velocity, in feet per second:

$$v = r\omega$$

$$= 6 \text{ ft}\left(\frac{3\pi}{10}\right) \frac{\text{rad}}{\text{sec}}$$

$$= \frac{9\pi}{5} \frac{\text{ft}}{\text{sec}}$$

$$= 5.65 \text{ ft/sec}$$

51.   Angular velocity, ω, is $\dfrac{1}{15}$ revolutions per minute. To convert to radians per minute, we multiply by 2π.

$$\omega = \frac{1}{15}(2\pi)$$

$$= \frac{2\pi}{15} \text{ radians per minute}$$

$$v = r\omega \qquad\qquad \text{Relationship between angular and linear velocities}$$

$$= \left(\frac{197}{2}\right)\left(\frac{2\pi}{15}\right) \qquad \text{Substitute known values}$$

This problem is continued on the next page

$$= \frac{197\pi}{15} \text{ ft/min}$$

To convert to miles, we divide by 5,280 ft and to convert to hours, we multiply by 60 min.

$$\frac{197\pi}{15} \cdot \frac{1}{5,280} \cdot \frac{60}{1} = 0.47 \frac{\text{mi}}{\text{hr}}$$

53.

$v = \dfrac{s}{t}$	Formula for linear velocity
$= \dfrac{16 \text{ km}}{1 \text{ hr}}$	Substitute known values
$= 16 \text{ km per hour}$	Simplify
$= 16,000 \text{ meters per hour}$	Change to meters per hour by multiplying by 1,000
$\omega = \dfrac{v}{r}$	Relationship between angular velocity and linear velocity
$= \dfrac{16,000 \text{ m/hr}}{0.3 \text{ m}}$	Substitute known values and change radius to meters
$= 53,300 \text{ radians per hour}$	Simplify

55. The angular velocity, ω_1, of the chain ring is 90 revolutions per minute. We want to convert this to radians per hour:

$$\omega_1 = 90 \frac{\text{rev}}{\text{min}} \cdot \frac{2\pi \text{ rad}}{1 \text{ rev}} \cdot \frac{60 \text{ min}}{1 \text{ hr}}$$

$$= 10,800\pi \frac{\text{rad}}{\text{hr}}$$

Next, we want the radii in kilometers:

$$\text{radius of the chain ring: } r_1 = \frac{1}{2}(150 \text{ mm}) \cdot \frac{1 \text{ km}}{1,000,000 \text{ mm}} = 0.000075 \text{ km}$$

$$\text{radius of the sprocket: } r_2 = \frac{1}{2}(95 \text{ mm}) \cdot \frac{1 \text{ km}}{1,000,000 \text{ mm}} = 0.0000475 \text{ km}$$

$$\text{radius of the wheel: } r_3 = \frac{1}{2}(700 \text{ mm}) \cdot \frac{1 \text{ km}}{1,000,000 \text{ mm}} = 0.000350 \text{ km}$$

We know that $r_1\omega_1 = r_2\omega_2$ because the linear velocity of the chain ring equals the linear velocity of the sprocket.

Substituting the given values, we can solve for ω_2:

$$0.000075(10,800\pi) = 0.0000475\omega_2$$

$$\frac{0.000075(10,800\pi)}{0.0000475} = \omega_2$$

$$\omega_2 = 53,572.42209 \text{ rad}$$

This problem is continued on the next page

We know that the angular velocity of the sprocket, ω_2, is equal to the angular velocity of the wheel, ω_3. Now we can find the linear velocity of the wheel, v_3:

$$v_3 = r_3\omega_3$$
$$= 0.000350(53,572.42209)$$
$$= 18.8\frac{km}{hr}$$

57. We use the same notation and formulas that we did in #55. In this problem:

$$r_1 = \frac{1}{2}(210\text{ mm}) = 105\text{ mm}$$

$$\omega_1 = 85\frac{rev}{min} \cdot \frac{2\pi \text{ rad}}{1 \text{ rev}} \cdot \frac{60 \text{ min}}{1 \text{ hr}} = 10,200\pi\frac{rad}{hr}$$

$$r_3 = 350\text{ mm}$$

$$v_3 = 45\frac{km}{hr} \cdot \frac{1,000,000 \text{ mm}}{1 \text{ km}} = 45,000,000\frac{mm}{hr}$$

We can find $\omega_2 = \omega_3$

$$\omega_2 = \omega_3 = \frac{v_3}{r_3} = \frac{45,000,000}{350} = 128,571.4286\frac{rad}{hr}$$

Now we can find r_2:

$$r_2\omega_2 = r_1\omega_1$$
$$r_2 = \frac{r_1\omega_1}{\omega_2}$$
$$= \frac{105(10,200\pi)}{128571.4286}$$
$$= 26.17\text{ mm}$$

The diameter is $2(26.17\text{ mm}) = 52.3\text{ mm}$

59. We will use the same notation and formulas used in problems 55 and 57.
We are given: $\quad r_1 = 75\text{ mm}$

$$r_2 = 40\text{ mm}$$
$$r_3 = 350\text{ mm}$$
$$v_3 = 20\frac{km}{hr}$$

We must convert v_3 to millimeters per minute:

$$20\frac{km}{hr} \cdot \frac{1,000,000 \text{ mm}}{1 \text{ km}} \cdot \frac{1 \text{ hr}}{60 \text{ min}} = \frac{1,000,000}{3}\frac{mm}{min}$$

Next we will find ω_3 which is equal to ω_2:

This problem is continued on the next page

$$\omega_3 = \omega_2 = \frac{v_3}{r_3} = \frac{\dfrac{1,000,000}{3}}{350} = \frac{20,000}{21} \frac{\text{rad}}{\text{min}}$$

Now we can solve for ω_1:

$$r_1\omega_1 = r_2\omega_2$$

$$\omega_1 = \frac{r_2\omega_2}{r_1}$$

$$= \frac{40\left(\dfrac{20,000}{21}\right)}{75} = 507.9365 \frac{\text{rad}}{\text{min}}$$

We must convert the angular velocity, ω_1, to $\dfrac{\text{rev}}{\text{min}}$:

$$507.9365 \frac{\text{rad}}{\text{min}} \cdot \frac{1\,\text{rev}}{2\pi\,\text{rad}} = 80.8 \frac{\text{rev}}{\text{min}}$$

61.   We can find x using the Pythagorean Theorem:

$$x = \sqrt{(9.50)^2 + (4.25)^2}$$

$$= \sqrt{108.3125}$$

$$= 10.4 \text{ mph}$$

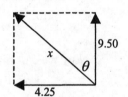

Now we'll find θ: $\tan\theta = \dfrac{4.25}{9.50}$

$$\tan\theta = 0.4474$$

$$\theta = \tan^{-1}(0.4474)$$

$$= 24.1°$$

The true course of the boat is 10.4 mph in the direction, N 24.1° W.

63. $|V| = 67 \text{ ft/sec}$ and $\theta = 37°$

$$|V_x| = |V|\cos\theta \qquad\qquad |V_y| = |V|\sin\theta$$

$$= 68\cos 37° \qquad\qquad\qquad = 68\sin 37°$$

$$= 54.3 \text{ ft/sec} \qquad\qquad\quad = 40.9 \text{ ft/sec}$$

65. $\theta = 180° + 32.7° = 212.7°$

$$|V_x| = |V|\cos\theta \qquad\qquad |V_y| = |V|\sin\theta$$

$$= 85.5\cos 212.7° \qquad\qquad = 85.5\sin 212.7°$$

$$= -71.9 \qquad\qquad\qquad\quad = -46.2$$

The ship has sailed 71.9 miles west and 46.2 miles south.

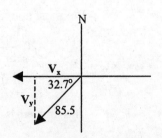

Chapter 3 Test

1. $\hat{\theta} = 235° - 180°$
  $= 55°$

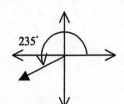

2. $\hat{\theta} = 180° - 117.8°$
  $= 62.2°$

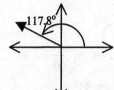

3. $\hat{\theta} = 410°20' - 360°$
  $= 50°20'$

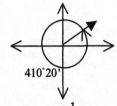

4. $\theta = -225° + 360°$
  $= 135°$

  $\hat{\theta} = 180° - 135°$
  $= 45°$

5. $\cot 320° = \dfrac{1}{\tan 320°}$ Reciprocal identity

  $= \dfrac{1}{-0.8391}$ Simplifying using a calculator

  $= -1.1918$

6. $\cot(-25°) = \dfrac{1}{\tan(-25°)}$ Reciprocal identity

  $= \dfrac{1}{-\tan 25°}$ Tangent is an odd function

  $= -\dfrac{1}{0.4663}$ Simplify, using a calculator

  $= -2.1445$

7. $\csc(-236.7°) = \dfrac{1}{\sin(-236.7°)}$ Reciprocal identity

  $= \dfrac{1}{-\sin 236.7°}$ Sine is an odd function

  $= -\dfrac{1}{0.8358}$ Simplify, using a calculator

  $= 1.1964$

8.
$$\sec(322.3°) = \frac{1}{\cos 322.3} \qquad \text{Reciprocal identity}$$

$$= \frac{1}{0.7912} \qquad \text{Simplify, using a calculator}$$

$$= 1.2639$$

9.
$$140°20' = 140° = \left(\frac{20}{60}\right)°$$

$$= 140.33°$$

$$\sec(140.33°) = \frac{1}{\cos 140.33°} \qquad \text{Reciprocal identity}$$

$$= \frac{1}{-0.7698} \qquad \text{Simplify, using a calculator}$$

$$= -1.2991$$

10.
$$188°50' = 188° + \left(\frac{50}{60}\right)°$$

$$= 188.833°$$

$$\csc(188.8.33°) = \frac{1}{\sin 188.833°} \qquad \text{Reciprocal identity}$$

$$= \frac{1}{-0.1536} \qquad \text{Simplify, using a calculator}$$

$$= -6.5121$$

11. $\sin\hat{\theta} = 0.1045 \qquad\qquad \theta = 180° - 6.0°$ **12.** $\cos\hat{\theta} = 0.4772 \qquad\qquad \theta = 180° - 61.5°$

$\qquad \hat{\theta} = 6.0° \qquad\qquad\qquad\quad = 174.0°$ $\qquad \hat{\theta} = 61.5° \qquad\qquad\qquad = 241.5°$

13. $\cot\hat{\theta} = 0.9659$

$\tan\hat{\theta} = \dfrac{1}{0.9659}$

Scientific Calculator: 0.9659 $\boxed{1/x}$ $\boxed{\text{inv}}$ $\boxed{\text{tan}}$

Graphing Calculator: $\boxed{\text{2nd}}$ $\boxed{\text{tan}}$ $\boxed{(}$ 0.9659 $\boxed{x^{-1}}$ $\boxed{)}$ $\boxed{\text{ENTER}}$

Answer: 46.0°

Since θ is in QIII, $\theta = 180° + \hat{\theta}$

$$= 180° + 46.0°$$

$$= 226.0°$$

14.

$\sec\hat{\theta} = 1.545$

$\cos\hat{\theta} = \dfrac{1}{1.545}$

Scientific Calculator: 1.545 $\boxed{1/x}$ $\boxed{\text{inv}}$ $\boxed{\cos}$

Graphing Calculator: $\boxed{\text{2nd}}$ $\boxed{\cos}$ $\boxed{(}$ 1.545 $\boxed{x^{-1}}$ $\boxed{)}$ $\boxed{\text{ENTER}}$

Answer: $49.7°$

Since θ is in QIV, $\theta = 360° - \hat{\theta}$

$= 360° - 49.7°$

$= 310.3°$

15.

$\hat{\theta} = 225° - 180°$

$\quad = 45°$

$\sin 225° = -\sin 45° \qquad \theta$ is in QIII

$\qquad = -\dfrac{1}{\sqrt{2}}$ or $-\dfrac{\sqrt{2}}{2}$

16.

$\hat{\theta} = 180° - 135°$

$\quad = 45°$

$\cos 135° = -\cos 45° \qquad \theta$ is in QII

$\qquad = -\dfrac{1}{\sqrt{2}}$ or $-\dfrac{\sqrt{2}}{2}$

17.

$\hat{\theta} = 360° - 330°$

$\quad = 30°$

$\tan 330° = -\tan 30° \qquad \theta$ is in QIV

$\qquad = -\dfrac{1}{\sqrt{3}}$

18.

$\hat{\theta} = 390° - 360°$

$\quad = 30°$

$\sec 390° = \sec 30° \qquad \theta$ is in QI

$\qquad = \dfrac{1}{\cos 30°}$

$\qquad = \dfrac{1}{\sqrt{3}/2}$

$\qquad = \dfrac{2}{\sqrt{3}}$

19. $\theta = 250 \cdot \dfrac{\pi}{180} = \dfrac{25\pi}{18}$ radians

20. $\theta = -390 \cdot \dfrac{\pi}{180} = -\dfrac{13\pi}{6}$ radians

21. $\theta = \dfrac{4\pi}{3}\left(\dfrac{180}{\pi}\right)°$

$\quad = 240°$

22. $\theta = \dfrac{7\pi}{12}\left(\dfrac{180}{\pi}\right)°$

$\quad = 105°$

23. $\hat{\theta} = \pi - \dfrac{2\pi}{3} = \dfrac{\pi}{3}$

$\sin\dfrac{2\pi}{3} = \sin\dfrac{\pi}{3}$ θ is in QII

$= \dfrac{\sqrt{3}}{2}$

24. $\hat{\theta} = \pi - \dfrac{2\pi}{3} = \dfrac{\pi}{3}$

$\cos\dfrac{2\pi}{3} = -\cos\dfrac{\pi}{3}$ θ is in QII

$= -\dfrac{1}{2}$

25. $4\cos\left(-\dfrac{3\pi}{4}\right) = 4\cos\dfrac{3\pi}{4}$ Cosine is an even function

$= 4\left(-\cos\dfrac{\pi}{4}\right)$ θ is in QII

$= 4\left(-\dfrac{1}{\sqrt{2}}\right)$ Substitute exact value

$= -\dfrac{4}{\sqrt{2}}$ Simplify

$= -2\sqrt{2}$ Rationalize the denominator

26. $2\cos\left(-\dfrac{5\pi}{3}\right) = 2\cos\dfrac{5\pi}{3}$ Cosine is an even function

$= 2\cos\dfrac{\pi}{3}$ θ is in QIV

$= 2\left(\dfrac{1}{2}\right)$ Substitute exact value

$= 1$ Simplify

27. $\sec\dfrac{5\pi}{6} = -\sec\dfrac{\pi}{6}$ θ is in QII

$= -\dfrac{2}{\sqrt{3}}$ Substitute exact value

28. $\csc\dfrac{5\pi}{6} = \csc\dfrac{\pi}{6}$ θ is in QII

$= 2$ Substitute exact value

29.

$$2\cos\left(3x - \frac{\pi}{2}\right) = 2\cos\left[3\left(\frac{\pi}{3}\right) - \frac{\pi}{2}\right]$$

$$= 2\cos\left(\pi - \frac{\pi}{2}\right)$$

$$= 2\cos\frac{\pi}{2}$$

$$= 2(0)$$

$$= 0$$

30.

$$4\sin\left(2x + \frac{\pi}{4}\right) = 4\sin\left[2\left(\frac{\pi}{4}\right) + \frac{\pi}{4}\right]$$

$$= 4\sin\frac{3\pi}{4}$$

$$= 4\left(\frac{\sqrt{2}}{2}\right)$$

$$= 2\sqrt{2}$$

31.

$$\cot(-x) = \frac{\cos(-x)}{\sin(-x)} \qquad \text{Ratio identity}$$

$$= \frac{\cos x}{-\sin x} \qquad \text{Cosine is an even function and sine is an odd function}$$

$$= -\frac{\cos x}{\sin x}$$

$$= -\cot x$$

Therefore, cotangent is an odd function.

32.

$$\sin(-\theta)\sec(-\theta)\cot(-\theta) = \sin(-\theta) \cdot \frac{1}{\cos(-\theta)} \cdot \frac{\cos(-\theta)}{\sin(-\theta)} = 1$$

33.

$$s = r\theta$$

$$= 12\left(\frac{\pi}{6}\right)$$

$$= 2\pi \quad \text{meters}$$

$$= 6.28 \quad \text{meters}$$

34.

$$s = r\theta$$

$$= 6\left[60\left(\frac{\pi}{180}\right)\right] \qquad \theta \text{ must be in radians}$$

$$= 6\left(\frac{\pi}{3}\right)$$

$$= 2\pi \text{ or } 6.28 \text{ ft}$$

35.

$$r = \frac{s}{\theta}$$

$$= \frac{\pi}{\pi/4}$$

$$= 4 \text{ cm}$$

36.

$$r = \frac{s}{\theta}$$

$$= \frac{\pi/4}{2\pi/3}$$

$$= \frac{3}{8} \text{ or } 0.375 \text{ cm}$$

37. $A = \dfrac{1}{2}r^2\theta$ where $\theta = 90\left(\dfrac{\pi}{180}\right) = \dfrac{\pi}{2}$ radians

$\quad = \dfrac{1}{2}(4)^2\left(\dfrac{\pi}{2}\right)$

$\quad = 4\pi \text{ in}^2$

$\quad = 12.6 \text{ in}^2$

38. $A = \dfrac{1}{2}r^2\theta$

$\quad = \dfrac{1}{2}(3)^2(2.4)$

$\quad = 10.8 \text{ cm}^2$

39. In 30 minutes, θ is π radians.

$s = r\theta$

$\quad = 2\pi \text{ or } 6.28 \text{ cm}$

40. $r = 5$ ft, $\omega = 0.5 \dfrac{\text{rev}}{\text{min}}$, and $t = 2$ min

$\omega = 0.5\dfrac{\text{rev}}{\text{min}} \cdot \dfrac{2\pi \text{ rad}}{1 \text{ rev}} = \pi \dfrac{\text{rad}}{\text{min}}$

$v = r\omega$

$\quad = 5 \text{ ft}\left(\dfrac{\pi \text{ rad}}{\text{min}}\right) = 5\pi \dfrac{\text{ft}}{\text{min}}$

$d = v \cdot t$

$\quad = (5\pi \text{ ft}/\text{min})(2 \text{ min})$

$\quad = 10\pi = 31.4 \text{ ft}$

41. $r = \dfrac{s}{\theta}$ $\qquad A = \dfrac{1}{2}r^2\theta$

$\quad = \dfrac{8}{4}$ $\qquad\quad = \dfrac{1}{2}(2)^2(4)$

$\quad = 2$ $\qquad\qquad = 8 \text{ in}^2$

42. $s = vt$

$\quad = 30(3)$

$\quad = 90 \text{ feet}$

43. $s = vt$

$\quad = 660(60)$ (Time must be in seconds)

$\quad = 3960 \text{ ft}$

44.

$\theta = \omega t$	Formula for angular velocity
$\quad = 4(6)$	Substitute known values
$\quad = 24 \text{ radians}$	Simplify
$s = r\theta$	Formula for arc length
$\quad = 3(24)$	Substitute known values
$\quad = 72 \text{ in}$	Simplify

45.

$\theta = \omega t$ Formula for angular velocity

$= \dfrac{3\pi}{4}(20)$ Substitute known values

$= 15\pi$ Simplify

$s = r\theta$ Formula for arc length

$= 8(15\pi)$ Substitute known values

$= 120\pi$ or 377 ft   Simplify

46.

$\omega = 6(2\pi)$

$= 12\pi \, \text{rad/min}$

$= 37.7 \, \text{rad/min}$

47.

$\omega = 2(2\pi)$

$= 4\pi \, \text{rad/min}$

$\doteq 12.6 \, \text{rad/min}$

48.

$\omega = \dfrac{v}{r}$ Relationship between angular and linear velocities

$= \dfrac{5}{10}$ Substitute known values

$= 0.5 \, \text{rad/sec}$ Simplify

49.

$\omega = \dfrac{v}{r}$ Relationship between angular and linear velocities

$= \dfrac{5}{3} \, \text{rad/sec}$ Substitute known values

50.

$\omega = 20(2\pi)$

$= 40\pi \, \text{rad/min}$

$v = r\omega$

$= 2(40\pi)$

$= 80\pi \, \text{ft/min}$

$= 251 \, \text{ft/min}$

51.

$\omega = 10(2\pi)$

$= 20\pi \, \text{rad/min}$

$v = r\omega$

$= 1(20\pi)$

$= 20\pi \, \text{ft/min}$

$= 62.8 \, \text{ft/min}$

52.

$\omega_1 = \dfrac{v}{r_1}$

$= \dfrac{24}{6}$

$= 4 \, \text{rad/sec} \, (\text{for 6 cm pulley})$

$\omega_2 = \dfrac{v}{r_2}$

$= \dfrac{24}{8}$

$= 3 \, \text{rad/sec} \, (\text{for 8 cm pulley})$

53. $\omega = 900(2\pi)$

$= 1,800\pi \text{ rad/min}$

$v = r\omega$

$= 1.5(1800\pi)$

$= 2,700\pi \text{ ft/min}$

$= 8,480 \text{ ft/min}$

54. $\omega = 300\dfrac{\text{rev}}{\text{min}}$ and $r = 1.5 \text{ in}$

$= 300\dfrac{\text{rev}}{\text{min}} \cdot \dfrac{2\pi \text{ rad}}{1 \text{ rev}} = 600\pi \dfrac{\text{rad}}{\text{min}}$

$v = r\omega$

$= (1.5 \text{ in})\left(600\pi \dfrac{\text{rad}}{\text{min}}\right)$

$= 900\pi \dfrac{\text{in}}{\text{min}}$

$= 2,830 \dfrac{\text{in}}{\text{min}}$

55. $\omega = 2941\dfrac{\text{rev}}{\text{min}}$

$= 2941\dfrac{\text{rev}}{\text{min}} \cdot \dfrac{2\pi \text{ rad}}{1 \text{ rev}} = 5882\pi \dfrac{\text{rad}}{\text{min}}$

$r = 1 \text{ in}$ $\qquad\qquad\qquad\qquad$ $r = 1.5 \text{ in}$

$v = r\omega$ $\qquad\qquad\qquad\qquad$ $v = r\omega$

$= 1(5882\pi)$ $\qquad\qquad\qquad$ $= 1.5(5882\pi)$

$= 5882\pi \dfrac{\text{in}}{\text{min}}$ $\qquad\qquad$ $= 8823\pi \dfrac{\text{in}}{\text{min}}$

We must convert this to miles per hour:

$5882\pi \dfrac{\text{in}}{\text{min}} \cdot \dfrac{60 \text{ min}}{1 \text{ hr}} \cdot \dfrac{1 \text{ ft}}{12 \text{ in}} \cdot \dfrac{1 \text{ mi}}{5280 \text{ ft}} = 175 \dfrac{\text{mi}}{\text{hr}}$ $\qquad$ $8823\pi \dfrac{\text{in}}{\text{min}} \cdot \dfrac{60 \text{ min}}{1 \text{ hr}} \cdot \dfrac{1 \text{ ft}}{12 \text{ in}} \cdot \dfrac{1 \text{ mi}}{5280 \text{ ft}} = 26.2 \dfrac{\text{mi}}{\text{hr}}$

56. $v = 11 \dfrac{\text{mi}}{\text{hr}} \cdot \dfrac{5,280\,\text{ft}}{1\,\text{mi}} \cdot \dfrac{1\,\text{hr}}{60\,\text{min}}$ and $r = \dfrac{1}{2}(14\,\text{ft}) = 7\,\text{ft}$

$\quad\quad = 968\,\text{ft}$

$\omega = \dfrac{v}{r}$

$\quad = \dfrac{968\,\text{ft}/\text{min}}{7\,\text{ft}}$

$\quad = \dfrac{968}{7}\dfrac{\text{rad}}{\text{min}}$

$\omega = \dfrac{968}{7}\dfrac{\text{rad}}{\text{min}} \cdot \dfrac{1\,\text{rev}}{2\pi\,\text{rad}}$

$\quad = 22.0 \dfrac{\text{rev}}{\text{min}}$

57. The radius of the chain ring, r_1, is $\dfrac{1}{2}(210)\,\text{mm} = 105\,\text{mm}$.

The radius of the sprocket, r_2, is $\dfrac{1}{2}(50)\,\text{mm} = 25\,\text{mm}$.

The radius of the wheel, r_3, is $\dfrac{1}{2}(700)\,\text{mm} = 350\,\text{mm}$.

Also, the angular velocity of the chain ring, ω_1, is $75\,\dfrac{\text{rev}}{\text{min}}$.

We must convert the radii to kilometers and the angular velocity to radians per hour:

$r_1 = 105\,\text{mm} \cdot \dfrac{1\,\text{km}}{1,000,000\,\text{mm}} = 0.000105$

$r_2 = 25\,\text{mm} \cdot \dfrac{1\,\text{km}}{1,000,000\,\text{mm}} = 0.000025$

$r_3 = 350\,\text{mm} \cdot \dfrac{1\,\text{km}}{1,000,000\,\text{mm}} = 0.000350$

$\omega_1 = 75\,\dfrac{\text{rev}}{\text{min}} \cdot \dfrac{2\pi\,\text{rad}}{1\,\text{rev}} \cdot \dfrac{60\,\text{min}}{1\,\text{hr}} = 9,000\pi\,\dfrac{\text{rad}}{\text{hr}}$

Now we can compute ω_2 and v_3:

$r_2\omega_2 = r_1\omega_1$

$\omega_2 = \dfrac{r_1\omega_1}{r_2} = \dfrac{(0.000105)(9,000\pi)}{0.000025} = 118,752.2023$

Remember $\omega_2 = \omega_3 : v_3 = r_3\omega_3$

$v_3 = 0.000350(118,752.2023)$

$\quad = 41.6\,\dfrac{\text{km}}{\text{hr}}$

CHAPTER 4   Graphing and Inverse Functions

Problem Set 4.1

13. Refer to the graphs in problem 1 and 8. Find the x-values corresponding to $y = 0$. You will notice the pattern that exists: $x = \dfrac{\pi}{2} + k\pi$, where k is any integer.

15. Refer to the graphs in problems 4 and 7. Find the x-values corresponding to $y = 1$. You will notice the pattern that exists: $x = \dfrac{\pi}{2} + 2k\pi$, where k is any integer.

17. Refer to the graphs in problems 5 and 12. Find the x-values corresponding to $y = 1$. You will notice the pattern that exists: $x = k\pi$, where k is any integer.

19. Refer to the graphs in problems 3 and 10. Find the x-values corresponding to $y = 1$. You will notice the pattern that exists: $x = \dfrac{\pi}{2} + 2k\pi$, where k is any integer.

21. Refer to the graphs in problems 5 and 12. Find the x-values where there is a vertical asymptote and the y-value is undefined. You will notice the pattern that exists: $x = \dfrac{\pi}{2} + k\pi$, where k is any integer.

23. Refer to the graphs in problems 3 and 10. Find the x-values where there is a vertical asymptote and the y-value is undefined. You will notice the pattern that exists: $x = k\pi$, where k is any integer.

25. The amplitude is 3 because the greatest y-value is 3 and the least y-value is -3. The period is π because the graph repeats itself every π units, or $f(x + \pi) = f(x)$.

27. The amplitude is 2 because the greatest y-value is 2 and the least y-value is -2. The period is 2 because the graph repeats itself every 2 units, or $f(x + 2) = f(x)$.

29. The amplitude is 3 because the greatest y-value is 3 and the least y-value is -3. The period is π because one complete curve goes from $-\dfrac{\pi}{4}$ to $\dfrac{3\pi}{4}$ or $\dfrac{3\pi}{4} - \left(-\dfrac{\pi}{4}\right) = \dfrac{3\pi}{4} + \dfrac{\pi}{4} = \dfrac{4\pi}{4} = \pi$.

31.

x	0	$\pi/2$	π	$3\pi/2$	2π
$y = 2\sin x$	0	2	0	–2	0

See example 3 in text for the graph. The amplitude of this curve is 2.

33. Look at example 1 in section 4.2 in your text for the table and graph of this curve. The amplitude of this curve is 1 and the period is π.

35. The value of A affects the amplitude of the graph. It stretches the graph away from the x-axis when $A>1$.

37. The value of A affects the amplitude of the graph. It compresses the graph closer to the x-axis when $0<A<1$.

39. The value of A compresses the graph closer to the x-axis when $0<A<1$.

41. The value of A stretches the graph away from the x-axis when $A>1$.

43. The value of B affects the period of the graph. When B is 2, there is one complete cycle of the graph between 0 and π. The period of this graph is π.

45. The value of B affects the period of the graph. When B is $\frac{1}{2}$, there is one complete cycle of the graph between 0 and 4π. The period of this graph is 4π.

47. The value of B affects the period of the graph. When B is 2, there is one complete cycle of the graph between 0 and $\frac{\pi}{2}$. The period of this graph is $\frac{\pi}{2}$.

49. The value of B affects the period of the graph. When B is $\frac{1}{2}$, there is one complete cycle of the graph between 0 and 4π. The period of this graph is 4π.

63. The negative sign reflects the graph across the x-axis.

65. The negative sign reflects the graph across the x-axis.

67. The negative sign reflects the graph across the x-axis.

69. $$\cos\theta\tan\theta = \cos\theta \cdot \frac{\sin\theta}{\cos\theta} \qquad \text{Ratio identity}$$
$$= \frac{\cos\theta\sin\theta}{\cos\theta} \qquad \text{Multiplication of fractions}$$
$$= \sin\theta \qquad \text{Division of common factor}$$

71. $$(1+\sin\theta)(1-\sin\theta) = 1-\sin^2\theta \qquad \text{Multiplication}$$
$$= \cos^2\theta \qquad \text{Pythagorean identity}$$

73. $$\csc\theta + \sin(-\theta) = \csc\theta - \sin\theta \qquad \text{Sine is an odd function}$$
$$= \frac{1}{\sin\theta} - \sin\theta \qquad \text{Reciprocal identity}$$
$$= \frac{1}{\sin\theta} - \sin\theta \cdot \frac{\sin\theta}{\sin\theta} \qquad \text{L.C.D. is } \sin\theta$$
$$= \frac{1}{\sin\theta} - \frac{\sin^2\theta}{\sin\theta} \qquad \text{Multiplication of fractions}$$
$$= \frac{1-\sin^2\theta}{\sin\theta} \qquad \text{Subtraction of fractions}$$
$$= \frac{\cos^2\theta}{\sin\theta} \qquad \text{Pythagorean identity}$$

75. $\dfrac{\pi}{3} = \dfrac{\pi}{3} \cdot \left(\dfrac{180}{\pi}\right)^{\circ} = 60^{\circ}$

77. $\dfrac{\pi}{4} = \dfrac{\pi}{4} \cdot \left(\dfrac{180}{\pi}\right)^{\circ} = 45^{\circ}$

79. $\dfrac{2\pi}{3} = \dfrac{2\pi}{3} \cdot \left(\dfrac{180}{\pi}\right)^{\circ} = 120^{\circ}$

81. $\dfrac{11\pi}{6} = \dfrac{11\pi}{6} \cdot \left(\dfrac{180}{\pi}\right)^{\circ} = 330^{\circ}$

Problem Set 4.2

1. $y = \sin 2x$

        Period $= \dfrac{2\pi}{2} = \pi$

3. $y = \cos \dfrac{1}{3}x$

        Period $= \dfrac{2\pi}{1/3} = 6\pi$

5. $y = \sin \pi x$

        Period $= \dfrac{2\pi}{\pi} = 2$

7. $y = \sin \dfrac{\pi}{2}x$

        Period $= \dfrac{2\pi}{\pi/2} = 4$

9. $y = \csc 3x$

        Period $= \dfrac{2\pi}{3}$

11. $y = \sec \dfrac{\pi}{4}x$

        Period $= \dfrac{2\pi}{\pi/4} = 8$

13. $y = \tan \dfrac{1}{2}x$

        Period $= \dfrac{\pi}{1/2} = 2\pi$

15. $y = \cot 4x$

        Period $= \dfrac{\pi}{4}$

17. $y = 4\sin 2x$ Amplitude $= 4$

                          Period $= \dfrac{2\pi}{2} = \pi$

19. $y = 3\sin \dfrac{1}{2}x$ Amplitude $= 3$

                          Period $= \dfrac{2\pi}{1/2} = 4\pi$

21. $y = \dfrac{1}{2}\cos 3x$ Amplitude $= \dfrac{1}{2}$

                          Period $= \dfrac{2\pi}{3}$

23. $y = \dfrac{1}{2}\sin \dfrac{\pi}{2}x$ Amplitude $= \dfrac{1}{2}$

                          Period $= \dfrac{2\pi}{\pi/2} = 4$

25. $y = \dfrac{1}{2}\csc 3x$ Period $= \dfrac{2\pi}{3}$

                          No amplitude

27. $y = 3\sec \dfrac{1}{2}x$ Period $= \dfrac{2\pi}{1/2} = 4\pi$

                          No amplitude

29. $y = 2\tan 3x$ $\quad$ Period $= \dfrac{\pi}{3}$

$\qquad\qquad\qquad\qquad$ No amplitude

31. $y = \dfrac{1}{2}\cot\dfrac{\pi}{2}x$ $\quad$ Period $= \dfrac{\pi}{\pi/2} = 2$

$\qquad\qquad\qquad\qquad$ No amplitude

33. $y = 4 + 4\sin 2x$ $\qquad$ Period $= \dfrac{2\pi}{2} = \pi$

$\qquad\qquad\qquad\qquad$ Amplitude $= 4$

$\qquad\qquad\qquad\qquad$ Vertical translation $= 4$

The graph in problem #17 is shifted up 4 units.

35. $y = -1 + \dfrac{1}{2}\cos 3x$ $\qquad$ Period $= \dfrac{2\pi}{3}$

$\qquad\qquad\qquad\qquad\qquad$ Amplitude $= \dfrac{1}{2}$

$\qquad\qquad\qquad\qquad\qquad$ Vertical translation $= -1$

The graph in problem #21 is shifted down 1 unit.

37. $y = -2 + 3\sec\dfrac{1}{2}x$ $\qquad$ Period $= \dfrac{2\pi}{1/2} = 4\pi$

$\qquad\qquad\qquad\qquad\qquad$ No amplitude

$\qquad\qquad\qquad\qquad\qquad$ Vertical translation $= -2$

The graph in problem #27 is shifted down 2 units.

39. $y = 3 + \dfrac{1}{2}\cot\dfrac{\pi}{2}x$ $\qquad$ Period $= \dfrac{\pi}{\pi/2} = 2$

$\qquad\qquad\qquad\qquad\qquad$ No amplitude

$\qquad\qquad\qquad\qquad\qquad$ Vertical translation $= 3$

The graph in problem #31 is shifted up 3 units.

41. $y = 2\sin \pi x$ $\quad$ Period $= \dfrac{2\pi}{\pi} = 2$

$\qquad\qquad\qquad\qquad$ Amplitude $= 2$

43. $y = 3\sin 2x$ $\quad$ Period $= \dfrac{2\pi}{2} = \pi$

$\qquad\qquad\qquad\qquad$ Amplitude $= 3$

45. $y = -3\cos\dfrac{1}{2}x$ $\quad$ Period $= \dfrac{2\pi}{1/2} = 4\pi$

$\qquad\qquad\qquad\qquad$ Amplitude $= |-3| = 3$

This graph is reflected across the x-axis.

47. $y = -2\sin(-3x)$ Period $= \dfrac{2\pi}{3}$

$\quad\quad = -(-2\sin 3x)$ Amplitude $= 2$

$\quad\quad = 2\sin 3x$

Note: $\sin(-x) = -\sin x$ because sine is an odd function.

49. $y = -2\csc 3x$ Period $= \dfrac{2\pi}{3}$

No amplitude

This graph is reflected across the x - axis.

51. $y = -\cot 2x$ Period $= \dfrac{\pi}{2}$

No amplitude

This graph is reflected across the x - axis.

55. $d = 100\tan\dfrac{\pi}{2}t$

t	0	1/2	1	3/2	2	5/2	3	7/2	4
d	0	100	N.D.	−100	0	100	N.D.	−100	0

63.   (a)   There is a horizontal shift to the left of $\dfrac{\pi}{4}$ units.

(b)   There is a horizontal shift to the right of $\dfrac{\pi}{4}$ units.

65.   (a)   There is a horizontal shift to the left of $\dfrac{\pi}{6}$ units.

(b)   There is a horizontal shift to the right of 6 units.

67. $\sin\left(x + \dfrac{\pi}{2}\right) = \sin\left(\dfrac{\pi}{2} + \dfrac{\pi}{2}\right)$ **69.** $\cos\left(y - \dfrac{\pi}{6}\right) = \cos\left(\dfrac{\pi}{6} - \dfrac{\pi}{6}\right)$

$\quad\quad\quad\quad\quad = \sin\pi$ $\quad\quad\quad\quad\quad = \cos 0$

$\quad\quad\quad\quad\quad = 0$ $\quad\quad\quad\quad\quad = 1$

71.
$$\sin(x+y) = \sin\left(\frac{\pi}{2} + \frac{\pi}{6}\right)$$
$$= \sin\left(\frac{2\pi}{3}\right)$$
$$= \sin\frac{\pi}{3} = \frac{\sqrt{3}}{2}$$

73.
$$\sin x + \sin y = \sin\frac{\pi}{2} + \sin\frac{\pi}{6}$$
$$= 1 + \frac{1}{2}$$
$$= \frac{3}{2}$$

75.
$$45° = 45\left(\frac{\pi}{180}\right)$$
$$= \frac{\pi}{4}$$

77.
$$60° = 60\left(\frac{\pi}{180}\right)$$
$$= \frac{\pi}{3}$$

79.
$$150° = 150\left(\frac{\pi}{180}\right)$$
$$= \frac{5\pi}{6}$$

81.
$$225° = 225\left(\frac{\pi}{180}\right)$$
$$= \frac{5\pi}{4}$$

Problem Set 4.3

11. Amplitude $= 1$ Period $= \dfrac{2\pi}{2} = \pi$ Phase Shift $= \dfrac{\pi}{2}$

Spacing $= \dfrac{1}{4}(\pi) = \dfrac{\pi}{4}$ $c = \dfrac{3\pi}{4} + \dfrac{\pi}{4} = \pi$

$a = $ starting point $= \dfrac{\pi}{2}$ $d = \pi + \dfrac{\pi}{4} = \dfrac{5\pi}{4}$

$b = \dfrac{\pi}{2} + \dfrac{\pi}{4} = \dfrac{3\pi}{4}$ $e = \dfrac{5\pi}{4} + \dfrac{\pi}{4} = \dfrac{6\pi}{4} = \dfrac{3\pi}{2}$

The 5 points we use on the x-axis are $\dfrac{\pi}{2}, \dfrac{3\pi}{4}, \pi, \dfrac{5\pi}{4}, \dfrac{3\pi}{2}$.

The 2 points we use on the y-axis are -1 and 1.

13. Amplitude $= 1$ Period $= \dfrac{2\pi}{\pi} = 2$ Phase Shift $= \dfrac{-\dfrac{\pi}{2}}{\pi} = -\dfrac{1}{2}$

Spacing $= \dfrac{1}{4}(2) = \dfrac{1}{2}$ $c = 0 + \dfrac{1}{2} = \dfrac{1}{2}$

This problem is continued on the next page.

$$a = \text{starting point} = -\frac{1}{2} \qquad\qquad d = \frac{1}{2} + \frac{1}{2} = 1$$

$$b = -\frac{1}{2} + \frac{1}{2} = 0 \qquad\qquad e = 1 + \frac{1}{2} = \frac{3}{2}$$

The 5 points we use on the x-axis are $-\frac{1}{2}, 0, \frac{1}{2}, 1, \frac{3}{2}$.

The 2 points we use on the y-axis are -1 and 1.

15. Amplitude $= |-1| = 1$ $\qquad$ Period $= \frac{2\pi}{2} = \pi$ $\qquad$ Phase Shift $= \dfrac{-\frac{\pi}{2}}{2} = -\frac{\pi}{4}$

$$\text{Spacing} = \frac{1}{4}(\pi) = \frac{\pi}{4} \qquad\qquad c = 0 + \frac{\pi}{4} = \frac{\pi}{4}$$

$$a = -\frac{\pi}{4} \qquad\qquad\qquad\qquad d = \frac{\pi}{4} + \frac{\pi}{4} = \frac{\pi}{2}$$

$$b = -\frac{\pi}{4} + \frac{\pi}{4} = 0 \qquad\qquad e = \frac{\pi}{2} + \frac{\pi}{4} = \frac{3\pi}{4}$$

The 5 points we use on the x-axis are $-\frac{\pi}{4}, 0, \frac{\pi}{4}, \frac{\pi}{2}, \frac{3\pi}{4}$.

The 2 points we use on the y-axis are -1 and 1. The graph is reflected across the x-axis.

17. Amplitude $= 2$ $\qquad$ Period $= \dfrac{2\pi}{\frac{1}{2}} = 4\pi$ $\qquad$ Phase Shift $= \dfrac{-\frac{\pi}{2}}{\frac{1}{2}} = -\pi$

$$\text{Spacing} = \frac{1}{4}(4\pi) = \pi \qquad\qquad c = 0 + \pi = \pi$$

$$a = -\pi \qquad\qquad\qquad\qquad d = \pi + \pi = 2\pi$$

$$b = -\pi + \pi = 0 \qquad\qquad\quad e = 2\pi + \pi = 3\pi$$

The 5 points we use on the x-axis are $-\pi, 0, \pi, 2\pi, 3\pi$.

The 2 points we use on the y-axis are -2 and 2.

19.   Amplitude $= \dfrac{1}{2}$ Period $= \dfrac{2\pi}{3}$ Phase Shift $= \dfrac{\frac{\pi}{2}}{3} = \dfrac{\pi}{6}$

Spacing $= \dfrac{1}{4}\left(\dfrac{2\pi}{3}\right) = \dfrac{\pi}{6}$ $c = \dfrac{\pi}{3} + \dfrac{\pi}{6} = \dfrac{2\pi}{6} + \dfrac{\pi}{6} = \dfrac{3\pi}{6} = \dfrac{\pi}{2}$

$a = \dfrac{\pi}{6}$ $d = \dfrac{\pi}{2} + \dfrac{\pi}{6} = \dfrac{3\pi}{6} + \dfrac{\pi}{6} = \dfrac{4\pi}{6} = \dfrac{2\pi}{3}$

$b = \dfrac{\pi}{6} + \dfrac{\pi}{6} = \dfrac{2\pi}{6} = \dfrac{\pi}{3}$ $e = \dfrac{2\pi}{3} + \dfrac{\pi}{6} = \dfrac{4\pi}{6} + \dfrac{\pi}{6} = \dfrac{5\pi}{6}$

The 5 points we use on the x-axis are $\dfrac{\pi}{6}, \dfrac{\pi}{3}, \dfrac{\pi}{2}, \dfrac{2\pi}{3}, \dfrac{5\pi}{6}$.

The 2 points we use on the y-axis are $-\dfrac{1}{2}$ and $\dfrac{1}{2}$.

21.   Amplitude $= 3$ Period $= \dfrac{2\pi}{\frac{\pi}{3}} = 6$ Phase Shift $= \dfrac{\frac{\pi}{3}}{\frac{\pi}{3}} = 1$

Spacing $= \dfrac{1}{4}(6) = \dfrac{3}{2}$ $c = \dfrac{5}{2} + \dfrac{3}{2} = \dfrac{8}{2} = 4$

$a = 1$ $d = 4 + \dfrac{3}{2} = \dfrac{11}{2}$

$b = 1 + \dfrac{3}{2} = \dfrac{5}{2}$ $e = \dfrac{11}{2} + \dfrac{3}{2} = \dfrac{14}{2} = 7$

The 5 points we use on the x-axis are $1, \dfrac{5}{2}, 4, \dfrac{11}{2}, 7$.

The 2 points we use on the y-axis are -3 and 3.

23.   The graph in problem #11 is shifted up 1 unit.

25.   The graph in problem #13 is shifted down 3 units.

27.   The graph in problem #15 is shifted up 2 units.

29.   The graph in problem #17 is shifted down 2 units.

31.   The graph in problem #19 is shifted up $\dfrac{3}{2}$ units.

33. Amplitude $= 4$ $\qquad$ Period $= \dfrac{2\pi}{2} = \pi$ $\qquad$ Phase Shift $= \dfrac{\dfrac{\pi}{2}}{2} = \dfrac{\pi}{4}$

Spacing $= \dfrac{1}{4}(\pi) = \dfrac{\pi}{4}$ $\qquad\qquad$ $c = \dfrac{\pi}{2} + \dfrac{\pi}{4} = \dfrac{3\pi}{4}$

$a = \dfrac{\pi}{4}$ $\qquad\qquad\qquad\qquad$ $d = \dfrac{3\pi}{4} + \dfrac{\pi}{4} = \pi$

$b = \dfrac{\pi}{4} + \dfrac{\pi}{4} = \dfrac{\pi}{2}$ $\qquad\qquad$ $e = \pi + \dfrac{\pi}{4} = \dfrac{5\pi}{4}$

For one complete cycle, the points we use on the x-axis are $\dfrac{\pi}{4}, \dfrac{\pi}{2}, \dfrac{3\pi}{4}, \pi, \dfrac{5\pi}{4}$.

The points we use on the y-axis are 4 and -4.

We must extend our graph from $-\dfrac{\pi}{4}$ to $\dfrac{3\pi}{2}$.

35. The graph in problem #33 is reflected across the x-axis.

37. Amplitude $= \dfrac{2}{3}$ $\qquad$ Period $= \dfrac{2\pi}{3}$ $\qquad$ Phase Shift $= \dfrac{-\dfrac{\pi}{2}}{3} = -\dfrac{\pi}{6}$

Spacing $= \dfrac{1}{4}\left(\dfrac{2\pi}{3}\right) = \dfrac{\pi}{6}$ $\qquad\qquad$ $c = 0 + \dfrac{\pi}{6} = \dfrac{\pi}{6}$

$a = -\dfrac{\pi}{6}$ $\qquad\qquad\qquad\qquad$ $d = \dfrac{\pi}{6} + \dfrac{\pi}{6} = \dfrac{\pi}{3}$

$b = -\dfrac{\pi}{6} + \dfrac{\pi}{6} = 0$ $\qquad\qquad$ $e = \dfrac{\pi}{3} + \dfrac{\pi}{6} = \dfrac{\pi}{2}$

For one complete cycle, the points we use on the x-axis are $-\dfrac{\pi}{6}, 0, \dfrac{\pi}{6}, \dfrac{\pi}{3}, \dfrac{\pi}{2}$.

The points we use on the y-axis are $\dfrac{2}{3}$ and $-\dfrac{2}{3}$.

We must extend our graph from $-\pi$ to π,

39. The graph in problem #37 is reflected across the x-axis.

41. First, we will sketch the reciprocal function $y = \sin\left(x + \dfrac{\pi}{4}\right)$ (from Problem 1).

The 5 points on the x-axis are $-\dfrac{\pi}{4}, \dfrac{\pi}{4}, \dfrac{3\pi}{4}, \dfrac{5\pi}{4}, \dfrac{7\pi}{4}$. The 2 points on the y-axis are 1 and -1.

We sketch this sine curve using a dotted line. Then we use the sine curve to graph

$y = \csc\left(x + \dfrac{\pi}{4}\right)$: the asymptotes will occur where $y = 0$, that is at $-\dfrac{\pi}{4}, \dfrac{3\pi}{4}$, and $\dfrac{7\pi}{4}$.

Using the asymptotes and the sine curve, we sketch the graph.

43. First, we will sketch the reciprocal function $y = 2\cos\left(2x - \dfrac{\pi}{2}\right)$ using a dotted line. (This is a very similar to problem #33, except the amplitude is 2, not 4.) The 5 points on the x-axis are $\dfrac{\pi}{4}, \dfrac{\pi}{2}, \dfrac{3\pi}{4}, \pi, \dfrac{5\pi}{4}$. The 2 points on the y-axis are -2 and 2. Then we use the cosine curve to graph $y = 2\sec\left(2x - \dfrac{\pi}{2}\right)$: the asymptotes will occur at $\dfrac{\pi}{2}$ and π. Using the asymptotes and the cosine curve, we sketch the graph.

45. First, we will sketch the reciprocal function $y = 3\sin\left(2x + \dfrac{\pi}{3}\right)$

Amplitude $= 3$ Period $= \dfrac{2\pi}{2} = \pi$ Phase Shift $= \dfrac{-\dfrac{\pi}{3}}{2} = -\dfrac{\pi}{6}$

The 5 points on the x-axis are $-\dfrac{\pi}{6}, \dfrac{\pi}{12}, \dfrac{\pi}{3}, \dfrac{7\pi}{12}, \dfrac{5\pi}{6}$.

The 2 points we use on the y-axis are 3 and -3.

We sketch this sine curve using a dotted line. Then we use the sine curve to graph $y = 3\csc\left(2x + \dfrac{\pi}{3}\right)$: the asymptotes will occur where $y = 0$, that is at $-\dfrac{\pi}{6}, \dfrac{\pi}{3},$ and $\dfrac{5\pi}{6}$.

Using the asymptotes and the sine curve, we sketch the graph.

47. Period $= \pi$ Phase Shift $= -\dfrac{\pi}{4}$ No amplitude

Asymptotes will occur at $-\dfrac{\pi}{2} - \dfrac{\pi}{4} + k\pi$ or $-\dfrac{3\pi}{4} + k\pi$, where k is an integer. For one complete cycle, the asymptotes will occur at $x = -\dfrac{3\pi}{4}$ and $x = \dfrac{\pi}{4}$ and the x-intercept will be at $x = -\dfrac{\pi}{4}$.

49. Period $= \pi$ Phase Shift $= \dfrac{\pi}{4}$ No amplitude

Asymptotes will occur at $\left(0 + \dfrac{\pi}{4}\right) + k\pi$ or at $\dfrac{\pi}{4} + k\pi$, where k is an integer. For one complete cycle, the asymptotes will occur at $\dfrac{\pi}{4}$ and $\dfrac{5\pi}{4}$ and the x-intercept will be at $\dfrac{3\pi}{4}$.

51. Period $=\dfrac{\pi}{2}$ Phase Shift $=\dfrac{\frac{\pi}{2}}{2}=\dfrac{\pi}{4}$ No amplitude

Since the period is $\dfrac{\pi}{2}$, without any phase shift, one complete cycle would go from

$-\dfrac{\pi}{4}$ to $\dfrac{\pi}{4}$. With a phase shift of $\dfrac{\pi}{4}$, the asymptotes will occur at $-\dfrac{\pi}{4}+\dfrac{\pi}{4}+\dfrac{\pi}{2}k$ or

$0+\dfrac{\pi}{2}k$ where k is an integer. For one complete cycle, the asymptotes will occur at

$x=0$ and $x=\dfrac{\pi}{2}$ and the x-intercept will be at $x=\dfrac{\pi}{4}$.

53. $s=r\,\theta$

$\quad=10\left(\dfrac{\pi}{6}\right)$

$\quad=\dfrac{5\pi}{3}$ cm

55. $s=r\,\theta$

$\quad s=2.6\left(\dfrac{30}{60}\cdot 2\pi\right)$

$\quad=2.6\pi$

$\quad=8.2$ cm

57. $r=\dfrac{s}{\theta}$

$\quad=\dfrac{4}{6}$

$\quad=\dfrac{2}{3}$ ft or 8 in

Problem Set 4.4

1. Since the line crosses the y-axis at 1, we know that $b=1$. The ratio of vertical change to horizontal change between any 2 points is $\dfrac{1}{2}$. Therefore $m=\dfrac{1}{2}$. The equation of the line must be $y=\dfrac{1}{2}x+1$.

3. Since the line crosses the y-axis at -3, we know that $b=-3$. The ratio of vertical change to horizontal change between any 2 points is $\dfrac{2}{1}$ or 2. Therefore $m=2$. The equation of the line must be $y=2x-3$.

5.   The graph is a sine curve with an amplitude of 1, period 2π, and no phase shift. The equation is $y = \sin x$.

7.   The graph is a cosine curve with an amplitude of 3, period 2π, and no phase shift. The equation is $y = 3 \cos x$.

9.   The graph is a cosine curve that has been reflected about the x-axis. The amplitude is 3, the period 2π, and no phase shift. The equation is $y = -3 \cos x$.

11.   The graph is a sine curve with an amplitude of 1, period $\dfrac{2\pi}{3}$, and no phase shift. To find B,

we set $\dfrac{2\pi}{3}$ equal to $\dfrac{2\pi}{B}$: $\dfrac{2\pi}{3} = \dfrac{2\pi}{B}$
$$B = 3$$
Therefore, the equation is $y = \sin 3x$

13.   The graph is a sine curve with an amplitude of 1, period 6π, and no phase shift. To find B,

we set 6π equal to $\dfrac{2\pi}{B}$:
$$6\pi = \dfrac{2\pi}{B}$$
$$6\pi B = 2\pi$$
$$B = \dfrac{1}{3}$$
Therefore, the equation is $y = \sin \dfrac{1}{3} x$

15.   The graph is a cosine curve with an amplitude of 2, period $\dfrac{2\pi}{3}$, and no phase shift. In problem #11, we found that $B = 3$. Therefore, the equation is $y = 2 \cos 3x$.

17.   The graph is a sine curve with an amplitude of 4, period 2, and no phase shift.

To find B, we use $2 = \dfrac{2\pi}{B}$
$$2B = 2\pi$$
$$B = \pi$$
Therefore, the equation is $y = 4 \sin \pi x$.

19. The graph is a sine curve that has been reflected across the *x*-axis. The amplitude is 4, the period 2 and no phase shift. In problem #17, we found that $B = \pi$. Therefore, the equation is $y = -4 \sin \pi x$.

21. The graph is the sine curve from #19 that has been moved up 2 units. Therefore, the equation is $y = 2 - 4 \sin \pi x$.

23. The graph is a cosine curve with an amplitude of 3. The period is $\dfrac{3\pi}{4} - \left(-\dfrac{\pi}{4}\right)$ or π and the phase shift is $-\dfrac{\pi}{4}$. To find *B*, we use $\pi = \dfrac{2\pi}{B}$

To find *C*, we use $-\dfrac{\pi}{4} = -\dfrac{C}{B}$

$$B\pi = 2\pi$$

$$-\dfrac{\pi}{4} = -\dfrac{C}{2}$$

$$B = 2$$

$$C = \dfrac{\pi}{2}$$

Therefore, the equation is $y = 3\cos\left(2x + \dfrac{\pi}{2}\right)$.

25. The graph is a cosine curve that has been reflected about the *x*-axis. The amplitude is 2. The period is $\dfrac{\pi}{2} - \left(-\dfrac{\pi}{6}\right) = \dfrac{4\pi}{6} = \dfrac{2\pi}{3}$. The phase shift is $-\dfrac{\pi}{6}$.

To find *B*, we use $\dfrac{2\pi}{3} = \dfrac{2\pi}{B}$

To find *C*, we use $-\dfrac{\pi}{6} = -\dfrac{C}{B}$

$$B = 3$$

$$-\dfrac{\pi}{6} = -\dfrac{C}{3}$$

$$C = \dfrac{\pi}{2}$$

Therefore, the equation is $y = -2\cos\left(3x + \dfrac{\pi}{2}\right)$

27. The graph is the cosine curve from problem #23 that has been moved down 3 units. Therefore the equation is $y = -3 + 3\cos\left(2x + \dfrac{\pi}{2}\right)$.

29. The graph is the cosine curve from problem #25 moved up 2 units. Therefore, the equation is $y = 2 - 2\cos\left(3x + \dfrac{\pi}{2}\right)$.

33.
$$\hat{\theta} = 360° - 321°$$
$$= 39°$$

35.
$$\hat{\theta} = 236° - 180°$$
$$= 56°$$

37.
$$\theta = -276° + 360°$$
$$= 84°$$
$$\hat{\theta} = 84°$$

39.
$$\sin\hat{\theta} = 0.7455$$
$$\hat{\theta} = \sin^{-1}(0.7455)$$
$$\hat{\theta} = 48.2°$$
Therefore, $\theta = 180° - 48.2°$
$$= 131.8°$$

41.
$$\csc\hat{\theta} = 2.3228$$
$$\sin\hat{\theta} = \frac{1}{2.3228}$$
$$\hat{\theta} = \sin^{-1}(0.4305)$$
$$\hat{\theta} = 25.5°$$
Therefore, $\theta = 180° + 25.5°$
$$= 205.5°$$

43.
$$\cot\hat{\theta} = 0.2089$$
$$\tan\hat{\theta} = \frac{1}{0.2089}$$
$$\hat{\theta} = \tan^{-1}(4.7870)$$
$$\hat{\theta} = 78.2°$$
Therefore, $\theta = 360° - 78.2°$
$$= 281.8°$$

Problem 4.5

1. We let $y_1 = 1$ and $y_2 = \sin x$ and graph y_1, y_2, and $y = y_1 + y_2$ on the same coordinate system.

3. We let $y_1 = 2$ and $y_2 = -\cos x$ (which is a cosine curve reflected across the x-axis). Then we graph y_1, y_2, and $y = y_1 + y_2$ on the same coordinate system.

5. We let $y_1 = 4$ and $y_2 = 2\sin x$ (which is a sine curve with an amplitude of 2 and a period of 2π). Then we graph y_1, y_2, and $y = y_1 + y_2$ on the same coordinate system.

7. We let $y_1 = \frac{1}{3}x$ and $y_2 = -\cos x$ Then we graph y_1, y_2, and $y = y_1 + y_2$ on the same coordinate system.

9. We let $y_1 = \frac{1}{2}x$ and $y_2 = -\cos x$ Then we graph y_1, y_2, and $y = y_1 + y_2$ on the same coordinate system.

11. We let $y_1 = x$ and $y_2 = \sin \pi x$ (which is a sine curve with a period of 2). Then we graph y_1, y_2, and $y = y_1 + y_2$ on the same coordinate system.

13. We let $y_1 = 3\sin x$ (which is a sine curve with an amplitude of 3 and a period of 2π) and $y_2 = \cos 2x$ (which is a cosine curve with an amplitude of 1 and a period of π). Then we graph y_1, y_2, and $y = y_1 + y_2$ on the same coordinate system.

15. We let $y_1 = 2\sin x$ (which is a sine curve with an amplitude of 2) and $y_2 = -\cos 2x$ (which is a cosine curve with a period of π and reflected across the x-axis). Then we graph y_1, y_2, and $y = y_1 + y_2$ on the same coordinate system.

17. We let $y_1 = \sin x$ and $y_2 = \sin \dfrac{x}{2}$ (which is a sine curve with an amplitude of 1 and a period of 4π). Then we graph y_1, y_2, and $y = y_1 + y_2$ on the same coordinate system.

19. We let $y_1 = \sin x$ and $y_2 = \sin 2x$ (which is a sine curve with a period of π). Then we graph y_1, y_2, and $y = y_1 + y_2$ on the same coordinate system.

21. We let $y_1 = \cos x$ and $y_2 = \dfrac{1}{2}\sin 2x$ (which is a sine curve with an amplitude of $\dfrac{1}{2}$ and a period of π). Then we graph y_1, y_2, and $y = y_1 + y_2$ on the same coordinate system.

23. We let $y_1 = \sin x$ and $y_2 = -\cos x$ (which is a cosine curve reflected across the x-axis). Then we graph y_1, y_2, and $y = y_1 + y_2$ on the same coordinate system.

25. $y = x \sin x$

x	0	$\pi/2$	π	$3\pi/2$	2π	$5\pi/2$	3π	$7\pi/2$	4π
$\sin x$	0	1	0	-1	0	1	0	-1	0
$x\sin x$	0	$\pi/2$	0	$-3\pi/2$	0	$5\pi/2$	0	$-7\pi/2$	0

37. $v = \dfrac{s}{t}$

$= \dfrac{5\,\text{ft}}{20\,\text{sec}} = \dfrac{1}{4}\,\text{ft/sec}$

39. $s = vt$

$= \left(20\dfrac{\text{ft}}{\text{sec}}\right)(60\,\text{sec})$

$= 1200\,\text{ft}$

Note: We change 1 minute to 60 seconds so that the units will agree.

41.
$$v = r\omega$$
$$= 6\,\text{m}\left(3\,\frac{\text{rad}}{\text{sec}}\right)$$
$$= 18\,\frac{\text{m}}{\text{sec}}$$

$$s = vt$$
$$= 18\,\frac{\text{m}}{\text{sec}} \cdot 10\,\text{sec}$$
$$= 180\,\text{m}$$

43.
$$120\,\frac{\text{rev}}{\text{min}} \cdot \frac{2\pi\,\text{rad}}{1\,\text{rev}} \cdot \frac{1\,\text{min}}{60\,\text{sec}} = 4\pi\,\frac{\text{rad}}{\text{sec}}$$

45.
$$\theta = 2(2\pi) = 4\pi$$

$$s = r\theta$$
$$= 10\,\text{cm}(4\pi)$$
$$= 40\pi\,\text{cm}$$

Problem Set 4.6

5. The angle between $-\dfrac{\pi}{2}$ and $\dfrac{\pi}{2}$ whose sine is $\dfrac{\sqrt{3}}{2}$ is $\dfrac{\pi}{3}$.

7. The angle between 0 and π whose cosine is -1 is π.

9. The angle between $-\dfrac{\pi}{2}$ and $\dfrac{\pi}{2}$ whose tangent is 1 is $\dfrac{\pi}{4}$.

11. The angle between 0 and π whose cosine is $-\dfrac{1}{\sqrt{2}}$ is $\dfrac{3\pi}{4}$.

13. The angle between $-\dfrac{\pi}{2}$ and $\dfrac{\pi}{2}$ whose sine is $-\dfrac{1}{2}$ is $-\dfrac{\pi}{6}$.

15. The angle between $-\dfrac{\pi}{2}$ and $\dfrac{\pi}{2}$ whose tangent is $\sqrt{3}$ is $\dfrac{\pi}{3}$.

17. The angle between $-\dfrac{\pi}{2}$ and $\dfrac{\pi}{2}$ whose sine is 0 is 0.

19. The angle between $-\dfrac{\pi}{2}$ and $\dfrac{\pi}{2}$ whose tangent is $-\dfrac{1}{\sqrt{3}}$ is $-\dfrac{\pi}{6}$.

21. The angle between 0 and π whose cosine is $-\dfrac{1}{2}$ is $\dfrac{2\pi}{3}$.

23. The angle between 0 and π whose cosine is $\dfrac{\sqrt{3}}{2}$ is $\dfrac{\pi}{6}$.

25. Scientific Calculator: 0.1702 $\boxed{\text{INV}}$ $\boxed{\text{SIN}}$
Graphing Calculator: $\boxed{\text{2nd}}$ $\boxed{\text{SIN}}$ $\boxed{(}$ 0.1702 $\boxed{)}$ $\boxed{\text{ENTER}}$
Answer: 9.8°

27. Scientific Calculator: 0.8425 $\boxed{+/-}$ $\boxed{\text{INV}}$ $\boxed{\text{COS}}$
Graphing Calculator: $\boxed{\text{2nd}}$ $\boxed{\text{COS}}$ $\boxed{(}$ $\boxed{-}$ 0.8425 $\boxed{)}$ $\boxed{\text{ENTER}}$
Answer: 147.4°

29. Scientific Calculator: 0.3799 $\boxed{\text{INV}}$ $\boxed{\text{TAN}}$
Graphing Calculator: $\boxed{\text{2nd}}$ $\boxed{\text{TAN}}$ $\boxed{(}$ 0.3799 $\boxed{)}$ $\boxed{\text{ENTER}}$
Answer: 20.8°

31. Scientific Calculator: 0.9627 $\boxed{\text{INV}}$ $\boxed{\text{SIN}}$
Graphing Calculator: $\boxed{\text{2nd}}$ $\boxed{\text{SIN}}$ $\boxed{(}$ 0.9627 $\boxed{)}$ $\boxed{\text{ENTER}}$
Answer: 74.3°

33. Scientific Calculator: 0.4664 $\boxed{+/-}$ $\boxed{\text{INV}}$ $\boxed{\text{COS}}$
Graphing Calculator: $\boxed{\text{2nd}}$ $\boxed{\text{COS}}$ $\boxed{(}$ $\boxed{-}$ 0.4664 $\boxed{)}$ $\boxed{\text{ENTER}}$
Answer: 117.8°

35. Scientific Calculator: 2.748 $\boxed{+/-}$ $\boxed{\text{INV}}$ $\boxed{\text{TAN}}$
Graphing Calculator: $\boxed{\text{2nd}}$ $\boxed{\text{TAN}}$ $\boxed{(}$ $\boxed{-}$ 2.748 $\boxed{)}$ $\boxed{\text{ENTER}}$
Answer: −70.0°

37.   Scientific Calculator: 0.7660 $\boxed{+/-}$ $\boxed{\text{INV}}$ $\boxed{\text{SIN}}$

Graphing Calculator: $\boxed{\text{2nd}}$ $\boxed{\text{SIN}}$ $\boxed{(}$ $\boxed{-}$ 0.7660 $\boxed{)}$ $\boxed{\text{ENTER}}$

Answer: $-50.0°$

39.   Using a TI-83 Plus: graph $y_1 = \sin^{-1}(x)$ in degree mode. Set the window to x between -1 and 1 with Xscl $= 0.5$ and y between -90 and 90 with Yscl $= 30$.

(a)   Then push $\boxed{\text{2nd}}$ $\boxed{\text{TRACE}}$ $\boxed{1: \text{VALUE}}$. At the cursor, enter $\boxed{(-)}$ 1 $\boxed{\div}$ 2 $\boxed{\text{ENTER}}$. The answer is $y = -30°$.

(b)   Then push $\boxed{\text{2nd}}$ $\boxed{\text{TRACE}}$ $\boxed{1: \text{VALUE}}$. At the cursor, enter $\boxed{\sqrt{}}$ $\boxed{(}$ 3 $\boxed{)}$ $\boxed{\div}$ 2 $\boxed{\text{ENTER}}$. The answer is $y = 60°$.

(c)   Then push $\boxed{\text{2nd}}$ $\boxed{\text{TRACE}}$ $\boxed{1: \text{VALUE}}$. At the cursor, enter $\boxed{(-)}$ 1 $\boxed{\div}$ $\boxed{\sqrt{}}$ $\boxed{(}$ 2 $\boxed{)}$ $\boxed{\text{ENTER}}$. The answer is $y = -45°$.

41.   Follow the directions in problem #39, except graph $y_1 = \tan^{-1}(x)$ with the window set to x between -10 and 10.

(a)   Then push $\boxed{\text{2nd}}$ $\boxed{\text{TRACE}}$ $\boxed{1: \text{VALUE}}$. At the cursor, enter $\boxed{(-)}$ 1 $\boxed{\text{ENTER}}$. The answer is $y = -45°$.

(b)   Then push $\boxed{\text{2nd}}$ $\boxed{\text{TRACE}}$ $\boxed{1: \text{VALUE}}$. At the cursor, enter $\boxed{\sqrt{}}$ $\boxed{(}$ 3 $\boxed{)}$ $\boxed{\text{ENTER}}$. The answer is $60°$.

(c)   Then push $\boxed{\text{2nd}}$ $\boxed{\text{TRACE}}$ $\boxed{1: \text{VALUE}}$. At the cursor, enter $\boxed{(-)}$ 1 $\boxed{\div}$ $\boxed{\sqrt{}}$ $\boxed{(}$ 3 $\boxed{)}$ $\boxed{\text{ENTER}}$. The answer is $-30°$.

43.   If $\theta = \cos^{-1}\dfrac{x}{2}$, then $\cos\theta = \dfrac{x}{2}$ and $0 \le \theta \le \pi$. For any value in this interval, $\sin\theta$ will be positive. Therefore $|\sin\theta| = \sin\theta$ and we can simplify: $2|\sin\theta| = 2\sin\theta$.

45.   Let $\theta = \sin^{-1}\dfrac{3}{5}$, then $\sin\theta = \dfrac{3}{5}$ and $-\dfrac{\pi}{2} \le \theta \le \dfrac{\pi}{2}$. We want to find $\sin\theta$ which is equal to $\dfrac{3}{5}$.

47.   Let $\theta = \cos^{-1}\dfrac{1}{2}$, then $\cos\theta = \dfrac{1}{2}$ and $0 \le \theta \le \pi$. We want to find $\cos\theta$ which is equal to $\dfrac{1}{2}$.

49. Let $\theta = \tan^{-1}\dfrac{1}{2}$, then $\tan\theta = \dfrac{1}{2}$ and $-\dfrac{\pi}{2} < \theta < \dfrac{\pi}{2}$. We want to find $\tan\theta$ which is equal to $\dfrac{1}{2}$.

51. First we want to find $\sin 225°$: $\quad \sin 225° = -\sin 45°$

$$= -\frac{\sqrt{2}}{2}$$

Next we want to find $\sin^{-1}\left(-\dfrac{\sqrt{2}}{2}\right)$:

Let $\theta = \sin^{-1}\left(-\dfrac{\sqrt{2}}{2}\right)$

Then $\sin\theta = -\dfrac{\sqrt{2}}{2}$

$$\sin\hat{\theta} = \frac{\sqrt{2}}{2}$$

$\hat{\theta} = 45°$ and $-90° \le \theta \le 90°$

Therefore, $\theta = -45°$.

53. First we want to find $\sin\dfrac{\pi}{3}$: $\quad \sin\dfrac{\pi}{3} = \dfrac{\sqrt{3}}{2}$

Next we want to find $\sin^{-1}\left(\dfrac{\sqrt{3}}{2}\right)$:

Let $\theta = \sin^{-1}\left(\dfrac{\sqrt{3}}{2}\right)$

Then $\sin\theta = \dfrac{\sqrt{3}}{2}$ and $-\dfrac{\pi}{2} \le \theta \le \dfrac{\pi}{2}$

Therefore, $\theta = \dfrac{\pi}{3}$.

55. First we want to find $\cos 120°$: $\quad \cos 120° = -\cos 60°$

$$= -\frac{1}{2}$$

Next we want to find $\cos^{-1}\left(-\dfrac{1}{2}\right)$:

Let $\theta = \cos^{-1}\left(-\dfrac{1}{2}\right)$

Then $\cos\theta = -\dfrac{1}{2}$

This problem is continued on the next page

$$\cos\hat{\theta} = \frac{1}{2}$$

$$\hat{\theta} = 60° \text{ and } 0° \le \theta \le 180°$$

Therefore, $\theta = 180° - 60° = 120°$.

57. First we want to find $\cos\dfrac{7\pi}{4}$: $\quad \cos\dfrac{7\pi}{4} = \cos\dfrac{\pi}{4}$

$$= \frac{\sqrt{2}}{2}$$

Next we want to find $\cos^{-1}\left(\dfrac{\sqrt{2}}{2}\right)$:

Let $\theta = \cos^{-1}\left(\dfrac{\sqrt{2}}{2}\right)$

Then $\cos\theta = \dfrac{\sqrt{2}}{2}$ and $0 \le \theta \le \pi$

Therefore, $\theta = \dfrac{\pi}{4}$.

59. First we want to find $\tan 45°$: $\quad \tan 45° = 1$

Next we want to find $\tan^{-1} 1$:

Let $\theta = \tan^{-1} 1$

Then $\tan\theta = 1$ and $-90° < \theta < 90°$

Therefore, $\theta = 45°$.

61. First we want to find $\tan\dfrac{5\pi}{6}$: $\quad \tan\dfrac{5\pi}{6} = -\tan\dfrac{\pi}{6}$

$$= -\frac{1}{\sqrt{3}}$$

Next we want to find $\tan^{-1}\left(-\dfrac{1}{\sqrt{3}}\right)$:

Let $\theta = \tan^{-1}\left(-\dfrac{1}{\sqrt{3}}\right)$

Then $\tan\theta = -\dfrac{1}{\sqrt{3}}$

$$\tan\hat{\theta} = \frac{1}{\sqrt{3}}$$

$$\hat{\theta} = \frac{\pi}{6} \text{ and } -\frac{\pi}{2} < \theta < \frac{\pi}{2} \qquad\qquad \text{Therefore, } \theta = -\frac{\pi}{6}.$$

63. Let $\theta = \tan^{-1}\dfrac{3}{4}$, then $\tan\theta = \dfrac{3}{4}$ and $-\dfrac{\pi}{2} < \theta < \dfrac{\pi}{2}$.

Next, we draw a triangle and find the hypotenuse:

$$\begin{aligned}\text{hypotenuse} &= \sqrt{3^2 + 4^2} \\ &= \sqrt{9 + 16} \\ &= \sqrt{25} \\ &= 5\end{aligned}$$

Then, we find $\cos\theta = \dfrac{4}{5}$.

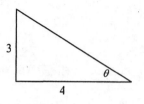

65. Let $\theta = \sin^{-1}\dfrac{3}{5}$, then $\sin\theta = \dfrac{3}{5}$ and $-\dfrac{\pi}{2} \le \theta \le \dfrac{\pi}{2}$.

Next, we draw a triangle and find the adjacent side:

$$\begin{aligned}\text{adjacent side} &= \sqrt{5^2 - 3^2} \\ &= \sqrt{25 - 9} \\ &= \sqrt{16} \\ &= 4\end{aligned}$$

Then, we find $\tan\theta = \dfrac{3}{4}$.

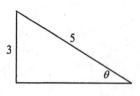

67. Let $\theta = \cos^{-1}\dfrac{1}{\sqrt{5}}$, then $\cos\theta = \dfrac{1}{\sqrt{5}}$ and $0 \le \theta \le \pi$.

We are looking for $\sec\theta$.

$$\begin{aligned}\sec\theta &= \dfrac{1}{\cos\theta} \\ &= \dfrac{1}{\dfrac{1}{\sqrt{5}}} \\ &= \sqrt{5}\end{aligned}$$

69. Let $\theta = \cos^{-1}\dfrac{1}{2}$, then $\cos\theta = \dfrac{1}{2}$ and $0 \le \theta \le \pi$.

Next, we draw a triangle and find the opposite side:

$$\begin{aligned}\text{Opposite side} &= \sqrt{2^2 - 1^2} \\ &= \sqrt{4 - 1} = \sqrt{3}\end{aligned}$$

Then we find $\sin\theta = \dfrac{\sqrt{3}}{2}$

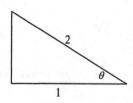

71. Let $\theta = \tan^{-1}\dfrac{1}{2}$, then $\tan\theta = \dfrac{1}{2}$ and $-\dfrac{\pi}{2} < \theta < \dfrac{\pi}{2}$.

We want to find $\cot\theta = \dfrac{1}{\tan\theta} = \dfrac{1}{\frac{1}{2}} = 2$

73. Since $y = \sin x$ and $y = \sin^{-1} x$ (for $-\dfrac{\pi}{2} \le x \le \dfrac{\pi}{2}$) are inverse functions, one function will "undo" the action performed by the other. Therefore, $\sin^{-1}(\sin x) = x$.

75. Let $\theta = \cos^{-1} x$, then $\cos\theta = x$ and $0 \le \theta \le \pi$. We want to find $\cos\theta$ which is equal to x.

77. Let $\theta = \sin^{-1} x$, then $\sin\theta = x$ and $-\dfrac{\pi}{2} \le \theta \le \dfrac{\pi}{2}$. We draw a triangle and find the adjacent

side: Adjacent side $= \sqrt{1^2 - x^2} = \sqrt{1 - x^2}$

From the figure, we find $\cos\theta = \dfrac{\sqrt{1-x^2}}{1} = \sqrt{1-x^2}$.

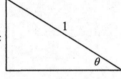

79. Let $\theta = \tan^{-1} x$, then $\tan\theta = x$ and $-\dfrac{\pi}{2} < \theta < \dfrac{\pi}{2}$. We draw a triangle and find the

hypotenuse: hypotenuse $= \sqrt{x^2 + 1^2} = \sqrt{x^2 + 1}$

From the figure, we find $\sin\theta = \dfrac{x}{\sqrt{x^2+1}}$.

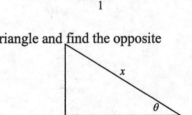

81. Let $\theta = \cos^{-1}\dfrac{1}{x}$, then $\cos\theta = \dfrac{1}{x}$ and $0 \le \theta \le \pi$. We draw a triangle and find the opposite

side: Opposite side $= \sqrt{x^2 - 1^2} = \sqrt{x^2 - 1}$

From the figure, we find $\sin\theta = \dfrac{\sqrt{x^2-1}}{x}$.

83. Let $\theta = \cos^{-1}\dfrac{1}{x}$, then $\cos\theta = \dfrac{1}{x}$ and $0 \le \theta \le \pi$.

We want to find $\sec\theta = \dfrac{1}{\cos\theta} = \dfrac{1}{1/x} = x$

85. The graph is a sine curve with amplitude of 4 and period of $\dfrac{2\pi}{2}$ or π.

87. The graph is a sine curve with amplitude of 2 and period of $\dfrac{2\pi}{\pi}$ or 2.

89. The graph is a cosine curve with amplitude of 3 and period of $\dfrac{2\pi}{1/2}$ or 4π.

The graph is reflected across the x-axis.

91. The graph is a standard sine curve with a phase shift of $\dfrac{\pi}{4}$.

93. Amplitude $= 1$ $\qquad$ Period $= \dfrac{2\pi}{2} = \pi$ $\qquad$ Phase shift $= \dfrac{-\dfrac{\pi}{2}}{2} = -\dfrac{\pi}{4}$

Spacing $= \dfrac{1}{4}(\pi) = \dfrac{\pi}{4}$ $\qquad\qquad$ $c = 0 + \dfrac{\pi}{4} = \dfrac{\pi}{4}$

$a = -\dfrac{\pi}{4}$ $\qquad\qquad\qquad\qquad$ $d = \dfrac{\pi}{4} + \dfrac{\pi}{4} = \dfrac{\pi}{2}$

$b = -\dfrac{\pi}{4} + \dfrac{\pi}{4} = 0$ $\qquad\qquad$ $e = \dfrac{\pi}{2} + \dfrac{\pi}{4} = \dfrac{3\pi}{4}$

The 5 points we use on the x-axis are $-\dfrac{\pi}{4}, 0, \dfrac{\pi}{4}, \dfrac{\pi}{2}, \dfrac{3\pi}{4}$.

The 2 points we use on the y-axis are -1 and 1.

95. Amplitude $= 3$ $\qquad$ Period $= \dfrac{2\pi}{2} = \pi$ $\qquad$ Phase shift $= \dfrac{\dfrac{\pi}{3}}{2} = \dfrac{\pi}{6}$

Spacing $= \dfrac{1}{4}(\pi) = \dfrac{\pi}{4}$ $\qquad\qquad$ $c = \dfrac{5\pi}{6} + \dfrac{\pi}{4} = \dfrac{2\pi}{3}$

$a = \dfrac{\pi}{6}$ $\qquad\qquad\qquad\qquad$ $d = \dfrac{2\pi}{3} + \dfrac{\pi}{4} = \dfrac{11\pi}{12}$

$b = \dfrac{\pi}{6} + \dfrac{\pi}{4} = \dfrac{5\pi}{12}$ $\qquad\qquad$ $e = \dfrac{11\pi}{12} + \dfrac{\pi}{4} = \dfrac{7\pi}{6}$

The 5 points we use on the x-axis are $\dfrac{\pi}{6}, \dfrac{5\pi}{12}, \dfrac{2\pi}{3}, \dfrac{11\pi}{12}, \dfrac{7\pi}{6}$.

The 2 points we use on the y-axis are -3 and 3.

Chapter 4 Test

8.   The value where $\sec x$ is undefined occur where $\cos x = 0$. These x-values are $-\dfrac{7\pi}{2}, -\dfrac{5\pi}{2}, -\dfrac{3\pi}{2}, -\dfrac{\pi}{2}, \ \dfrac{\pi}{2}, \dfrac{3\pi}{2}, \ \dfrac{5\pi}{2}$ and $\dfrac{7\pi}{2}$.

9.   Amplitude $= 1$ Period $= \dfrac{2\pi}{\pi} = 2$

10.   Amplitude $= |-3| = 3$ Period $= 2\pi$
    The graph is reflected across the x-axis.

11.   Amplitude $= 3$ Period $= \dfrac{2\pi}{2} = \pi$ Vertical translation $= 2$

12.   Amplitude $= 2$ Period $= \dfrac{2\pi}{\pi} = 2$

13.   Amplitude $= 1$ Period $= 2\pi$ Phase shift $= -\dfrac{\pi}{4}$

      Spacing $= \dfrac{1}{4}(2\pi) = \dfrac{\pi}{2}$ $c = \dfrac{\pi}{4} + \dfrac{\pi}{2} = \dfrac{3\pi}{4}$

      $a = \dfrac{\pi}{4}$ $d = \dfrac{3\pi}{4} + \dfrac{\pi}{2} = \dfrac{5\pi}{4}$

      $b = -\dfrac{\pi}{4} + \dfrac{\pi}{2} = \dfrac{\pi}{4}$ $e = \dfrac{5\pi}{4} + \dfrac{\pi}{2} = \dfrac{7\pi}{4}$

      Points we use on the x-axis are $-\dfrac{\pi}{4}, \dfrac{\pi}{4}, \dfrac{3\pi}{4}, \dfrac{5\pi}{4}, \dfrac{7\pi}{4}$.
      Points we use on the y-axis are -1 and 1.

14.   Amplitude $= 1$ Period $= 2\pi$ Phase shift $= \dfrac{\pi}{2}$

      The problem is continued on the next page

$$\text{Spacing} = \frac{1}{4}(2\pi) = \frac{\pi}{2} \qquad\qquad c = \pi + \frac{\pi}{2} = \frac{3\pi}{2}$$

$$a = \frac{\pi}{2} \qquad\qquad\qquad\qquad\qquad d = \frac{3\pi}{2} + \frac{\pi}{2} = 2\pi$$

$$b = \frac{\pi}{2} + \frac{\pi}{2} = \pi \qquad\qquad\qquad e = 2\pi + \frac{\pi}{2} = \frac{5\pi}{2}$$

Points we use on the x-axis are $\dfrac{\pi}{2}, \pi, \dfrac{3\pi}{2}, 2\pi, \dfrac{5\pi}{2}$.

Points we use on the y-axis are -1 and 1.

15. Amplitude $= 3$ $\qquad$ Period $= \dfrac{2\pi}{2} = \pi$ $\qquad$ Phase shift $= \dfrac{\frac{\pi}{3}}{2} = \dfrac{\pi}{6}$

$$\text{Spacing} = \frac{1}{4}(\pi) = \frac{\pi}{4} \qquad\qquad c = \frac{5\pi}{12} + \frac{\pi}{6} = \frac{2\pi}{3}$$

$$a = \frac{\pi}{6} \qquad\qquad\qquad\qquad\qquad d = \frac{2\pi}{3} + \frac{\pi}{6} = \frac{11\pi}{12}$$

$$b = \frac{\pi}{6} + \frac{\pi}{4} = \frac{5\pi}{12} \qquad\qquad e = \frac{11\pi}{12} + \frac{\pi}{6} = \frac{7\pi}{6}$$

Points we use on the x-axis are $\dfrac{\pi}{6}, \dfrac{5\pi}{12}, \dfrac{2\pi}{3}, \dfrac{11\pi}{12}, \dfrac{7\pi}{6}$.

Points we use on the y-axis are -3 and 3.

16. Amplitude $= 3$ $\qquad$ Period $= \dfrac{2\pi}{\frac{\pi}{3}} = 6$ $\qquad$ Phase shift $= \dfrac{\frac{\pi}{3}}{\frac{\pi}{3}} = 1$

Vertical translation $= -3$

$$\text{Spacing} = \frac{1}{4}(6) = \frac{3}{2} \qquad\qquad c = \frac{5}{2} + \frac{3}{2} = 4$$

$$a = 1 \qquad\qquad\qquad\qquad\qquad d = 4 + \frac{3}{2} = \frac{11}{2}$$

$$b = 1 + \frac{3}{2} = \frac{5}{2} \qquad\qquad\qquad e = \frac{11}{2} + \frac{3}{2} = 7$$

Points we use on the x-axis are $1, \dfrac{5}{2}, 4, \dfrac{11}{2}, 7$.

Points we use on the y-axis are 0 and -6. We move the graph of $y = 3\sin\left(\dfrac{\pi}{3}x - \dfrac{\pi}{3}\right)$ down 3 units.

17. Using Problem 13 above, draw in the asymptotes where $y = 0$, at $-\dfrac{\pi}{4}, \dfrac{3\pi}{4}$, and $\dfrac{7\pi}{4}$. Now

sketch the reciprocal function using the asymptotes and the curve $y = \sin\left(x + \dfrac{\pi}{4}\right)$ as your guides.

18. $\text{Period} = \dfrac{\pi}{2}$ $\text{Phase shift} = \dfrac{\dfrac{\pi}{2}}{2} = \dfrac{\pi}{4}$

Considering first the period of $\dfrac{\pi}{2}$, the asymptotes would be at $-\dfrac{\pi}{4}$ and $\dfrac{\pi}{4}$.

Since the graph has phase shift of $\dfrac{\pi}{4}$, the asymptotes are at $-\dfrac{\pi}{4} + \dfrac{\pi}{4} = 0$ and $\dfrac{\pi}{4} + \dfrac{\pi}{4} = \dfrac{\pi}{2}$.

For one complete cycle, the asymptotes will be at 0 and $\dfrac{\pi}{2}$. The x-intercept will be at $\dfrac{\pi}{4}$.

19. $\text{Amplitude} = 2$ $\text{Period} = \dfrac{2\pi}{3}$ $\text{Phase shift} = \dfrac{\pi}{3}$

$\text{Spacing} = \dfrac{1}{4}\left(\dfrac{2\pi}{3}\right) = \dfrac{\pi}{6}$ $c = \dfrac{\pi}{2} + \dfrac{\pi}{6} = \dfrac{2\pi}{3}$

$a = \dfrac{\pi}{3}$ $d = \dfrac{2\pi}{3} + \dfrac{\pi}{6} = \dfrac{5\pi}{6}$

$b = \dfrac{\pi}{3} + \dfrac{\pi}{6} = \dfrac{\pi}{2}$ $e = \dfrac{5\pi}{6} + \dfrac{\pi}{6} = \pi$

The 5 points we use on the x-axis are $\dfrac{\pi}{3}, \dfrac{\pi}{2}, \dfrac{2\pi}{3}, \dfrac{5\pi}{6}, \pi$.

The 2 points we use on the y-axis are -2 and 2. Then extend the graph from $-\dfrac{\pi}{3}$ to $\dfrac{5\pi}{3}$.

20. $\text{Amplitude} = 2$ $\text{Period} = \dfrac{2\pi}{\dfrac{\pi}{2}} = 4$ $\text{Phase shift} = \dfrac{\dfrac{\pi}{4}}{\dfrac{\pi}{2}} = \dfrac{1}{2}$

$\text{Spacing} = \dfrac{1}{4}(4) = 1$ $c = \dfrac{3}{2} + 1 = \dfrac{5}{2}$

$a = \dfrac{1}{2}$ $d = \dfrac{5}{2} + 1 = \dfrac{7}{2}$

$b = \dfrac{1}{2} + 1 = \dfrac{3}{2}$ $e = \dfrac{7}{2} + 1 = \dfrac{9}{2}$

The 5 points we use on the x-axis are $\dfrac{1}{2}, \dfrac{3}{2}, \dfrac{5}{2}, \dfrac{7}{2}, \dfrac{9}{2}$.

The 2 points we use on the y-axis are -2 and 2. Then extend the graph from $-\dfrac{1}{2}$ to $\dfrac{13}{2}$.

21. The graph is a sine curve with amplitude of 2, period of 4π, and phase shift of $-\pi$.

To find B, we use $\qquad 4\pi = \dfrac{2\pi}{B}$ $\qquad$ Then to find C, we use $\qquad -\pi = -\dfrac{C}{B}$

$$4\pi B = 2\pi$$ $$-\pi = -\dfrac{C}{1/2}$$

$$B = \dfrac{1}{2}$$ $$\dfrac{\pi}{2} = C$$

Therefore, the equation is $y = 2\sin\left(\dfrac{1}{2}x + \dfrac{\pi}{2}\right)$.

22. The graph is a sine curve with amplitude of $\dfrac{1}{2}$ and period of 4, that has been translated up $\dfrac{1}{2}$ unit. There is no phase shift.

To find B, we use $\qquad 4 = \dfrac{2\pi}{B}$

$$4B = 2\pi$$

$$B = \dfrac{\pi}{2}$$

Therefore, the equation is $y = \dfrac{1}{2} + \dfrac{1}{2}\sin\dfrac{\pi}{2}x$.

23. Let $y_1 = \dfrac{1}{2}x$ (a straight line) and $y_2 = -\sin x$ (a sine curve reflected about the x-axis). Then graph y_1, y_2 and $y = y_1 + y_2$ on the same coordinate system.

24. Let $y_1 = \sin x$ and $y_2 = \cos 2x$ (a cosine curve with period of π). Then graph y_1, y_2 and $y = y_1 + y_2$ on the same coordinate system.

25. The graph of $y = \cos^{-1} x$ is equivalent to $x = \cos y$ and $0 \le y \le \pi$. We make a table of values to aid in the graphing.

x	y
1	0
0	$\pi/2$
-1	π

26. The graph of $y = \arcsin x$ is equivalent to $x = \sin y$ and $-\dfrac{\pi}{2} \le y \le \dfrac{\pi}{2}$. We make a table of values to aid in graphing:

x	y
-1	$-\pi/2$
0	0
1	$\pi/2$

27. The angle between $-\dfrac{\pi}{2}$ and $\dfrac{\pi}{2}$ whose sine is $\dfrac{1}{2}$ is $\dfrac{\pi}{6}$.

28. The angle between 0 and π whose cosine is $-\dfrac{\sqrt{3}}{2}$ is $\dfrac{5\pi}{6}$.

29. The angle between $-\dfrac{\pi}{2}$ and $\dfrac{\pi}{2}$ whose tangent is -1 is $-\dfrac{\pi}{4}$.

30. The angle between $-\dfrac{\pi}{2}$ and $\dfrac{\pi}{2}$ whose sine is 1 is $\dfrac{\pi}{2}$.

31. Scientific Calculator: 0.5934 $\boxed{\text{INV}}$ $\boxed{\text{SIN}}$
Graphing Calculator: $\boxed{\text{2nd}}$ $\boxed{\text{SIN}}$ $\boxed{(}$ 0.5934 $\boxed{)}$ $\boxed{\text{ENTER}}$
Answer: 36.4°

32. Scientific Calculator: 0.8302 $\boxed{\text{+/-}}$ $\boxed{\text{INV}}$ $\boxed{\text{TAN}}$
Graphing Calculator: $\boxed{\text{2nd}}$ $\boxed{\text{TAN}}$ $\boxed{(}$ $\boxed{-}$ 0.8302 $\boxed{)}$ $\boxed{\text{ENTER}}$
Answer: −39.7°

33. Scientific Calculator: 0.6981 $\boxed{\text{+/-}}$ $\boxed{\text{INV}}$ $\boxed{\text{COS}}$
Graphing Calculator: $\boxed{\text{2nd}}$ $\boxed{\text{COS}}$ $\boxed{(}$ $\boxed{-}$ 0.6981 $\boxed{)}$ $\boxed{\text{ENTER}}$
Answer: 134.3°

34. Scientific Calculator: 0.2164 $\boxed{+/-}$ $\boxed{\text{INV}}$ $\boxed{\text{SIN}}$

Graphing Calculator: $\boxed{\text{2nd}}$ $\boxed{\text{SIN}}$ $\boxed{(}$ $\boxed{-}$ 0.2164 $\boxed{)}$ $\boxed{\text{ENTER}}$

Answer: $-12.5°$

35. Let $\theta = \cos^{-1}\dfrac{2}{3}$, then $\cos\theta = \dfrac{2}{3}$ and $0 \leq \theta \leq \pi$.

Next, we draw a triangle and find the opposite side:

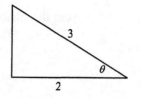

$$\begin{aligned} \text{opposite side} &= \sqrt{3^2 - 2^2} \\ &= \sqrt{9 - 4} \\ &= \sqrt{5} \end{aligned}$$

From the figure, we find $\tan\theta = \dfrac{\sqrt{5}}{2}$.

36. Let $\theta = \tan^{-1}\dfrac{2}{3}$, then $\tan\theta = \dfrac{2}{3}$ and $-\dfrac{\pi}{2} < \theta < \dfrac{\pi}{2}$.

Next, we draw a triangle and find the hypotenuse:

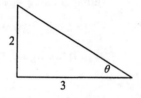

$$\begin{aligned} \text{hypotenuse} &= \sqrt{2^2 + 3^2} \\ &= \sqrt{4 + 9} \\ &= \sqrt{13} \end{aligned}$$

From the figure, we find $\cos\theta = \dfrac{3}{\sqrt{13}}$.

37. Since $y = \cos x$ and $y = \cos^{-1} x$ are inverse functions when $-90° \leq x \leq 90°$, one function undoes the other. Therefore, $\cos^{-1}(\cos 30°) = 30°$.

38. We cannot use inverse functions because $\dfrac{7\pi}{6}$ is not between $-\dfrac{\pi}{2}$ and $\dfrac{\pi}{2}$.

First, we find $\tan\dfrac{7\pi}{6}$: $\qquad \tan\dfrac{7\pi}{6} = \tan\dfrac{\pi}{6}$

$$= \dfrac{1}{\sqrt{3}}$$

Next, we find $\tan^{-1}\dfrac{1}{\sqrt{3}}$:

Let $\theta = \tan^{-1}\dfrac{1}{\sqrt{3}}$, then $\tan\theta = \dfrac{1}{\sqrt{3}}$ and $-\dfrac{\pi}{2} < \theta < \dfrac{\pi}{2}$.

Therefore, $\theta = \dfrac{\pi}{6}$.

39. Let $\theta = \cos^{-1} x$, then $\cos\theta = x$ and $0 \le \theta \le \pi$
Next, we draw a triangle and label the adjacent side and the hypotenuse.
Now we find the opposite side using the Pythagorean Theorem.

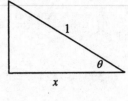

$$\text{Opposite side} = \sqrt{1^2 - x^2}$$
$$= \sqrt{1 - x^2}$$

From the figure, we find $\sin\theta = \dfrac{\sqrt{1-x^2}}{1}$
$$= \sqrt{1-x^2}$$

40. Let $\theta = \sin^{-1} x$, then $\sin\theta = x$ and $-\dfrac{\pi}{2} \le \theta \le \dfrac{\pi}{2}$.

Next we draw a triangle and label the opposite side and the hypotenuse.

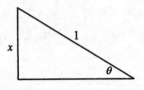

Now we find the adjacent side using the Pythagorean Theorem.

$$\text{adjacent side} = \sqrt{1^2 - x^2}$$
$$= \sqrt{1 - x^2}$$

From the figure, we find $\tan\theta = \dfrac{x}{\sqrt{1-x^2}}$.

CHAPTER 5   Identities and Formulas

Problem Set 5.1

1. $\cos\theta\tan\theta = \cos\theta \cdot \dfrac{\sin\theta}{\cos\theta}$ Ratio identity

$\qquad\qquad = \dfrac{\cos\theta\sin\theta}{\cos\theta}$ Multiply

$\qquad\qquad = \sin\theta$ Reduce

3. $\csc\theta\tan\theta = \dfrac{1}{\sin\theta} \cdot \dfrac{\sin\theta}{\cos\theta}$ Ratio and reciprocal identities

$\qquad\qquad = \dfrac{1}{\cos\theta}$ Reduce

$\qquad\qquad = \sec\theta$ Reciprocal identity

5. $\dfrac{\tan A}{\sec A} = \dfrac{\dfrac{\sin A}{\cos A}}{\dfrac{1}{\cos A}}$ Ratio and reciprocal identities

$\qquad\quad = \dfrac{\sin A\cos A}{\cos A}$ Divide

$\qquad\quad = \sin A$ Reduce

7. $\sec\theta\cot\theta\sin\theta = \dfrac{1}{\cos\theta} \cdot \dfrac{\cos\theta}{\sin\theta} \cdot \sin\theta$ Ratio and reciprocal identities

$\qquad\qquad\quad = 1$ Reduce

9. $\cos x(\csc x + \tan x) = \cos x\csc x + \cos x\tan x$ Distributive property

$\qquad\qquad = \cos x \cdot \dfrac{1}{\sin x} + \cos x \cdot \dfrac{\sin x}{\cos x}$ Reciprocal and ratio identities

$\qquad\qquad = \dfrac{\cos x}{\sin x} + \dfrac{\cos x\sin x}{\cos x}$ Multiply

$\qquad\qquad = \cot x + \sin x$ Ratio identity and reduce second fraction

11. $\cos x(\csc x - \sec x) = \cos x\left(\dfrac{1}{\sin x} - \dfrac{1}{\cos x}\right)$ Reciprocal identities

$$= \dfrac{\cos x}{\sin x} - 1 \qquad \text{Multiply}$$

$$= \cot x - 1 \qquad \text{Ratio identity}$$

13. $\cos^2 x\left(1 + \tan^2 x\right) = \cos^2 x\left(\sec^2 x\right)$ Pythagorean identity

$$= \cos^2 x\left(\dfrac{1}{\cos^2 x}\right) \qquad \text{Reciprocal identity}$$

$$= 1 \qquad \text{Multiply and reduce}$$

15. $(1 - \sin x)(1 + \sin x) = 1 - \sin^2 x$ Multiply

$$= \cos^2 x \qquad \text{Pythagorean identity}$$

17. $\dfrac{\cos^4 t - \sin^4 t}{\sin^2 t} = \dfrac{\left(\cos^2 t + \sin^2 t\right)\left(\cos^2 t - \sin^2 t\right)}{\sin^2 t}$ Factor

$$= \dfrac{1\left(\cos^2 t - \sin^2 t\right)}{\sin^2 t} \qquad \text{Pythagorean identity}$$

$$= \dfrac{\cos^2 t}{\sin^2 t} - \dfrac{\sin^2 t}{\sin^2 t} \qquad \text{Separate into 2 fractions}$$

$$= \cot^2 t - 1 \qquad \text{Ratio identity and reduce second fraction}$$

19. $\dfrac{\cos^2 \theta}{1 - \sin \theta} = \dfrac{1 - \sin^2 \theta}{1 - \sin \theta}$ Pythagorean identity

$$= \dfrac{(1 - \sin \theta)(1 + \sin \theta)}{1 - \sin \theta} \qquad \text{Factor numerator}$$

$$= 1 + \sin \theta \qquad \text{Reduce}$$

21. $\dfrac{1 - \sin^4 \theta}{1 + \sin^2 \theta} = \dfrac{\left(1 - \sin^2 \theta\right)\left(1 + \sin^2 \theta\right)}{1 + \sin^2 \theta}$ Factor

$$= 1 - \sin^2 \theta \qquad \text{Reduce}$$

$$= \cos^2 \theta \qquad \text{Pythagorean identity}$$

23. $\sec^2 \theta - \tan^2 \theta = \left(\tan^2 \theta + 1\right) - \tan^2 \theta$ Pythagorean identity

$$= 1 \qquad \text{Combine}$$

25.
$$\sec^4 - \tan^4\theta = \left(\sec^2\theta - \tan^2\theta\right)\left(\sec^2\theta + \tan^2\theta\right) \qquad \text{Factor}$$
$$= 1\left(\sec^2\theta + \tan^2\theta\right) \qquad \text{Pythagorean identity (problem \#23)}$$
$$= \frac{1}{\cos^2\theta} + \frac{\sin^2\theta}{\cos^2\theta} \qquad \text{Reciprocal and ratio identities}$$
$$= \frac{1 + \sin^2\theta}{\cos^2\theta} \qquad \text{Add fractions}$$

27.
$$\frac{\sin^2\theta - \cos^2\theta}{\sin\theta\cos\theta} = \frac{\sin^2\theta}{\sin\theta\cos\theta} - \frac{\cos^2\theta}{\sin\theta\cos\theta} \qquad \text{Separate into 2 fractions}$$
$$= \frac{\sin\theta}{\cos\theta} - \frac{\cos\theta}{\sin\theta} \qquad \text{Reduce}$$
$$= \tan\theta - \cot\theta \qquad \text{Ratio identities}$$

29.
$$\csc B - \sin B = \frac{1}{\sin B} - \sin B \qquad \text{Reciprocal identity}$$
$$= \frac{1}{\sin B} - \sin B \cdot \frac{\sin B}{\sin B} \qquad \text{LCD is } \sin B$$
$$= \frac{1 - \sin^2 B}{\sin B} \qquad \text{Subtract fractions}$$
$$= \frac{\cos^2 B}{\sin B} \qquad \text{Pythagorean identity}$$
$$= \frac{\cos B}{\sin B} \cdot \cos B \qquad \text{Separate fraction}$$
$$= \cot B \cos B \qquad \text{Ratio identity}$$

31.
$$\cot\theta\cos\theta + \sin\theta = \frac{\cos\theta}{\sin\theta} \cdot \cos\theta + \sin\theta \qquad \text{Ratio identity}$$
$$= \frac{\cos^2\theta}{\sin\theta} + \sin\theta \cdot \frac{\sin\theta}{\sin\theta} \qquad \text{LCD is } \sin\theta$$
$$= \frac{\cos^2\theta}{\sin\theta} + \frac{\sin^2\theta}{\sin\theta} \qquad \text{Multiply}$$
$$= \frac{\cos^2\theta + \sin^2\theta}{\sin\theta} \qquad \text{Add fractions}$$
$$= \frac{1}{\sin\theta} \qquad \text{Pythagorean identity}$$
$$= \csc\theta \qquad \text{Reciprocal identity}$$

33.
$$\frac{\cos x}{1+\sin x}+\frac{1+\sin x}{\cos x}=\frac{\cos x}{1+\sin x}\cdot\frac{\cos x}{\cos x}+\frac{1+\sin x}{\cos x}\cdot\frac{1+\sin x}{1+\sin x} \qquad \text{LCD}$$

$$=\frac{\cos^2 x}{\cos x(1+\sin x)}+\frac{1+2\sin x+\sin^2 x}{\cos x(1+\sin x)} \qquad \text{Multiply fractions}$$

$$=\frac{(\cos^2 x+\sin^2 x)+1+2\sin x}{\cos x(1+\sin x)} \qquad \text{Add fractions}$$

$$=\frac{1+1+2\sin x}{\cos x(1+\sin x)} \qquad \text{Pythagorean identity}$$

$$=\frac{2+2\sin x}{\cos x(1+\sin x)} \qquad \text{Combine}$$

$$=\frac{2(1+\sin x)}{\cos x(1+\sin x)} \qquad \text{Factor out a 2}$$

$$=\frac{2}{\cos x} \qquad \text{Reduce}$$

$$=2\sec x \qquad \text{Reciprocal identity}$$

35.
$$\frac{1}{1+\cos x}+\frac{1}{1-\cos x}=\frac{1}{1+\cos x}\cdot\frac{1-\cos x}{1-\cos x}+\frac{1}{1-\cos x}\cdot\frac{1+\cos x}{1+\cos x} \qquad \text{LCD}$$

$$=\frac{1-\cos x}{1-\cos^2 x}+\frac{1+\cos x}{1-\cos^2 x} \qquad \text{Multiply}$$

$$=\frac{1-\cos x+1+\cos x}{1-\cos^2 x} \qquad \text{Add fractions}$$

$$=\frac{2}{1-\cos^2 x} \qquad \text{Combine numerator}$$

$$=\frac{2}{\sin^2 x} \qquad \text{Pythagorean identity}$$

$$=2\csc^2 x \qquad \text{Reciprocal identity}$$

37.
$$\frac{1-\sec x}{1+\sec x}=\frac{1-\dfrac{1}{\cos x}}{1+\dfrac{1}{\cos x}} \qquad \text{Reciprocal identity}$$

$$=\frac{\cos x\left(1-\dfrac{1}{\cos x}\right)}{\cos x\left(1+\dfrac{1}{\cos x}\right)} \qquad \text{Multiply numerator and denominator by LCD}$$

$$=\frac{\cos x-1}{\cos x+1} \qquad \text{Distributive property}$$

39.

$$\frac{\cos t}{1 + \sin t} = \frac{\cos t}{1 + \sin t} \cdot \frac{1 - \sin t}{1 - \sin t} \qquad \text{Multiply numerator and denominator by } 1 - \sin t$$

$$= \frac{\cos t (1 - \sin t)}{1 - \sin^2 t} \qquad \text{Multiply fractions}$$

$$= \frac{\cos t (1 - \sin t)}{\cos^2 t} \qquad \text{Pythagorean identity}$$

$$= \frac{1 - \sin^2 t}{\cos t} \qquad \text{Reduce}$$

41.

$$\frac{1 - \sin t}{1 + \sin t} = \frac{1 - \sin t}{1 + \sin t} \cdot \frac{1 - \sin t}{1 - \sin t} \qquad \text{Multiply numerator and denominator by } 1 - \sin t$$

$$= \frac{(1 - \sin t)^2}{1 - \sin^2 t} \qquad \text{Multiply fractions}$$

$$= \frac{(1 - \sin t)^2}{\cos^2 t} \qquad \text{Pythagorean identity}$$

43.

$$\frac{\sec \theta + 1}{\tan \theta} = \frac{\sec \theta + 1}{\tan \theta} \cdot \frac{\sec \theta - 1}{\sec \theta - 1} \qquad \text{Multiply numerator and denominator by } \sec \theta - 1$$

$$= \frac{\sec^2 \theta - 1}{\tan \theta (\sec \theta - 1)} \qquad \text{Multiply fractions}$$

$$= \frac{\tan^2 \theta}{\tan \theta (\sec \theta - 1)} \qquad \text{Pythagorean identity}$$

$$= \frac{\tan \theta}{\sec \theta - 1} \qquad \text{Reduce}$$

45.

$$(\sec x - \tan x)^2 = \left(\frac{1}{\cos x} - \frac{\sin x}{\cos x} \right)^2 \qquad \text{Reciprocal and ratio identities}$$

$$= \left(\frac{1 - \sin x}{\cos x} \right)^2 \qquad \text{Subtract fractions}$$

$$= \frac{(1 - \sin x)^2}{\cos^2 x} \qquad \text{Property of exponents}$$

$$= \frac{(1 - \sin x)^2}{1 - \sin^2 x} \qquad \text{Pythagorean identity}$$

$$= \frac{(1 - \sin x)(1 - \sin x)}{(1 - \sin x)(1 + \sin x)} \qquad \text{Factor}$$

$$= \frac{1 - \sin x}{1 + \sin x} \qquad \text{Reduce}$$

47. $\sec x + \tan x = \dfrac{\sec x + \tan x}{1} \cdot \dfrac{\sec x - \tan x}{\sec x - \tan x}$ Multiply numerator and denominator by $\sec x - \tan x$

$$= \frac{\sec^2 x - \tan^2 x}{\sec x - \tan x} \qquad \text{Multiply fractions}$$

$$= \frac{1}{\sec x - \tan x} \qquad \text{Pythagorean identity}$$

49. $\dfrac{\sin x + 1}{\cos x + \cot x} = \dfrac{\sin x + 1}{\cos x + \dfrac{\cos x}{\sin x}}$ Ratio identity

$$= \frac{\sin x}{\sin x} \cdot \frac{(\sin x + 1)}{\left(\cos x + \dfrac{\cos x}{\sin x}\right)} \qquad \text{Multiply numerator and denominator by LCD}$$

$$= \frac{\sin x (\sin x + 1)}{\sin x \cos x + \cos x} \qquad \text{Distributive property}$$

$$= \frac{\sin x (\sin x + 1)}{\cos x (\sin x + 1)} \qquad \text{Factor}$$

$$= \frac{\sin x}{\cos x} \qquad \text{Reduce}$$

$$= \tan x \qquad \text{Ratio identity}$$

51. $\sin^4 A - \cos^4 A = \left(\sin^2 A + \cos^2 A\right)\left(\sin^2 A - \cos^2 A\right)$ Factor

$$= 1\left(\sin^2 A - \cos^2 A\right) \qquad \text{Pythagorean identity}$$

$$= 1 - \cos^2 A - \cos^2 A \qquad \text{Pythagorean identity}$$

$$= 1 - 2\cos^2 A \qquad \text{Combine}$$

53. $\dfrac{\sin^2 B - \tan^2 B}{1 - \sec^2 B} = \dfrac{\sin^2 B - \dfrac{\sin^2 B}{\cos^2 B}}{1 - \dfrac{1}{\cos^2 B}}$ Ratio identity and reciprocal identity

$$= \frac{\cos^2 B \left(\sin^2 B - \dfrac{\sin^2 B}{\cos^2 B}\right)}{\cos^2 B \left(1 - \dfrac{1}{\cos^2 B}\right)} \qquad \text{Multiply numerator and denominator by LCD}$$

$$= \frac{\cos^2 B \sin^2 B - \sin^2 B}{\cos^2 B - 1} \qquad \text{Distributive property}$$

This problem is continued on the next page

$$= \frac{\sin^2 B \left(\cos^2 B - 1\right)}{\cos^2 B - 1}$$ Factor

$$= \sin^2 B$$ Reduce

55. $\dfrac{\sec^4 y - \tan^4 y}{\sec^2 y + \tan^2 y} = \dfrac{\left(\sec^2 y - \tan^2 y\right)\left(\sec^2 y + \tan^2 y\right)}{\sec^2 y + \tan^2 y}$ Factor

$$= \sec^2 y - \tan^2 y$$ Reduce

$$= 1$$ Pythagorean identity

57. $\dfrac{\sin^3 A - 8}{\sin A - 2} = \dfrac{\left(\sin A - 2\right)\left(\sin^2 A + 2\sin A + 4\right)}{\sin A - 2}$ Factor as difference of 2 cubes

$$= \sin^2 A + 2\sin A + 4$$ Reduce

59. $\dfrac{1 - \tan^3 t}{1 - \tan t} = \dfrac{\left(1 - \tan t\right)\left(1 + \tan t + \tan^2 t\right)}{1 - \tan t}$ Factor

$$= \left(1 + \tan^2 t\right) + \tan t$$ Reduce and regroup numerator

$$= \sec^2 t + \tan t$$ Pythagorean identity

61. $\dfrac{\sin^2 x + \sin x \cos x}{\cos x - 2\cos^3 x} = \dfrac{\sin x \left(\sin x + \cos x\right)}{\cos x \left(1 - 2\cos^2 x\right)}$ Factor

$$= \frac{\sin x \left(\sin x + \cos x\right)}{\cos x \left[\left(1 - \cos^2 x\right) - \cos^2 x\right]}$$ Regrouping denominator

$$= \frac{\sin x \left(\sin x + \cos x\right)}{\cos x \left(\sin^2 x - \cos^2 x\right)}$$ Pythagorean identity

$$= \frac{\sin x \left(\sin x + \cos x\right)}{\cos x \left(\sin x + \cos x\right)\left(\sin x - \cos x\right)}$$ Factor denominator

$$= \frac{\sin x}{\cos x \left(\sin x - \cos x\right)}$$ Reduce

$$= \frac{\tan x}{\sin x - \cos x}$$ Ratio identity

63. $\left(\tan \theta + \cot \theta\right)^2 = \tan^2 \theta + 2\tan \theta \cot \theta + \cot^2 \theta$ Multiply

$$= \sec^2 \theta - 1 + 2\tan \theta \cdot \frac{1}{\tan \theta} + \csc^2 \theta - 1$$ Reciprocal and Pythagorean identities

$$= \sec^2 \theta + \csc^2 \theta + 2 - 2$$ Reduce and combine

$$= \sec^2 \theta + \csc^2 \theta$$ Combine

65.
$$\frac{1+\sin\phi}{1-\sin\phi} - \frac{1-\sin\phi}{1+\sin\phi} = \frac{(1+\sin\phi)(1+\sin\phi)}{(1-\sin\phi)(1+\sin\phi)} - \frac{(1-\sin\phi)(1-\sin\phi)}{(1+\sin\phi)(1-\sin\phi)} \qquad \text{LCD}$$

$$= \frac{(1+2\sin\phi+\sin^2\phi)-(1-2\sin\phi+\sin^2\phi)}{(1-\sin\phi)(1+\sin\phi)} \qquad \text{Subtract fractions}$$

$$= \frac{4\sin\phi}{1-\sin^2\phi} \qquad \text{Simplify}$$

$$= \frac{4\sin\phi}{\cos^2\phi} \qquad \text{Pythagorean identity}$$

$$= 4 \cdot \frac{\sin\phi}{\cos\phi} \cdot \frac{1}{\cos\phi} \qquad \text{Separate fraction}$$

$$= 4\tan\phi\sec\phi \qquad \text{Ratio and reciprocal identities}$$

75. If $\theta = 210°$, $\sin 210° \neq \sqrt{1-\cos^2(210°)}$

$$-\frac{1}{2} \neq \sqrt{1-\left(-\sqrt{3}/2\right)^2}$$

$$-\frac{1}{2} \neq \sqrt{1-\frac{3}{4}}$$

$$-\frac{1}{2} \neq \sqrt{\frac{1}{4}}$$

$$-\frac{1}{2} \neq \frac{1}{2}$$

77. If $\theta = 30°$, then $\sin 30° \neq \dfrac{1}{\cos 30°}$

$$\frac{1}{2} \neq \frac{1}{\sqrt{3}/2}$$

$$\frac{1}{2} \neq \frac{2}{\sqrt{3}}$$

79. If $\theta = 30°$, $\sqrt{\sin^2(30°)+\cos^2(30°)} = \sqrt{\left(\frac{1}{2}\right)^2 + \left(\frac{\sqrt{3}}{2}\right)^2}$

$$= \sqrt{\frac{1}{4}+\frac{3}{4}} = \sqrt{1} = 1$$

Also, $\sin 30° + \cos 30° = \dfrac{1}{2} + \dfrac{\sqrt{3}}{2}$

$$= \frac{1+\sqrt{3}}{2} \qquad \text{Since } 1 \neq \frac{1+\sqrt{3}}{2}, \text{ this statement is false.}$$

81. $\sin(30° + 60°) = \sin 90°$ and $\sin 30° + \sin 60° = \dfrac{1}{2} + \dfrac{\sqrt{3}}{2}$

$$= 1$$

$$= \dfrac{1 + \sqrt{3}}{2}$$

Since $1 \neq \dfrac{1 + \sqrt{3}}{2}$, this statement is false.

83. $\cos A = \sqrt{1 - \sin^2 A}$ $\qquad\qquad \tan A = \dfrac{\sin A}{\cos A}$

$$= \sqrt{1 - \left(\dfrac{3}{5}\right)^2} \qquad\qquad = \dfrac{\dfrac{3}{5}}{\dfrac{4}{5}}$$

$$= \sqrt{\dfrac{16}{25}} \qquad\qquad\qquad = \dfrac{3}{4}$$

$$= \dfrac{4}{5}$$

85. $\sin \dfrac{\pi}{3} = \dfrac{\sqrt{3}}{2}$ $\qquad\qquad$ **87.** $\cos \dfrac{\pi}{6} = \dfrac{\sqrt{3}}{2}$

89. $\dfrac{\pi}{12} = \dfrac{\pi}{12} \cdot \dfrac{180}{\pi}$ $\qquad\qquad$ **91.** $\dfrac{7\pi}{12} = \dfrac{7\pi}{12} \cdot \dfrac{180}{\pi}$

$\qquad\quad = 15°$ $\qquad\qquad\qquad\qquad = 105°$

Problem Set 5.2

1. $\sin 15° = \sin(45° - 30°)$ $\qquad$ **3.** $\tan 15° = \tan(45° - 30°)$

$$= \sin 45° \cos 30° - \cos 45° \sin 30° \qquad = \dfrac{\tan 45° - \tan 30°}{1 + \tan 45° \tan 30°}$$

$$= \left(\dfrac{\sqrt{2}}{2}\right)\left(\dfrac{\sqrt{3}}{2}\right) - \left(\dfrac{\sqrt{2}}{2}\right)\left(\dfrac{1}{2}\right) \qquad = \dfrac{1 - \dfrac{1}{\sqrt{3}}}{1 + 1\left(\dfrac{1}{\sqrt{3}}\right)}$$

$$= \dfrac{\sqrt{6}}{4} - \dfrac{\sqrt{2}}{4} \qquad\qquad\qquad = \dfrac{\sqrt{3} - 1}{\sqrt{3} + 1} \qquad \text{Multiply numerator and}$$

$$= \dfrac{\sqrt{6} - \sqrt{2}}{4} \qquad\qquad\qquad\qquad\qquad\qquad \text{denominator by } \sqrt{3}$$

5. $\sin\dfrac{7\pi}{12} = \sin\left(\dfrac{3\pi}{12}+\dfrac{4\pi}{12}\right)$

$\qquad\qquad = \sin\left(\dfrac{\pi}{4}+\dfrac{\pi}{3}\right)$

$\qquad\qquad = \sin\dfrac{\pi}{4}\cos\dfrac{\pi}{3}+\cos\dfrac{\pi}{4}\sin\dfrac{\pi}{3}$

$= \left(\dfrac{\sqrt{2}}{2}\right)\left(\dfrac{1}{2}\right)+\left(\dfrac{\sqrt{2}}{2}\right)\left(\dfrac{\sqrt{3}}{2}\right)$

$= \dfrac{\sqrt{2}}{4}+\dfrac{\sqrt{6}}{4}$

$= \dfrac{\sqrt{2}+\sqrt{6}}{4}$

7. $\cos 105° = \cos\left(60°+45°\right)$

$\qquad\qquad = \cos 60°\cos 45° - \sin 60°\sin 45°$

$\qquad\qquad = \dfrac{1}{2}\left(\dfrac{\sqrt{2}}{2}\right)-\dfrac{\sqrt{3}}{2}\left(\dfrac{\sqrt{2}}{2}\right)$

$= \dfrac{\sqrt{2}}{4}-\dfrac{\sqrt{6}}{4}$

$= \dfrac{\sqrt{2}-\sqrt{6}}{4}$

9. $\sin\left(x+2\pi\right) = \sin x\cos 2\pi + \cos x\sin 2\pi$

$\qquad\qquad = \left(\sin x\right)\left(1\right)+\left(\cos x\right)\left(0\right)$

$\qquad\qquad = \sin x$

11. $\cos\left(x-\dfrac{\pi}{2}\right) = \cos x\cos\dfrac{\pi}{2}+\sin x\sin\dfrac{\pi}{2}$

$\qquad\qquad = \cos x\left(0\right)+\sin x\left(1\right)$

$\qquad\qquad = \sin x$

13. $\cos\left(180°-\theta\right) = \cos 180°\cos\theta + \sin 180°\sin\theta$

$\qquad\qquad = -1\left(\cos\theta\right)+0\left(\sin\theta\right)$

$\qquad\qquad = -\cos\theta$

15. $\sin\left(90°+\theta\right) = \sin 90°\cos\theta + \cos 90°\sin\theta$

$\qquad\qquad = 1\left(\cos\theta\right)+0\left(\sin\theta\right)$

$\qquad\qquad = \cos\theta$

17. $\tan\left(x+\dfrac{\pi}{4}\right) = \dfrac{\tan x + \tan\dfrac{\pi}{4}}{1-\tan x\tan\dfrac{\pi}{4}}$

$\qquad\qquad = \dfrac{\tan x+1}{1-\left(\tan x\right)\left(1\right)}$

$\qquad\qquad = \dfrac{1+\tan x}{1-\tan x}$

19. $\sin\left(\dfrac{3\pi}{2}-x\right) = \sin\dfrac{3\pi}{2}\cos x - \cos\dfrac{3\pi}{2}\sin x$

$\qquad\qquad = -1\left(\cos x\right)-0\left(\sin x\right)$

$\qquad\qquad = -\cos x$

21. $\sin 3x\cos 2x + \cos 3x\sin 2x = \sin\left(3x+2x\right) = \sin 5x$

23. $\cos 5x\cos x - \sin 5x\sin x = \cos\left(5x+x\right) = \cos 6x$

25. $\cos 15° \cos 75° - \sin 15° \sin 75° = \cos\left(15° + 75°\right) = \cos 90° = 0$

27. $y = \sin 5x \cos 3x - \cos 5x \sin 3x$

$y = \sin\left(5x - 3x\right) = \sin 2x$

The graph is a sine curve with amplitude

of 1 and period of $\dfrac{2\pi}{2} = \pi$.

29. $y = 3\cos 7x \cos 5x + 3\sin 7x \sin 5x$

$= 3\left[\cos 7x \cos 5x + \sin 7x \sin 5x\right]$

$y = 3\cos\left(7x - 5x\right) = 3\cos 2x$

The graph is a cosine curve with an

amplitude of 3 and a period of $\dfrac{2\pi}{2} = \pi$.

31. $y = \sin x \cos\dfrac{\pi}{4} + \cos x \sin\dfrac{\pi}{4}$

$y = \sin\left(x + \dfrac{\pi}{4}\right)$

The graph is a sine curve with: Amplitude $= 1$ Period $= 2\pi$ Phase shift $= -\dfrac{\pi}{4}$

Spacing $= \dfrac{1}{4}\left(2\pi\right) = \dfrac{\pi}{2}$ $c = \dfrac{\pi}{4} + \dfrac{\pi}{2} = \dfrac{3\pi}{4}$

$a = -\dfrac{\pi}{4}$ $d = \dfrac{3\pi}{4} + \dfrac{\pi}{2} = \dfrac{5\pi}{4}$

$b = -\dfrac{\pi}{4} + \dfrac{\pi}{2} = \dfrac{\pi}{4}$ $e = \dfrac{5\pi}{4} + \dfrac{\pi}{2} = \dfrac{7\pi}{4}$

The 5 points on the x-axis we use are: $-\dfrac{\pi}{4}, \dfrac{\pi}{4}, \dfrac{3\pi}{4}, \dfrac{5\pi}{4}, \dfrac{7\pi}{4}$.

The 2 points on the y-axis we use are: -1 and 1.

33. $y = 2\left(\sin x \cos\dfrac{\pi}{3} + \cos x \sin\dfrac{\pi}{3}\right)$

$y = 2\sin\left(x + \dfrac{\pi}{3}\right)$

The graph is a sine curve with: Amplitude $= 2$ Period $= 2\pi$ Phase shift $= -\dfrac{\pi}{3}$

Spacing $= \dfrac{1}{4}\left(2\pi\right) = \dfrac{\pi}{2}$ $c = \dfrac{\pi}{6} + \dfrac{\pi}{2} = \dfrac{2\pi}{3}$

$a = -\dfrac{\pi}{3}$ $d = \dfrac{2\pi}{3} + \dfrac{\pi}{2} = \dfrac{7\pi}{6}$

$b = -\dfrac{\pi}{3} + \dfrac{\pi}{2} = \dfrac{\pi}{6}$ $e = \dfrac{7\pi}{6} + \dfrac{\pi}{2} = \dfrac{5\pi}{3}$

The 5 points on the x-axis we use are: $-\dfrac{\pi}{3}, \dfrac{\pi}{6}, \dfrac{2\pi}{3}, \dfrac{7\pi}{6}, \dfrac{5\pi}{3}$.

The 2 points on the y-axis we use are: 2 and -2 .

35.   If $\sin A = \dfrac{3}{5}$ with A in QII, then

$$\cos A = -\sqrt{1 - \sin^2 A}$$

$$= -\sqrt{1 - \dfrac{9}{25}}$$

$$= -\sqrt{\dfrac{16}{25}}$$

$$= -\dfrac{4}{5}$$

Also, if $\sin B = -\dfrac{5}{13}$ B in QIII, then

$$\cos B = -\sqrt{1 - \dfrac{25}{169}}$$

$$= -\sqrt{\dfrac{144}{169}}$$

$$= -\dfrac{12}{13}$$

$$\sin(A + B) = \sin A \cos B + \cos A \sin B$$

$$= \dfrac{3}{5}\left(-\dfrac{12}{13}\right) + \left(-\dfrac{4}{5}\right)\left(-\dfrac{5}{13}\right)$$

$$= -\dfrac{36}{65} + \dfrac{20}{65} = -\dfrac{16}{65}$$

$$\cos(A + B) = \cos A \cos B - \sin A \sin B$$

$$= -\dfrac{4}{5}\left(-\dfrac{12}{13}\right) - \dfrac{3}{5}\left(-\dfrac{5}{13}\right)$$

$$= \dfrac{48}{65} + \dfrac{15}{65} = \dfrac{63}{65}$$

$$\tan A = \dfrac{\sin A}{\cos A}$$

$$= \dfrac{\dfrac{3}{5}}{-\dfrac{4}{5}} = -\dfrac{3}{4}$$

$$\tan B = \dfrac{\sin B}{\cos B}$$

$$= \dfrac{-\dfrac{5}{13}}{-\dfrac{12}{13}} = \dfrac{5}{12}$$

$$\tan(A + B) = \dfrac{\tan A + \tan B}{1 - \tan A \tan B}$$

$$= \dfrac{-\dfrac{3}{4} + \dfrac{5}{12}}{1 - \left(-\dfrac{3}{4}\right)\left(\dfrac{5}{12}\right)} = \dfrac{-\dfrac{1}{3}}{1 + \dfrac{5}{16}} = \dfrac{-\dfrac{1}{3}}{\dfrac{21}{16}} = -\dfrac{16}{63}$$

The angle $(A + B)$ must terminate in QIV because the cosine is positive and the sine and tangent are negative.

37.   If $\sin A = \dfrac{1}{\sqrt{5}}$ with A in QI, then

$$\cos A = \sqrt{1 - \sin^2 A}$$

$$= \sqrt{1 - \dfrac{1}{5}}$$

$$= \sqrt{\dfrac{4}{5}}$$

$$= \dfrac{2}{\sqrt{5}}$$

Also, $\tan A = \dfrac{\sin A}{\cos A}$

$$= \dfrac{1/\sqrt{5}}{2\sqrt{5}}$$

$$= \dfrac{1}{2}$$

We have $\tan A = \dfrac{1}{2}$ and $\tan B = \dfrac{3}{4}$

This problem is continued on the next page.

Therefore, $\tan(A+B) = \dfrac{\dfrac{1}{2}+\dfrac{3}{4}}{1-\left(\dfrac{1}{2}\right)\left(\dfrac{3}{4}\right)}$ $\qquad\qquad \cot(A+B) = \dfrac{1}{\tan(A+B)}$

$\qquad\qquad\qquad\quad = \dfrac{5/4}{5/8}$ $\qquad\qquad\qquad\qquad\qquad = \dfrac{1}{2}$

$\qquad\qquad\qquad\quad = 2$

The angle $(A + B)$ terminates in QI because its tangent is positive. (If its tangent were negative, it would terminate in QII.)

39. $\quad \tan(A+B) = \dfrac{\tan A + \tan B}{1 - \tan A \tan B}$

$\qquad\qquad 3 = \dfrac{\tan A + \dfrac{1}{2}}{1 - (\tan A)\left(\dfrac{1}{2}\right)}$

$\qquad\qquad \dfrac{3}{1} = \dfrac{2\tan A + 1}{2 - \tan A}$

$\qquad 6 - 3\tan A = 2\tan A + 1$

$\qquad\quad -5\tan A = -5$

$\qquad\qquad \tan A = 1$

41. $\quad \sin 2x = \sin(x + x)$

$\qquad\qquad = \sin x \cos x + \cos x \sin x$

$\qquad\qquad = 2\sin x \cos x$

43. $\quad \sin(90° + x) + \sin(90° - x)$

$\quad = \left(\sin 90° \cos x + \cos 90° \sin x\right) + \left(\sin 90° \cos x - \cos 90° \sin x\right)$ $\qquad$ Sum and difference formulas

$\quad = 2\sin 90° \cos x$ $\qquad\qquad$ Combine

$\quad = 2(1)\cos x$ $\qquad\qquad$ Substitute exact value

$\quad = 2\cos x$ $\qquad\qquad$ Multiply

45. $\quad \cos(x - 90°) - \cos(x + 90°)$

$\quad = \left[\cos x \cos 90° + \sin x \sin 90°\right] - \left[\cos x \cos 90° - \sin x \sin 90°\right]$ $\qquad$ Sum and difference formulas

$\quad = \left[(\cos x)(0) + (\sin x)(1)\right] - \left[(\cos x)(0) - (\sin x)(1)\right]$ $\qquad$ Substitute exact values

$\quad = \sin x - (-\sin x)$ $\qquad\qquad$ Multiply

$\quad = 2\sin x$ $\qquad\qquad$ Subtract

47. $\sin\left(\dfrac{\pi}{6}+x\right)+\sin\left(\dfrac{\pi}{6}-x\right)$

$=\left(\sin\dfrac{\pi}{6}\cos x+\cos\dfrac{\pi}{6}\sin x\right)+\left(\sin\dfrac{\pi}{6}\cos x-\cos\dfrac{\pi}{6}\sin x\right)$ Sum and difference formulas

$=2\sin\dfrac{\pi}{6}\cos x$ Combine

$=2\left(\dfrac{1}{2}\right)\cos x$ Substitute exact values

$=\cos x$ Multiply

49. $\cos\left(x+\dfrac{\pi}{4}\right)+\cos\left(x-\dfrac{\pi}{4}\right)$

$=\left(\cos x\cos\dfrac{\pi}{4}-\sin x\sin\dfrac{\pi}{4}\right)+\left(\cos x\cos\dfrac{\pi}{4}+\sin x\sin\dfrac{\pi}{4}\right)$ Sum and difference formulas

$=(\cos x)\dfrac{\sqrt{2}}{2}-(\sin x)\dfrac{\sqrt{2}}{2}+(\cos x)\dfrac{\sqrt{2}}{2}+(\sin x)\dfrac{\sqrt{2}}{2}$ Substitute exact values

$=\sqrt{2}\cos x$ Combine

51. $\sin\left(\dfrac{3\pi}{2}+x\right)+\sin\left(\dfrac{3\pi}{2}-x\right)$

$=\left(\sin\dfrac{3\pi}{2}\cos x+\cos\dfrac{3\pi}{2}\sin x\right)+\left(\sin\dfrac{3\pi}{2}\cos x-\cos\dfrac{3\pi}{2}\sin x\right)$ Sum and difference formulas

$=2\sin\dfrac{3\pi}{2}\cos x$ Combine

$=2(-1)\cos x$ Substitute exact value

$=-2\cos x$ Multiply

53. $\sin(A+B)+\sin(A-B)$

$=(\sin A\cos B+\cos A\sin B)+(\sin A\cos B-\cos A\sin B)$ Sum and difference formulas

$=2\sin A\cos B$ Combine

55. $\dfrac{\sin(A-B)}{\cos A\cos B}=\dfrac{\sin A\cos B-\cos A\sin B}{\cos A\cos B}$ Difference formula

$=\dfrac{\sin A\cos B}{\cos A\cos B}-\dfrac{\cos A\sin B}{\cos A\cos B}$ Separate into 2 fractions

$=\dfrac{\sin A}{\cos A}-\dfrac{\sin B}{\cos B}$ Reduce

$=\tan A-\tan B$ Ratio identity

57.

$$\sec(A+B)$$

$$= \frac{1}{\cos(A+B)}$$

Reciprocal identity

$$= \frac{1}{\cos(A+B)} \cdot \frac{\cos(A-B)}{\cos(A-B)}$$

Multiply numerator and

denominator by $\cos(A-B)$

$$= \frac{\cos(A-B)}{(\cos A \cos B - \sin A \sin B)(\cos A \cos B + \sin A \sin B)}$$

Sum and differences formulas

$$= \frac{\cos(A-B)}{\cos^2 A \cos^2 B - \sin^2 A \sin^2 B}$$

Multiply

$$= \frac{\cos(A-B)}{\cos^2 A \cos^2 B + \sin^2 A \cos^2 B - \sin^2 A \sin^2 B - \sin^2 A \cos^2 B}$$

Add & subtract $\sin^2 A \cos^2 B$

in denominator

$$= \frac{\cos(A-B)}{\cos^2 B(\cos^2 A + \sin^2 A) - \sin^2 A(\sin^2 B + \cos^2 B)}$$

Factor by grouping

$$= \frac{\cos(A-B)}{\cos^2 B(1) - \sin^2 A(1)}$$

Pythagorean identity

$$= \frac{\cos(A-B)}{\cos^2 B - \sin^2 A}$$

Multiply

65. Amplitude $= 4$

Period $= \dfrac{2\pi}{2} = \pi$

67. Amplitude $= 3$

Period $= \dfrac{2\pi}{1/2} = 4\pi$

69. Amplitude $= 2$

Period $= \dfrac{2\pi}{\pi} = 2$

71. Period $= \dfrac{2\pi}{3}$

Asymptotes occur where $y = \sin 3x$ is zero, or at 0, $\dfrac{\pi}{3}$, and $\dfrac{2\pi}{3}$.

We use the graph of $y = \sin 3x$ and the asymptotes to sketch the graph of $y = \csc 3x$.

73. Amplitude $= \dfrac{1}{2}$

Period $= \dfrac{2\pi}{3}$

75. Amplitude $= \dfrac{1}{2}$

Period $= \dfrac{2\pi}{\pi/2} = 4$

Problem Set 5.3

1. If $\sin A = -\dfrac{3}{5}$ with A in QIII, then

$$\cos A = -\sqrt{1 - \sin^2 A}$$

$$= -\sqrt{1 - \frac{9}{25}}$$

$$= -\sqrt{\frac{16}{25}}$$

$$= -\frac{4}{5}$$

$$\sin 2A = 2 \sin A \cos A$$

Therefore,

$$= 2\left(-\frac{3}{5}\right)\left(\frac{4}{5}\right) = \frac{24}{25}$$

3. $\cos 2A = \cos^2 A - \sin^2 A$

$$= \left(-\frac{4}{5}\right)^2 - \left(-\frac{3}{5}\right)^2$$

$$= \frac{16}{25} - \frac{9}{25} = \frac{7}{25}$$

$$\tan 2A = \frac{\sin 2A}{\cos 2A} = \frac{\dfrac{24}{25}}{\dfrac{7}{25}} = \frac{24}{7}$$

5. If $\cos x = \dfrac{1}{\sqrt{10}}$ with x in QIV, then

$$\cos 2x = 2\cos^2 x - 1$$

$$= 2\left(\frac{1}{\sqrt{10}}\right)^2 - 1$$

$$= 2\left(\frac{1}{10}\right) - 1$$

$$= \frac{1}{5} - 1 = -\frac{4}{5}$$

7. $\sin 2x = -\sqrt{1 - \cos^2(2x)}$

$$= -\sqrt{1 - \left(\frac{16}{25}\right)} = -\sqrt{\frac{9}{25}}$$

$$= -\frac{3}{5}$$

$$\cot 2x = \frac{\cos 2x}{\sin 2x}$$

$$= \frac{-\dfrac{4}{5}}{-\dfrac{3}{5}} = \frac{4}{3}$$

9. If $\tan\theta = \dfrac{5}{12}$ with θ in QI, we can draw the triangle at the right and

find the hypotenuse using the Pythagorean Theorem.

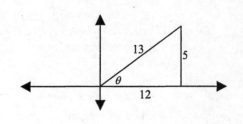

$$\text{hypotenuse} = \sqrt{5^2 + 12^2}$$
$$= \sqrt{25 + 144}$$
$$= \sqrt{169} = 13$$

Then, $\sin\theta = \dfrac{5}{13}$ and $\cos\theta = \dfrac{12}{13}$.

Therefore, $\sin 2\theta = 2\sin\theta\cos\theta$

$$= 2\left(\dfrac{5}{13}\right)\left(\dfrac{12}{13}\right) = \dfrac{120}{169}$$

11. $\csc 2\theta = \dfrac{1}{\sin 2\theta} = \dfrac{1}{\dfrac{120}{169}} = \dfrac{169}{120}$

13. If $\csc t = \sqrt{5}$ with t in QII, then $\sin t = \dfrac{1}{\sqrt{5}}$.

Therefore, $\cos 2t = 1 - 2\sin^2 t$

$$= 1 - 2\left(\dfrac{1}{\sqrt{5}}\right)^2$$

$$= 1 - \dfrac{2}{5} = \dfrac{3}{5}$$

15. $\sec 2t = \dfrac{1}{\cos 2t}$

$$= \dfrac{1}{3/5} = \dfrac{5}{3}$$

17. $y = 4 - 8\sin^2 x$

$\quad = 4\left(1 - 2\sin^2 x\right)$

$\quad = 4\cos 2x$

The graph is a cosine curve with amplitude $= 4$ and period $= \dfrac{2\pi}{2} = \pi$.

19. $y = 6\cos^2 x - 3$

$\quad = 3\left(2\cos^2 x - 1\right)$

$\quad = 3\cos 2x$

The graph is a cosine curve with amplitude $= 3$ and period $= \dfrac{2\pi}{2} = \pi$.

21. $y = 1 - 2\sin^2 2x$

$\quad = \cos 2(2x)$

$\quad = \cos 4x$

The graph is a cosine curve with amplitude $= 1$ and period $= \dfrac{2\pi}{4} = \dfrac{\pi}{2}$.

23. $\sin 60° = \dfrac{\sqrt{3}}{2}$ $\qquad\qquad$ $2\sin 30° \cos 30° = 2\left(\dfrac{1}{2}\right)\left(\dfrac{\sqrt{3}}{2}\right)$

$$= \dfrac{\sqrt{3}}{2}$$

Therefore, they are equal.

25. $\cos 120° = -\cos 60°$ $\qquad\qquad$ $\cos^2 60° - \sin^2 60° = \left(\dfrac{1}{2}\right)^2 - \left(\dfrac{\sqrt{3}}{2}\right)^2$

$$= -\dfrac{1}{2}$$

$$= \dfrac{1}{4} - \dfrac{3}{4}$$

$$= -\dfrac{1}{2}$$

Therefore, they are equal.

27. $\tan 2A = \dfrac{2\tan A}{1 - \tan^2 A}$ $\qquad\qquad$ **29.** $2\sin 15° \cos 15° = \sin 2\left(15°\right) = \sin 30° = \dfrac{1}{2}$

$$= \dfrac{2\left(\dfrac{3}{4}\right)}{1 - \left(\dfrac{3}{4}\right)^2} = \dfrac{\dfrac{3}{2}}{1 - \dfrac{9}{16}}$$

$$= \dfrac{\dfrac{3}{2}}{\dfrac{7}{16}} = \dfrac{24}{7}$$

31. $1 - 2\sin^2 75° = \cos 2\left(75°\right)$ $\qquad\qquad$ **33.** $\sin\dfrac{\pi}{12}\cos\dfrac{\pi}{12} = \dfrac{1}{2}\left(2\sin\dfrac{\pi}{12}\cos\dfrac{\pi}{12}\right)$

$$= \cos 150°$$

$$= \dfrac{1}{2}\sin 2\left(\dfrac{\pi}{12}\right)$$

$$= -\cos 30°$$

$$= \dfrac{1}{2}\sin\dfrac{\pi}{6}$$

$$= -\dfrac{\sqrt{3}}{2}$$

$$= \dfrac{1}{2}\left(\dfrac{1}{2}\right)$$

$$= \dfrac{1}{4}$$

35.
$$\frac{\tan 22.5°}{1 - \tan^2 22.5°} = \frac{1}{2}\tan 2(22.5°)$$
$$= \frac{1}{2}\tan 45°$$
$$= \frac{1}{2}(1)$$
$$= \frac{1}{2}$$

37.

$(\sin x - \cos x)^2 = \sin^2 x - 2\sin x \cos x + \cos^2 x$	Expand
$= (\sin^2 x + \cos^2 x) - 2\sin x \cos x$	Commutative property
$= 1 - 2\sin x \cos x$	Pythagorean identity
$= 1 - \sin 2x$	Double-angle identity

39.

$\dfrac{1 + \cos 2\theta}{2} = \dfrac{1 + 2\cos^2\theta - 1}{2}$	Double-angle identity
$= \dfrac{2\cos^2\theta}{2}$	Combine
$= \cos^2\theta$	Reduce

41.

$\dfrac{\sin 2\theta}{1 - \cos 2\theta} = \dfrac{2\sin\theta\cos\theta}{1 - (1 - 2\sin^2\theta)}$	Double-angle identities
$= \dfrac{2\sin\theta\cos\theta}{2\sin^2\theta}$	Subtract
$= \dfrac{\cos\theta}{\sin\theta}$	Reduce
$= \cot\theta$	Ratio identity

43.

$\tan x + \cot x = \dfrac{\sin x}{\cos x} + \dfrac{\cos x}{\sin x}$	Ratio identities
$= \dfrac{\sin x}{\cos x} \cdot \dfrac{\sin x}{\sin x} + \dfrac{\cos x}{\sin x} \cdot \dfrac{\cos x}{\cos x}$	LCD is $\sin x \cos x$
$= \dfrac{\sin^2 x + \cos^2 x}{\sin x \cos x}$	Multiply and add fractions
$= \dfrac{1}{\sin x \cos x}$	Pythagorean identity
$= \dfrac{1}{\frac{1}{2}\sin 2x}$	Double-angle identity
$= 2\csc 2x$	Reciprocal identity

45.
$$\sin 3\theta = \sin(2\theta + \theta)$$ Addition
$$= \sin 2\theta \cos \theta + \cos 2\theta \sin \theta$$ Sum formula
$$= (2 \sin \theta \cos \theta) \cos \theta + (1 - 2 \sin^2 \theta) \sin \theta$$ Double-angle identities
$$= 2 \sin \theta \cos^2 \theta + \sin \theta - 2 \sin^3 \theta$$ Multiply
$$= 2 \sin \theta (1 - \sin^2 \theta) + \sin \theta - 2 \sin^3 \theta$$ Pythagorean identity
$$= 2 \sin \theta - 2 \sin^3 \theta + \sin \theta - 2 \sin^3 \theta$$ Multiply
$$= 3 \sin \theta - 4 \sin^3 \theta$$ Combine

47.
$$\cos^4 x - \sin^4 x = (\cos^2 x + \sin^2 x)(\cos^2 x - \sin^2 x)$$ Factor
$$= 1(\cos^2 x - \sin^2 x)$$ Pythagorean identity
$$= \cos 2x$$ Double-angle identity

49.
$$\frac{\cos 2\theta}{\sin \theta \cos \theta} = \frac{\cos^2 \theta - \sin^2 \theta}{\sin \theta \cos \theta}$$ Double-angle identity
$$= \frac{\cos^2 \theta}{\sin \theta \cos \theta} - \frac{\sin^2 \theta}{\sin \theta \cos \theta}$$ Separate into 2 fractions
$$= \frac{\cos \theta}{\sin \theta} - \frac{\sin \theta}{\cos \theta}$$ Reduce
$$= \cot \theta - \tan \theta$$ Ratio identities

51.
$$\sin 4A = 2 \sin 2A \cos 2A$$ Double-angle identity
$$= 2(2 \sin A \cos A)(\cos^2 A - \sin^2 A)$$ Double-angle identities
$$= 4 \sin A \cos^3 A - 4 \sin^3 A \cos A$$ Distributive property

53.
$$\frac{1 - \tan x}{1 + \tan x} = \frac{1 - \dfrac{\sin x}{\cos x}}{1 + \dfrac{\sin x}{\cos x}}$$ Ratio identity
$$= \frac{\cos x \left(1 - \dfrac{\sin x}{\cos x}\right)}{\cos x \left(1 + \dfrac{\sin x}{\cos x}\right)}$$ Multiply numerator and denominator by LCD
$$= \frac{\cos x - \sin x}{\cos x + \sin x}$$ Distributive property
$$= \frac{\cos x - \sin x}{\cos x + \sin x} \cdot \frac{\cos x - \sin x}{\cos x - \sin x}$$ Multiply by a fraction equal to 1

This problem is continued on the next page

$$= \frac{\cos^2 x - 2\sin x \cos x + \sin^2 x}{\cos^2 x - \sin^2 x} \qquad \text{Multiply fractions}$$

$$= \frac{1 - 2\sin x \cos x}{\cos^2 x - \sin^2 x} \qquad \text{Pythagorean identity}$$

$$= \frac{1 - \sin 2x}{\cos 2x} \qquad \text{Double-angle identities}$$

61. First, we find θ: $\quad x = 5\tan\theta$

$$\tan\theta = \frac{x}{5}$$

$$\theta = \tan^{-1}\frac{x}{5}$$

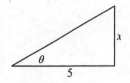

Next, we find $\sin 2\theta$ by drawing the triangle at the right and using the Pythagorean Theorem:

$$\text{hypotenuse} = \sqrt{x^2 + 5^2} \qquad\qquad \text{Then } \sin 2\theta = 2\sin\theta\cos\theta$$

$$= \sqrt{x^2 + 25} \qquad\qquad\qquad = 2\left(\frac{x}{\sqrt{x^2 + 25}}\right)\left(\frac{5}{\sqrt{x^2 + 25}}\right)$$

$$= \frac{10x}{x^2 + 25}$$

Last, we evaluate: $\dfrac{\theta}{2} - \dfrac{\sin 2\theta}{4} = \dfrac{1}{2}\tan^{-1}\dfrac{x}{5} - \dfrac{1}{4}\left(\dfrac{10x}{x^2 + 25}\right)$

$$= \frac{1}{2}\tan^{-1}\frac{x}{5} - \frac{1}{2}\left(\frac{5x}{x^2 + 25}\right)$$

$$= \frac{1}{2}\left(\tan^{-1}\frac{x}{5} - \frac{5x}{x^2 + 25}\right)$$

63. First we find θ: $\quad x = 3\sin\theta$

$$\sin\theta = \frac{x}{3}$$

$$\theta = \sin^{-1}\frac{x}{3}$$

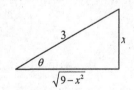

Next, we find $\sin 2\theta$ by drawing the triangle at the right and using the Pythagorean Theorem:

This problem is continued on the next page

$$\text{adjacent side} = \sqrt{3^2 - x^2}$$
$$= \sqrt{9 - x^2}$$

$$\text{Then, } \sin 2\theta = 2\sin\theta\cos\theta$$
$$= 2\left(\frac{x}{3}\right)\left(\frac{\sqrt{9-x^2}}{3}\right)$$
$$= \frac{2}{9}x\sqrt{9-x^2}$$

Last, we evaluate: $\dfrac{\theta}{2} - \dfrac{\sin 2\theta}{4} = \dfrac{1}{2}\sin^{-1}\dfrac{x}{3} - \dfrac{1}{4}\left(\dfrac{2}{9}x\sqrt{9-x^2}\right)$

$$= \frac{1}{2}\sin^{-1}\frac{x}{3} - \frac{1}{18}x\sqrt{9-x^2}$$

$$= \frac{1}{2}\left(\sin^{-1}\frac{x}{3} - \frac{x\sqrt{9-x^2}}{9}\right)$$

65. Let $y_1 = 2$ (a horizontal line) and $y_2 = -2\cos x$ (a cosine curve reflected across the x-axis with amplitude of 2). Graph y_1, y_2 and $y = y_1 + y_2$ on the same coordinate system.

67. Let $y_1 = 3$ (a horizontal line) and $y_2 = -3\cos x$ (a cosine curve reflected across the x-axis with amplitude of 3). Graph y_1, y_2 and $y = y_1 + y_2$ on the same coordinate system.

69. Let $y_1 = \cos x$ and $y_2 = \dfrac{1}{2}\sin 2x$ (a sine curve with amplitude of $\dfrac{1}{2}$ and a period of $\dfrac{2\pi}{2}$ or π). Graph y_1, y_2 and $y = y_1 + y_2$ on the same coordinate system.

71. Let $y_1 = \dfrac{1}{2}x$ (a line) and $y_2 = \sin \pi x$ (a sine curve with amplitude of 1 and period $= \dfrac{2\pi}{\pi} = 2$). Graph y_1, y_2 and $y = y_1 + y_2$ on the same coordinate system.

Problem Set 5.4

1. If A is in QIV, then $270° < A < 360°$ and $135° < \dfrac{A}{2} < 180°$. Therefore $\dfrac{A}{2}$ is in QII.

$$\sin\frac{A}{2} = \sqrt{\frac{1-\cos A}{2}}$$

$$= \sqrt{\frac{1 - \frac{1}{2}}{2}} = \sqrt{\frac{1}{4}} = \frac{1}{2}$$

3.
$$\csc \frac{A}{2} = \frac{1}{\sin \frac{A}{2}} = \frac{1}{\frac{1}{2}} = 2$$

5. If A is in QIII, then $180° < A < 270°$ and $90° < \frac{A}{2} < 135°$. Therefore, $\frac{A}{2}$ is in QII.

If $\sin A = -\frac{3}{5}$, then $\cos A = -\sqrt{1 - \sin^2 A}$

$$= -\sqrt{1 - \frac{9}{25}}$$

$$= -\sqrt{\frac{16}{25}}$$

$$= -\frac{4}{5}$$

Therefore, $\cos \frac{A}{2} = -\sqrt{\frac{1 + \cos A}{2}}$

$$= -\sqrt{\frac{1 - \frac{4}{5}}{2}}$$

$$= -\sqrt{\frac{\frac{1}{5}}{2}}$$

$$= -\sqrt{\frac{1}{10}} = -\frac{1}{\sqrt{10}}$$

7.
$$\sec \frac{A}{2} = \frac{1}{\cos \frac{A}{2}}$$

$$= \frac{1}{-\frac{1}{\sqrt{10}}}$$

$$= -\sqrt{10}$$

9. If B is in QIII, then $\frac{B}{2}$ is in QII. (See Problem 5)

If $\sin B = -\frac{1}{3}$, then $\cos B = -\sqrt{1 - \sin^2 B}$

$$= -\sqrt{1 - \frac{1}{9}}$$

$$= -\sqrt{\frac{8}{9}}$$

$$= -\frac{2\sqrt{2}}{3}$$

Therefore $\sin \frac{B}{2} = -\sqrt{\frac{1 - \cos B}{2}}$

$$= \sqrt{\frac{1 + \frac{2\sqrt{2}}{3}}{2}}$$

$$= \sqrt{\frac{3 + 2\sqrt{2}}{6}}$$ (Multiply numerator

and denominator by 3)

Use the information from problem 9 to solve #11 and #13:

11. $\cos\dfrac{B}{2} = -\sqrt{\dfrac{1+\cos B}{2}}$

$= -\sqrt{\dfrac{1-\dfrac{2\sqrt{2}}{3}}{2}}$

$= -\sqrt{\dfrac{3-2\sqrt{2}}{6}}$

13. $\tan\dfrac{B}{2} = \dfrac{1-\cos A}{\sin A}$

$= \dfrac{1+\dfrac{2\sqrt{2}}{3}}{-\dfrac{1}{3}}$

$= \dfrac{3+2\sqrt{2}}{-1}$

$= -3 - 2\sqrt{2}$

15. $\cos A = -\sqrt{1-\sin^2 A}$

$= -\sqrt{1-\dfrac{16}{25}}$

$= -\sqrt{\dfrac{9}{25}} = -\dfrac{3}{5}$

$\sin\dfrac{A}{2} = \sqrt{\dfrac{1-\cos A}{2}}$ (in QI)

$= \sqrt{\dfrac{1-\left(-\dfrac{3}{5}\right)}{2}} = \sqrt{\dfrac{\dfrac{8}{5}}{2}} = \sqrt{\dfrac{4}{5}} = \dfrac{2}{\sqrt{5}}$

17. $\cos 2A = 1 - 2\sin^2 A$

$= 1 - 2\left(\dfrac{4}{5}\right)^2$

$= 1 - \dfrac{32}{25}$

$= -\dfrac{7}{25}$

19. $\sec 2A = \dfrac{1}{\cos 2A}$

$= \dfrac{1}{-\dfrac{7}{25}} = -\dfrac{25}{7}$

21. If B is in QI, then $\dfrac{B}{2}$ must be in QI.

If $\sin B = \dfrac{3}{5}$, then $\cos B = \sqrt{1 - \sin^2 B}$

$= \sqrt{1 - \dfrac{9}{25}} = \sqrt{\dfrac{16}{25}} = \dfrac{4}{5}$

Therefore, $\cos\dfrac{B}{2} = \sqrt{\dfrac{1+\cos B}{2}}$

$= \sqrt{\dfrac{1+\dfrac{4}{5}}{2}} = \sqrt{\dfrac{9}{10}} = \dfrac{3}{\sqrt{10}}$

23. $\sin(A+B) = \sin A \cos B + \cos A \sin B$

$$= \left(\frac{4}{5}\right)\left(\frac{4}{5}\right) + \left(-\frac{3}{5}\right)\left(\frac{3}{5}\right)$$

$$= \frac{16}{25} - \frac{9}{25} = \frac{7}{25}$$

25. If $\sin A = \frac{4}{5}$, then $\cos A = -\sqrt{1 - \sin^2 A}$

$$= -\sqrt{1 - \frac{16}{25}} = -\sqrt{\frac{9}{25}} = -\frac{3}{5}$$

Therefore, $\cos(A-B) = \cos A \cos B + \sin A \sin B$

$$= -\frac{3}{5}\left(\frac{4}{5}\right) + \left(\frac{4}{5}\right)\left(\frac{3}{5}\right)$$

$$= -\frac{12}{25} + \frac{12}{25}$$

$$= 0$$

27. $y = 4\sin^2 \frac{x}{2}$

$$= 4\left(\pm\sqrt{\frac{1-\cos x}{2}}\right)^2$$

$$= 4\left(\frac{1-\cos x}{2}\right)$$

$$= 2 - 2\cos x$$

This graph is a cosine curve with amplitude of 2. It has been reflected across the x-axis and has a vertical translation of 2.

29. $y = 2\cos^2 \frac{x}{2}$

$$= 2\left(\pm\sqrt{\frac{1+\cos x}{2}}\right)^2$$

$$= 2\left(\frac{1+\cos x}{2}\right)$$

$$= 1 + \cos x \qquad \text{This graph is the standard cosine curve}$$

with a vertical translation of 1.

31. $\cos 15° = \cos\left(\frac{30°}{2}\right)$

$$= \sqrt{\frac{1+\cos 30°}{2}}$$

$$= \sqrt{\frac{1 + \frac{\sqrt{3}}{2}}{2}}$$

$$= \sqrt{\frac{2+\sqrt{3}}{4}}$$

$$= \frac{\sqrt{2+\sqrt{3}}}{2}$$

33.

$$\sin 75° = \sin\left(\frac{150°}{2}\right)$$

$$= -\sqrt{\frac{1-\cos 150°}{2}}$$

$$= \sqrt{\frac{1-\left(-\frac{\sqrt{3}}{2}\right)}{2}}$$

$$= \sqrt{\frac{2+\sqrt{3}}{4}}$$

$$= \frac{\sqrt{2+\sqrt{3}}}{2}$$

35.

$$\cos 105° = \cos\left(\frac{210°}{2}\right)$$

$$= -\sqrt{\frac{1+\cos 210°}{2}}$$

$$= -\sqrt{\frac{1+\left(-\frac{\sqrt{3}}{2}\right)}{2}}$$

$$= -\sqrt{\frac{2-\sqrt{3}}{4}}$$

$$= -\frac{\sqrt{2-\sqrt{3}}}{2}$$

37.

$$\frac{\csc\theta - \cot\theta}{2\csc\theta} = \frac{\frac{1}{\sin\theta} - \frac{\cos\theta}{\sin\theta}}{\frac{2}{\sin\theta}}$$ — Reciprocal and ratio identities

$$= \frac{\frac{1-\cos\theta}{\sin\theta}}{\frac{2}{\sin\theta}}$$ — Subtract

$$= \frac{1-\cos\theta}{2}$$ — Divide

$$= \sin^2\frac{\theta}{2}$$ — Half-angle formula

39.

$$\sec^2\frac{A}{2} = \frac{1}{\cos^2\frac{A}{2}}$$ — Reciprocal identity

$$= \frac{1}{\left(\pm\sqrt{\frac{1+\cos A}{2}}\right)^2}$$ — Half-angle identity

$$= \frac{1}{\frac{1+\cos A}{2}}$$ — Simplify

$$= \frac{2}{1+\cos A}$$ — Divide

$$= \frac{2}{1+\frac{1}{\sec A}}$$ — Reciprocal identity

$$= \frac{2\sec A}{\sec A + 1}$$ — Multiply numerator and denominator by sec$A$

41. $\tan\dfrac{B}{2} = \dfrac{1-\cos B}{\sin B}$ Half-angle identity

$\qquad = \dfrac{1}{\sin B} - \dfrac{\cos B}{\sin B}$ Separate fractions

$\qquad = \csc B - \cot B$ Reciprocal and ratio identities

43. $\tan\dfrac{x}{2} + \cot\dfrac{x}{2} = \dfrac{1-\cos x}{\sin x} + \dfrac{1}{\dfrac{\sin x}{1+\cos x}}$ Half-angle identities and reciprocal identity

$\qquad = \dfrac{1-\cos x}{\sin x} + \dfrac{1+\cos x}{\sin x}$ Divide second fraction

$\qquad = \dfrac{1-\cos x + 1 + \cos x}{\sin x}$ Add

$\qquad = \dfrac{2}{\sin x}$ Combine

$\qquad = 2\csc x$ Reciprocal identity

45. $\dfrac{\tan\theta + \sin\theta}{2\tan\theta} = \dfrac{\dfrac{\sin\theta}{\cos\theta} + \sin\theta}{\dfrac{2\sin\theta}{\cos\theta}}$ Ratio identity

$\qquad = \dfrac{\sin\theta + \sin\theta\cos\theta}{2\sin\theta}$ Multiply numerator and denominator by $\cos\theta$

$\qquad = \dfrac{\sin\theta(1+\cos\theta)}{2\sin\theta}$ Factor

$\qquad = \dfrac{1+\cos\theta}{2}$ Reduce

$\qquad = \cos^2\dfrac{\theta}{2}$ Half-angle identity

47. $\dfrac{1}{4} + \dfrac{\cos 2\theta}{2} + \dfrac{\cos^2 2\theta}{4} = \dfrac{1}{4} + \dfrac{2\cos^2\theta - 1}{2} + \dfrac{\left(2\cos^2\theta - 1\right)^2}{4}$ Double-angle identity

$\qquad = \dfrac{1}{4} + \dfrac{2\left(2\cos^2\theta - 1\right)}{4} + \dfrac{4\cos^4\theta - 4\cos^2\theta + 1}{4}$ LCD is 4

$\qquad = \dfrac{1 + 4\cos^2\theta - 2 + 4\cos^4\theta - 4\cos^2\theta + 1}{4}$ Combine

$\qquad = \dfrac{4\cos^4\theta}{4}$ Simplify

$\qquad = \cos^4\theta$ Reduce

49. Let $\theta = \arcsin\dfrac{3}{5}$. Then $\sin\theta = \dfrac{3}{5}$ and $-\dfrac{\pi}{2} \leq \theta \leq \dfrac{\pi}{2}$.

We want to find $\sin\theta$ which is equal to $\dfrac{3}{5}$.

51. Let $\beta = \arctan 2$. Then $\tan\beta = \dfrac{2}{1}$ and $-\dfrac{\pi}{2} < \beta < \dfrac{\pi}{2}$.

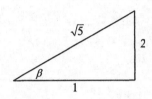

We draw a triangle and label the opposite and adjacent sides. Then we find the hypotenuse using the Pythagorean Theorem:

$$\text{hypotenuse} = \sqrt{1^2 + 2^2}$$
$$= \sqrt{1+4} = \sqrt{5}$$

Using the figure, $\cos\beta = \dfrac{1}{\sqrt{5}}$.

53. Let $\theta = \tan^{-1} x$. Then $\tan\theta = \dfrac{x}{1}$ and $-\dfrac{\pi}{2} < \theta < \dfrac{\pi}{2}$.

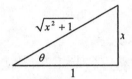

We draw a triangle and label the opposite and adjacent sides. Then we find the hypotenuse using the Pythagorean Theorem:

$$\text{hypotenuse} = \sqrt{x^2 + 1^2} = \sqrt{x^2 + 1}$$

Using the figure, $\sin\theta = \dfrac{x}{\sqrt{x^2+1}}$.

55. Let $\theta = \sin^{-1} x$. Then $\sin\theta = \dfrac{x}{1}$ and $-\dfrac{\pi}{2} \leq \theta \leq \dfrac{\pi}{2}$.

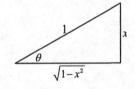

We draw a triangle, label the sides, and then find the missing side:

$$\text{adjacent side} = \sqrt{1^2 - x^2}$$
$$= \sqrt{1-x^2}$$

Using the figure, $\tan\theta = \dfrac{x}{\sqrt{1-x^2}}$.

57. The graph of $y = \sin^{-1} x$ is equivalent to $x = \sin y$. Choose values for y and find the corresponding x-values.

Problem Set 5.5

1.   Let $\alpha = \arcsin\dfrac{3}{5}$ and $\beta = \arctan 2$.

Then $\sin\alpha = \dfrac{3}{5}$ and $-90° \le \alpha \le 90°$ and $\tan\beta = \dfrac{2}{1}$ and $-90° < \beta < 90°$.

Also, $\sin\left(\arcsin\dfrac{3}{5} - \arctan 2\right) = \sin(\alpha - \beta)$

$$= \sin\alpha\cos\beta - \cos\alpha\sin\beta$$

We draw and label a triangle for α and another for β:

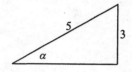   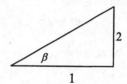

Using the Pythagorean Theorem we find the missing sides:

adjacent side $= \sqrt{5^2 - 3^2}$ hypotenuse $= \sqrt{1^2 + 2^2}$

$\qquad\qquad\quad = \sqrt{25 - 9}$ $\qquad\quad = \sqrt{1 + 4}$

$\qquad\qquad\quad = \sqrt{16} = 4$ $\qquad\quad = \sqrt{5}$

Then, $\sin\alpha = \dfrac{3}{5}$ $\sin\beta = \dfrac{2}{\sqrt{5}}$

$\qquad \cos\alpha = \dfrac{4}{5}$ $\cos\beta = \dfrac{1}{\sqrt{5}}$

Substituting these above, we get: $\sin\alpha\cos\beta - \cos\alpha\sin\beta = \dfrac{3}{5}\left(\dfrac{1}{\sqrt{5}}\right) - \dfrac{4}{5}\left(\dfrac{2}{\sqrt{5}}\right)$

$$= \dfrac{3}{5\sqrt{5}} - \dfrac{8}{5\sqrt{5}}$$

$$= -\dfrac{5}{5\sqrt{5}}$$

$$= -\dfrac{1}{\sqrt{5}}$$

3.   Let $\alpha = \tan^{-1}\dfrac{1}{2}$ and $\beta = \sin^{-1}\dfrac{1}{2}$.

Then $\tan\alpha = \dfrac{1}{2}$ and $-90° < \alpha < 90°$ and $\sin\beta = \dfrac{1}{2}$ and $-90° \le \beta \le 90°$.

This problem is continued on the next page

Also, $\cos\left(\tan^{-1}\dfrac{1}{2}+\sin^{-1}\dfrac{1}{2}\right)=\cos(\alpha+\beta)$

$$=\cos\alpha\cos\beta-\sin\alpha\sin\beta$$

We draw and label a triangle for α and another for β :

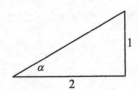 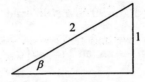

Using the Pythagorean Theorem we find the missing sides:

hypotenuse $=\sqrt{1^2+2^2}$ $\qquad$ adjacent side $=\sqrt{2^2-1^2}$

$\qquad\qquad =\sqrt{1+4}=\sqrt{5}$ $\qquad\qquad\qquad =\sqrt{4-1}=\sqrt{3}$

Then, $\sin\alpha=\dfrac{1}{\sqrt{5}}$ $\qquad\qquad\qquad$ $\sin\beta=\dfrac{1}{2}$

$\qquad\quad \cos\alpha=\dfrac{2}{\sqrt{5}}$ $\qquad\qquad\qquad$ $\cos\beta=\dfrac{\sqrt{3}}{2}$

Substituting these above we get:

$$\cos\alpha\cos\beta-\sin\alpha\sin\beta=\left(\dfrac{2}{\sqrt{5}}\right)\left(\dfrac{\sqrt{3}}{2}\right)-\left(\dfrac{1}{\sqrt{5}}\right)\left(\dfrac{1}{2}\right)$$

$$=\dfrac{2\sqrt{3}}{2\sqrt{5}}-\dfrac{1}{2\sqrt{5}}$$

$$=\dfrac{2\sqrt{3}-1}{2\sqrt{5}}$$

5. $\qquad$ Let $\alpha=\cos^{-1}\dfrac{1}{\sqrt{5}}$. Then $\cos\alpha=\dfrac{1}{\sqrt{5}}$ and $0°\le\alpha\le180°$.

$\qquad$ Also, $\sin\left(2\cos^{-1}\dfrac{1}{\sqrt{5}}\right)=\sin2\alpha$

$$=2\sin\alpha\cos\alpha$$

We draw and label the sides of a triangle and using the Pythagorean Theorem, we find the opposite side:

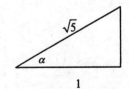

$\qquad$ opposite side $=\sqrt{\left(\sqrt{5}\right)^2-1^2}=\sqrt{4}=2$

From the figure, we have

$\qquad \sin\alpha=\dfrac{2}{\sqrt{5}}$ and $\cos\alpha=\dfrac{1}{\sqrt{5}}$

Substituting these above, we get

$\qquad 2\sin\alpha\cos\alpha=2\left(\dfrac{2}{\sqrt{5}}\right)\left(\dfrac{1}{\sqrt{5}}\right)=\dfrac{4}{5}$

7. Let $\alpha = \sin^{-1} x$. Then $\sin \alpha = \dfrac{x}{1}$ and $-90° \le \alpha \le 90°$.

We draw and label the sides of a triangle and find the adjacent side:

$$\text{adjacent side} = \sqrt{1^2 - x^2} = \sqrt{1 - x^2}$$

From the figure, we find $\tan \alpha = \dfrac{x}{\sqrt{1-x^2}}$

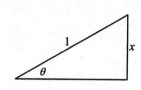

9. Let $\alpha = \sin^{-1} x$. Then $\sin \alpha = \dfrac{x}{1}$ and $-90° \le \alpha \le 90°$.

Also, $\sin\left(2\sin^{-1} x\right) = \sin 2\alpha$

$$= 2\sin\alpha\cos\alpha$$

From the figure in problem #7:

$$\sin\alpha = x \quad \text{and} \quad \cos\alpha = \sqrt{1-x^2}$$

Substituting these above, we get

$$2\sin\alpha\cos\alpha = 2x\sqrt{1-x^2}$$

11. Let $\alpha = \cos^{-1} x$. Then $\cos\alpha = x$ and $0° \le \alpha \le 180°$.
We want to find $\cos\left(2\cos^{-1} x\right) = \cos 2\alpha$

$$= 2\cos^2\alpha - 1$$
$$= 2x^2 - 1$$

13.
$$\sin 30° \sin 120° = \frac{1}{2}\left[\cos\left(-90°\right) - \cos 150°\right]$$
$$\frac{1}{2} \cdot \frac{\sqrt{3}}{2} = \frac{1}{2}\left[0 - \left(-\frac{\sqrt{3}}{2}\right)\right]$$
$$\frac{\sqrt{3}}{4} = \frac{\sqrt{3}}{4}$$

15.
$$10\sin 5x \cos 3x$$
$$= 10\left(\frac{1}{2}\right)\left[\sin\left(5x+3x\right) + \sin\left(5x-3x\right)\right]$$
$$= 5\left(\sin 8x + \sin 2x\right)$$

17.
$$\cos 8x \cos 2x = \frac{1}{2}\left[\cos\left(8x+2x\right) + \cos\left(8x-2x\right)\right]$$
$$= \frac{1}{2}\left(\cos 10x + \cos 6x\right)$$

19.
$$\sin 60° \cos 30°$$
$$= \frac{1}{2}\left[\sin\left(60° + 30°\right) + \sin\left(60° - 30°\right)\right]$$
$$= \frac{1}{2}\left(\sin 90° + \sin 30°\right)$$
$$= \frac{1}{2}\left(1 + \frac{1}{2}\right) = \frac{1}{2}\left(\frac{3}{2}\right) = \frac{3}{4}$$

21.
$$\sin 4\pi \sin 2\pi = \frac{1}{2}\left[\cos\left(4\pi - 2\pi\right) - \cos\left(4\pi + 2\pi\right)\right]$$
$$= \frac{1}{2}\left[\cos 2\pi - \cos 6\pi\right]$$
$$= \frac{1}{2}\left[1 - 1\right]$$
$$= 0$$

23.
$$\sin\alpha - \sin\beta = \sin 30° - \sin 90° \qquad\qquad 2\cos\frac{\alpha+\beta}{2}\sin\frac{\alpha-\beta}{2}$$
$$= \frac{1}{2} - 1 \qquad\qquad\qquad\qquad\qquad = 2\cos\frac{30° + 90°}{2}\sin\frac{30° - 90°}{2}$$
$$= -\frac{1}{2} \qquad\qquad\qquad\qquad\qquad = 2\cos 60°\sin\left(-30°\right)$$
$$\qquad\qquad\qquad\qquad\qquad = 2\left(\frac{1}{2}\right)\left(-\frac{1}{2}\right) = -\frac{1}{2}$$

25.
$$\sin 7x + \sin 3x = 2\sin\frac{7x + 3x}{2}\cos\frac{7x - 3x}{2} \qquad\quad \textbf{27.} \quad \cos 45° + \cos 15°$$
$$= 2\sin 5x\cos 2x \qquad\qquad\qquad\qquad\qquad = 2\cos\frac{45° + 15°}{2}\cos\frac{45° - 15°}{2}$$
$$\qquad\qquad\qquad\qquad\qquad\qquad\qquad = 2\cos 30°\cos 15°$$
$$\qquad\qquad\qquad\qquad\qquad\qquad\qquad = 2\left(\frac{\sqrt{3}}{2}\right)\cos 15° = \sqrt{3}\cos 15°$$

29.
$$\sin\frac{7\pi}{12} - \sin\frac{\pi}{12} = 2\cos\frac{\dfrac{7\pi}{12} + \dfrac{\pi}{12}}{2}\sin\frac{\dfrac{7\pi}{12} - \dfrac{\pi}{12}}{2}$$
$$= 2\cos\frac{\pi}{3}\sin\frac{\pi}{4}$$
$$= 2\left(\frac{1}{2}\right)\left(\frac{1}{\sqrt{2}}\right)$$
$$= \frac{1}{\sqrt{2}}$$

31.
$$\frac{\sin 3x + \sin x}{\cos 3x - \cos x} = \frac{2\sin\dfrac{3x + x}{2}\cos\dfrac{3x - x}{2}}{-2\sin\dfrac{3x + x}{2}\sin\dfrac{3x - x}{2}} \qquad \text{Sum to product formulas}$$
$$= \frac{2\sin 2x\cos x}{-2\sin 2x\sin x} \qquad\qquad \text{Simplify}$$

This problem is continued on the next page

$$= -\frac{\cos x}{\sin x}$$ Reduce

$$= -\cot x$$ Ratio identity

33. $$\frac{\sin 4x + \sin 6x}{\cos 4x - \cos 6x} = \frac{2\sin 5x \cos(-x)}{-2\sin 5x \cos(-x)}$$ Sum to product formulas

$$= -\frac{\cos(-x)}{\sin(-x)}$$ Reduce

$$= -\frac{\cos x}{-\sin x}$$ Cosine is an even function; sine is an odd function

$$= \cot x$$ Ratio identity

35. $$\frac{\sin 5x + \sin 3x}{\cos 3x + \cos 5x} = \frac{2\sin 4x \cos x}{2\cos 4x \cos x}$$ Sum to product formulas

$$= \frac{\sin 4x}{\cos 4x}$$ Reduce

$$= \tan 4x$$ Ratio identity

37.   The graph is a sine curve with a phase shift of $-\dfrac{\pi}{4}$.

39.   The graph is a cosine curve with a phase shift of $\dfrac{\pi}{3}$.

41.   Amplitude $= 1$ Period $= \dfrac{2\pi}{2} = \pi$ Phase shift $= \dfrac{\frac{\pi}{2}}{2} = \dfrac{\pi}{4}$

Spacing $= \dfrac{1}{4}(\pi) = \dfrac{\pi}{4}$ $c = \dfrac{\pi}{2} + \dfrac{\pi}{4} = \dfrac{3\pi}{4}$

$a = \dfrac{\pi}{4}$ $d = \dfrac{3\pi}{4} + \dfrac{\pi}{4} = \pi$

$b = \dfrac{\pi}{4} + \dfrac{\pi}{4} = \dfrac{\pi}{2}$ $e = \pi + \dfrac{\pi}{4} = \dfrac{5\pi}{4}$

The 5 points we use on the x-axis are: $\dfrac{\pi}{4}, \dfrac{\pi}{2}, \dfrac{3\pi}{4}, \pi, \dfrac{5\pi}{4}$.

The 2 points we use on the y-axis are: 1 and -1.

43.   Amplitude $=\dfrac{1}{2}$ Period $=\dfrac{2\pi}{3}$ Phase shift $=\dfrac{\frac{\pi}{2}}{3}=\dfrac{\pi}{6}$

     Spacing $=\dfrac{1}{4}\left(\dfrac{2\pi}{3}\right)=\dfrac{\pi}{6}$ $c=\dfrac{\pi}{3}+\dfrac{\pi}{6}=\dfrac{\pi}{2}$

     $a=\dfrac{\pi}{6}$ $d=\dfrac{\pi}{2}+\dfrac{\pi}{6}=\dfrac{2\pi}{3}$

     $b=\dfrac{\pi}{6}+\dfrac{\pi}{6}=\dfrac{\pi}{3}$ $e=\dfrac{2\pi}{3}+\dfrac{\pi}{6}=\dfrac{5\pi}{6}$

     The 5 points we use on the x-axis are: $\dfrac{\pi}{6},\dfrac{\pi}{3},\dfrac{\pi}{2},\dfrac{2\pi}{3},\dfrac{5\pi}{6}$.

     The 2 points we use on the y-axis are: $-\dfrac{1}{2}$ and $\dfrac{1}{2}$.

45.   Amplitude $=3$ Period $=\dfrac{2\pi}{\pi}=2$ Phase shift $=\dfrac{\frac{\pi}{2}}{\pi}=\dfrac{1}{2}$

     Spacing $=\dfrac{1}{4}(2)=\dfrac{1}{2}$ $c=1+\dfrac{1}{2}=\dfrac{3}{2}$

     $a=\dfrac{1}{2}$ $d=\dfrac{3}{2}+\dfrac{1}{2}=2$

     $b=\dfrac{1}{2}+\dfrac{1}{2}=1$ $e=2+\dfrac{1}{2}=\dfrac{5}{2}$

     The 5 points we use on the x-axis are: $\dfrac{1}{2},1,\dfrac{3}{2},2,\dfrac{5}{2}$.

     The 2 points we use on the y-axis are: -3 and 3.

Chapter 5 Test

1. $\sin\theta\sec\theta = \sin\theta \cdot \dfrac{1}{\cos\theta}$ Reciprocal identity

    $= \dfrac{\sin\theta}{\cos\theta}$ Multiply

    $= \tan\theta$ Ratio identity

2. $\dfrac{\cot\theta}{\csc\theta} = \dfrac{\dfrac{\cos\theta}{\sin\theta}}{\dfrac{1}{\sin\theta}}$ Ratio and reciprocal identities

    $= \cos\theta$ Divide

3. $(\sec x - 1)(\sec x + 1) = \sec^2 x - 1$ Multiply

    $= \tan^2 x$ Pythagorean identity

4. $\tan\theta\sin\theta = \dfrac{\sin\theta}{\cos\theta}\sin\theta$ Ratio identity

    $= \dfrac{\sin^2\theta}{\cos\theta}$ Multiply

    $= \dfrac{1 - \cos^2\theta}{\cos\theta}$ Pythagorean identity

    $= \dfrac{1}{\cos\theta} - \dfrac{\cos^2\theta}{\cos\theta}$ Separate into 2 fractions

    $= \sec\theta - \cos\theta$ Reciprocal identity and reduce second fraction

5. $\dfrac{\cos t}{1 - \sin t} = \dfrac{\cos t}{1 - \sin t} \cdot \dfrac{1 + \sin t}{1 + \sin t}$ Multiply by a fraction equal to 1

    $= \dfrac{\cos t(1 + \sin t)}{1 - \sin^2 t}$ Multiply

    $= \dfrac{\cos t(1 + \sin t)}{\cos^2 t}$ Pythagorean identity

    $= \dfrac{1 + \sin t}{\cos t}$ Reduce

6.

$$\frac{1}{1-\sin t} + \frac{1}{1+\sin t} = \frac{1(1+\sin t)+1(1-\sin t)}{(1-\sin t)(1+\sin t)} \qquad \text{Add fractions}$$

$$= \frac{2}{1-\sin^2 t} \qquad \text{Simplify}$$

$$= \frac{2}{\cos^2 t} \qquad \text{Pythagorean identity}$$

$$= 2\sec^2 t \qquad \text{Reciprocal identity}$$

7.

$$\sin(\theta-90°) = \sin\theta\cos 90° - \cos\theta\sin 90° \qquad \text{Difference identity}$$

$$= (\sin\theta)(0) - (\cos\theta)(1) \qquad \text{Substitute exact values}$$

$$= -\cos\theta \qquad \text{Simplify}$$

8.

$$\cos\left(\frac{\pi}{2}+\theta\right) = \cos\frac{\pi}{2}\cos\theta - \sin\frac{\pi}{2}\sin\theta \qquad \text{Sum identity}$$

$$= 0(\cos\theta) - 1(\sin\theta) \qquad \text{Substitute exact values}$$

$$= -\sin\theta \qquad \text{Simplify}$$

9.

$$\cos^4 A - \sin^4 A = (\cos^2 A + \sin^2 A)(\cos^2 A - \sin^2 A) \qquad \text{Factor}$$

$$= 1(\cos^2 A - \sin^2 A) \qquad \text{Pythagorean identity}$$

$$= \cos 2A \qquad \text{Double-angle identity}$$

10.

$$\frac{\sin 2A}{1-\cos 2A} = \frac{2\sin A\cos A}{1-(1-2\sin^2 A)} \qquad \text{Double-angle identities}$$

$$= \frac{2\sin A\cos A}{2\sin^2 A} \qquad \text{Subtract}$$

$$= \frac{\cos A}{\sin A} \qquad \text{Reduce}$$

$$= \cot A \qquad \text{Ratio identity}$$

11.

$$\frac{\cos 2x}{\sin x\cos x} = \frac{\cos^2 x - \sin^2 x}{\sin x\cos x} \qquad \text{Double-angle identity}$$

$$= \frac{\cos^2 x}{\sin x\cos x} - \frac{\sin^2 x}{\sin x\cos x} \qquad \text{Separate into 2 fractions}$$

$$= \frac{\cos x}{\sin x} - \frac{\sin x}{\cos x} \qquad \text{Reduce}$$

$$= \cot x - \tan x \qquad \text{Ratio identities}$$

12. $\dfrac{\tan x}{\sec x + 1} = \dfrac{\dfrac{\sin x}{\cos x}}{\dfrac{1}{\cos x} + 1}$ Ratio and reciprocal identities

$\qquad\qquad\quad = \dfrac{\sin x}{1 + \cos x}$ Multiply numerator and denominator by $\cos x$

$\qquad\qquad\quad = \tan\dfrac{x}{2}$ Half-angle identity

19. A is in QIV. Then, $\cos A = \sqrt{1 - \sin^2 A}$ B is in QII. Then, $\cos B = -\sqrt{1 - \sin^2 A}$

$\qquad\qquad\qquad\qquad = \sqrt{1 - \dfrac{9}{25}} \qquad\qquad\qquad\qquad\qquad\qquad = -\sqrt{1 - \dfrac{144}{169}}$

$\qquad\qquad\qquad\qquad = \sqrt{\dfrac{16}{25}} \qquad\qquad\qquad\qquad\qquad\qquad\quad = -\sqrt{\dfrac{25}{169}}$

$\qquad\qquad\qquad\qquad = \dfrac{4}{5} \qquad\qquad\qquad\qquad\qquad\qquad\qquad\quad = -\dfrac{5}{13}$

We have $\sin A = -\dfrac{3}{5} \qquad\qquad \sin B = \dfrac{12}{13}$

$\qquad\quad\ \cos A = \dfrac{4}{5} \qquad\qquad \cos B = -\dfrac{5}{13}$

Therefore, $\sin(A + B) = \sin A \cos B + \cos A \sin B$

$\qquad\qquad\qquad\quad = -\dfrac{3}{5}\left(-\dfrac{5}{13}\right) + \dfrac{4}{5}\left(\dfrac{12}{13}\right)$

$\qquad\qquad\qquad\quad = \dfrac{15}{65} + \dfrac{48}{65}$

$\qquad\qquad\qquad\quad = \dfrac{63}{65}$

20. $\cos(A - B) = \cos A \cos B + \sin A \sin B$

$\qquad\qquad\quad = \dfrac{4}{5}\left(-\dfrac{5}{13}\right) + \left(-\dfrac{3}{5}\right)\left(\dfrac{12}{13}\right)$ (See Problem 19)

$\qquad\qquad\quad = -\dfrac{20}{65} - \dfrac{36}{65}$

$\qquad\qquad\quad = -\dfrac{56}{65}$

21. $\cos 2B = 1 - 2\sin^2 B$

$$= 1 - 2\left(\frac{12}{13}\right)^2$$

$$= 1 - 2\left(\frac{144}{169}\right)$$

$$= 1 - \frac{288}{169}$$

$$= -\frac{119}{169}$$

22. $\sin 2B = 2\sin B \cos B$

$$= 2\left(\frac{12}{13}\right)\left(-\frac{5}{13}\right) \quad \text{(See Problem 19)}$$

$$= -\frac{120}{169}$$

23. Using the information from Problem 19:

If $270° \le A \le 360°$, then $135° \le \dfrac{A}{2} \le 180°$.

Therefore, $\dfrac{A}{2}$ is in QII.

$$\sin\frac{A}{2} = \sqrt{\frac{1 - \cos A}{2}}$$

$$= \sqrt{\frac{1 - \dfrac{4}{5}}{2}}$$

$$= \sqrt{\frac{1}{10}}$$

$$= \frac{1}{\sqrt{10}}$$

24. $\cos\dfrac{A}{2} = -\sqrt{\dfrac{1 + \cos A}{2}}$ (See Problem 23)

$$= -\sqrt{\frac{1 + \dfrac{4}{5}}{2}}$$

$$= -\sqrt{\frac{9}{10}}$$

$$= -\frac{3}{\sqrt{10}}$$

25. $\sin 75° = \sin\left(30° + 45°\right)$

$$= \sin 30° \cos 45° + \cos 30° \sin 45°$$

$$= \frac{1}{2}\left(\frac{\sqrt{2}}{2}\right) + \frac{\sqrt{3}}{2}\left(\frac{\sqrt{2}}{2}\right)$$

$$= \frac{\sqrt{2}}{4} + \frac{\sqrt{6}}{4}$$

$$= \frac{\sqrt{2} + \sqrt{6}}{4}$$

26. $\cos 15° = \cos\left(45° - 30°\right)$

$$= \cos 45° \cos 30° + \sin 45° \sin 30°$$

$$= \frac{\sqrt{2}}{2}\left(\frac{\sqrt{3}}{2}\right) + \frac{\sqrt{2}}{2}\left(\frac{1}{2}\right)$$

$$= \frac{\sqrt{6}}{4} + \frac{\sqrt{2}}{4}$$

$$= \frac{\sqrt{6} + \sqrt{2}}{4}$$

27. $\tan\dfrac{\pi}{12} = \tan\left(\dfrac{\pi}{3} - \dfrac{\pi}{4}\right)$

$$= \dfrac{\tan\dfrac{\pi}{3} - \tan\dfrac{\pi}{4}}{1 + \tan\dfrac{\pi}{3}\tan\dfrac{\pi}{4}}$$

$$= \dfrac{\sqrt{3} - 1}{1 + \left(\sqrt{3}\right)(1)} = \dfrac{\sqrt{3} - 1}{\sqrt{3} + 1}$$

28. $\cot\dfrac{\pi}{12} = \dfrac{1}{\tan\dfrac{\pi}{12}}$

$$= \dfrac{1}{\left[\dfrac{\sqrt{3} - 1}{\sqrt{3} + 1}\right]} \qquad \text{(See Problem 27)}$$

$$= \dfrac{\sqrt{3} + 1}{\sqrt{3} - 1}$$

29. $\cos 4x \cos 5x - \sin 4x \sin 5x = \cos\left(4x + 5x\right) = \cos 9x$

30. $\sin 15° \cos 75° + \cos 15° \sin 75° = \sin\left(15° + 75°\right) = \sin 90° = 1$

31. $\cos 2A = 1 - 2\sin^2 A$

$$= 1 - 2\left(-\dfrac{1}{\sqrt{5}}\right)^2$$

$$= 1 - \dfrac{2}{5}$$

$$= \dfrac{3}{5}$$

For $\cos\dfrac{A}{2}$ we must find $\cos A$.

$\cos A = -\sqrt{1 - \sin^2 A} \qquad (A \text{ is in QIII})$

$$= -\sqrt{1 - \dfrac{1}{5}} = -\sqrt{\dfrac{4}{5}} = -\dfrac{2}{\sqrt{5}} = -\dfrac{2\sqrt{5}}{5}$$

If $180° \le A \le 270°$, then $90° \le \dfrac{A}{2} \le 135°$, and $\dfrac{A}{2}$ is in QII.

$\cos\dfrac{A}{2} = -\sqrt{\dfrac{1 + \cos A}{2}}$

$$= -\sqrt{\dfrac{1 + \left(-\dfrac{2\sqrt{5}}{5}\right)}{2}}$$

$$= \sqrt{\dfrac{5 - 2\sqrt{5}}{10}}$$

32. If $\sec A = \sqrt{10}$, then $\cos A = \dfrac{1}{\sqrt{10}}$.

$$\sin A = \sqrt{1 - \cos^2 A}$$

$$= \sqrt{1 - \frac{1}{10}} = \sqrt{\frac{9}{10}} = \frac{3}{\sqrt{10}}$$

Then, $\sin 2A = 2 \sin A \cos A$

$$= 2 \left(\frac{3}{\sqrt{10}} \right) \left(\frac{1}{\sqrt{10}} \right) = \frac{6}{10} = \frac{3}{5}$$

Also, $\sin \dfrac{A}{2} = \sqrt{\dfrac{1 - \cos A}{2}}$

$$= \sqrt{\frac{1 - \dfrac{1}{\sqrt{10}}}{2}}$$

$$= \sqrt{\frac{\sqrt{10} - 1}{2\sqrt{10}}} \ \text{ or } \ \sqrt{\frac{10 - \sqrt{10}}{20}}$$

33. $\tan(A + B) = \dfrac{\tan A + \tan B}{1 - \tan A \tan B}$

$$3 = \frac{\tan A + \dfrac{1}{2}}{1 - (\tan A)\left(\dfrac{1}{2}\right)}$$

$$3 = \frac{2\tan A + 1}{2 - \tan A}$$

$$3(2 - \tan A) = 2\tan A + 1$$

$$6 - 3\tan A = 2\tan A + 1$$

$$-5\tan A = -5$$

$$\tan A = 1$$

34.

$\cos 2x = 2\cos^2 x - 1$	Double-angle identity
$\dfrac{1}{2} = 2\cos^2 x - 1$	Substitute given value
$\dfrac{3}{2} = 2\cos^2 x$	Subtract 1 from both sides
$\cos^2 x = \dfrac{3}{4}$	Multiply both sides by $\dfrac{1}{2}$
$\cos x = \pm \dfrac{\sqrt{3}}{2}$	Take square root of both sides

35. Let $\alpha = \arcsin \dfrac{4}{5}$ and $\beta = \arctan 2$.

Then $\sin \alpha = \dfrac{4}{5}$ and $\tan \beta = \dfrac{2}{1}$.

We can draw 2 triangles and label the sides accordingly:

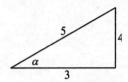 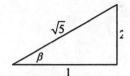

From the figures, we have

$\sin \alpha = \dfrac{4}{5}$ $\qquad$ $\sin \beta = \dfrac{2}{\sqrt{5}}$

$\cos \alpha = \dfrac{3}{5}$ $\qquad$ $\cos \beta = \dfrac{1}{\sqrt{5}}$

Therefore, $\cos(\alpha - \beta) = \cos \alpha \cos \beta + \sin \alpha \sin \beta$

$$= \frac{3}{5}\left(\frac{1}{\sqrt{5}}\right) + \frac{4}{5}\left(\frac{2}{\sqrt{5}}\right)$$

$$= \frac{3}{5\sqrt{5}} + \frac{8}{5\sqrt{5}}$$

$$= \frac{11}{5\sqrt{5}}$$

36. Let $\alpha = \arccos \dfrac{4}{5}$ and $\beta = \arctan 2$. Then $\cos \alpha = \dfrac{4}{5}$ and $\tan \beta = \dfrac{2}{1}$.

We can draw 2 triangles and label the sides accordingly:

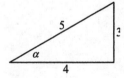 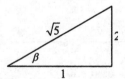

From the figures, we have: $\qquad \sin \alpha = \dfrac{3}{5} \qquad\qquad \sin \beta = \dfrac{2}{\sqrt{5}}$

$\qquad\qquad\qquad\qquad\qquad\qquad\quad \cos \alpha = \dfrac{4}{5} \qquad\qquad \cos \beta = \dfrac{1}{\sqrt{5}}$

This problem is continued on the next page

Therefore, $\cos(\alpha - \beta) = \cos\alpha\cos\beta + \sin\alpha\sin\beta$

$$= \frac{3}{5}\left(\frac{1}{\sqrt{5}}\right) + \frac{4}{5}\left(\frac{2}{\sqrt{5}}\right)$$

$$= \frac{3}{5\sqrt{5}} + \frac{8}{5\sqrt{5}}$$

$$= \frac{11}{5\sqrt{5}}$$

37. Let $\alpha = \sin^{-1} x$. Then $\sin\alpha = x$

$\cos 2\alpha = 1 - 2\sin^2\alpha$

$\qquad = 1 - 2x^2$

38. Let $\alpha = \cos^{-1} x$. Then $\cos\alpha = \dfrac{x}{1}$. We can draw a triangle and

use the Pythagorean Theorem to find the opposite side:

$\quad$ opposite side $= \sqrt{1^2 - x^2} = \sqrt{1 - x^2}$

$\quad$ Therefore, $\sin\alpha = \dfrac{\sqrt{1 - x^2}}{1} = \sqrt{1 - x^2}$

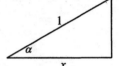

$\qquad \sin 2\alpha = 2\sin\alpha\cos\alpha$

$\qquad\qquad = 2\sqrt{1 - x^2}\,(x) = 2x\sqrt{1 - x^2}$

39. $\sin 6x \sin 4x = \dfrac{1}{2}\Big[\cos(6x - 4x) - \cos(6x + 4x)\Big]$

$\qquad\qquad\qquad = \dfrac{1}{2}(\cos 2x - \cos 10x)$

40. $\cos 15° + \cos 75° = 2\cos\dfrac{15° + 75°}{2}\cos\dfrac{15° - 75°}{2}$

$\qquad\qquad\qquad = 2\cos 45°\cos(-30°)$

$\qquad\qquad\qquad = 2\cos 45°\cos 30°$

$\qquad\qquad\qquad = 2\left(\dfrac{\sqrt{2}}{2}\right)\left(\dfrac{\sqrt{3}}{2}\right)$

$\qquad\qquad\qquad = \dfrac{\sqrt{6}}{2}$

CHAPTER 6   Equations

Problem Set 6.1

1. $2\sin\theta = 1$

$\sin\theta = \dfrac{1}{2}$

$\hat{\theta} = 30°$ and θ is in QI or QII

$\theta = 30°$ or $150°$

3. $2\cos\theta - \sqrt{3} = 0$

$2\cos\theta = \sqrt{3}$

$\cos\theta = \dfrac{\sqrt{3}}{2}$

$\hat{\theta} = 30°$ and θ is in QI or QIV

$\theta = 30°$ or $330°$

5. $2\tan\theta + 2 = 0$

$2\tan\theta = -2$

$\tan\theta = -1$

$\hat{\theta} = 45°$ and θ in in QII or QIV

$\theta = 135°$ or $315°$

7. $4\sin t - \sqrt{3} = 2\sin t$

$2\sin t = \sqrt{3}$

$\sin t = \dfrac{\sqrt{3}}{2}$

$\hat{t} = \dfrac{\pi}{3}$ and t is in QI or QII

$t = \dfrac{\pi}{3}$ or $\dfrac{2\pi}{3}$

9. $2\cos t = 6\cos t - \sqrt{12}$

$-4\cos t = -2\sqrt{3}$

$\cos t = \dfrac{\sqrt{3}}{2}$

$\hat{t} = \dfrac{\pi}{6}$ and t is in QI or QIV

$t = \dfrac{\pi}{6}$ or $\dfrac{11\pi}{6}$

11. $3\sin t + 5 = -2\sin t$

$5\sin t = -5$

$\sin t = -1$

$t = \dfrac{3\pi}{2}$

13. $4\sin\theta - 3 = 0$

$4\sin\theta = 3$

$\sin\theta = 0.75$

$\hat{\theta} = 48.6°$ and θ is in QI or QII

$\theta = 48.6°$ or $131.4°$

15. $2\cos\theta - 5 = 3\cos\theta - 2$

$-\cos\theta = 3$

$\cos\theta = -3$

There is no solution because $\cos\theta$ must be between -1 and 1.

17.
$$\sin\theta - 3 = 5\sin\theta$$
$$-3 = 4\sin\theta$$
$$\sin\theta = -0.75$$

$$\hat{\theta} = 48.6° \text{ and } \theta \text{ is in QIII or QIV}$$

$$\theta = 228.6° \text{ or } 311.4°$$

19.
$$(\sin x - 1)(2\sin x - 1) = 0$$
$$\sin x - 1 = 0 \text{ or } 2\sin x - 1 = 0$$
$$\sin x = 1 \qquad 2\sin x = 1$$
$$x = \frac{\pi}{2} \qquad \sin x = \frac{1}{2}$$
$$\hat{x} = \frac{\pi}{6}$$
$$x = \frac{\pi}{6} \text{ or } \frac{5\pi}{6}$$

21.
$$\tan x(\tan x - 1) = 0$$
$$\tan x = 0 \text{ or } \tan x - 1 = 0$$
$$x = 0, \pi \qquad \tan x = 1$$
$$x = \frac{\pi}{4} \text{ or } \frac{5\pi}{4}$$

23.
$$\sin x + 2\sin x \cos x = 0$$
$$\sin x(1 + 2\cos x) = 0$$
$$\sin x = 0 \text{ or } 1 + 2\cos x = 0$$
$$x = 0, \pi \qquad 2\cos x = -1$$
$$\cos x = -\frac{1}{2}$$
$$\hat{x} = \frac{\pi}{3}$$
$$x = \frac{2\pi}{3} \text{ or } \frac{4\pi}{3}$$

25.
$$2\sin^2 x - \sin x - 1 = 0$$
$$(2\sin x + 1)(\sin x - 1) = 0$$
$$2\sin x + 1 = 0 \text{ or } \sin x - 1 = 0$$
$$2\sin x = -1 \qquad \sin x = 1$$
$$\sin x = -\frac{1}{2} \qquad x = \frac{\pi}{2}$$
$$x = \frac{7\pi}{6} \text{ or } \frac{11\pi}{6}$$

27.
$$(2\cos\theta + \sqrt{3})(2\cos\theta + 1) = 0$$
$$2\cos\theta + \sqrt{3} = 0 \text{ or } 2\cos\theta + 1 = 0$$
$$2\cos\theta = -\sqrt{3} \qquad 2\cos\theta = -1$$
$$\cos\theta = -\frac{\sqrt{3}}{2} \qquad \cos\theta = -\frac{1}{2}$$
$$\hat{\theta} = 30° \qquad \hat{\theta} = 60°$$
$$\theta = 150° \text{ or } 210° \quad \theta = 120° \text{ or } 240°$$

29.
$$\sqrt{3}\tan\theta - 2\sin\theta\tan\theta = 0$$
$$\tan\theta(\sqrt{3} - 2\sin\theta) = 0$$

$$\tan\theta = 0 \text{ or } \quad \sqrt{3} - 2\sin\theta = 0$$
$$\theta = 0° \text{ or } 180° \quad -2\sin\theta = -\sqrt{3}$$
$$\sin\theta = \frac{\sqrt{3}}{2}$$
$$\theta = 60° \text{ or } 120°$$

31.
$$2\cos^2\theta + 11\cos\theta + 5 = 0$$
$$(2\cos\theta + 1)(\cos\theta + 5) = 0$$
$$2\cos\theta + 1 = 0 \quad \text{or} \quad \cos\theta + 5 = 0$$
$$2\cos\theta = -1 \qquad\qquad \cos\theta = -5$$
$$\cos\theta = -\frac{1}{2} \qquad \text{No solution}$$
$$\hat{\theta} = 60°$$
$$\theta = 120° \text{ or } 240°$$

33.
$$2\sin^2\theta - 2\sin\theta - 1 = 0 \qquad\qquad \text{where } a = 2, \, b = -2, \, c = -1$$
$$\sin\theta = \frac{-(-2) \pm \sqrt{(-2)^2 - 4(2)(-1)}}{2(2)}$$
$$= \frac{2 \pm \sqrt{12}}{4}$$
$$\sin\theta = 1.3666 \quad \text{or} \quad \sin\theta = -0.3660$$
$$\text{No solution} \qquad\qquad \hat{\theta} = 21.5° \text{ and } \theta \text{ is in QIII or QIV}$$
$$\theta = 201.5° \text{ or } 338.5°$$

35.
$$\cos^2\theta + \cos\theta - 1 = 0 \qquad \text{where } a = 1, \quad b = 1, \quad c = -1$$
$$\cos\theta = \frac{-1 \pm \sqrt{1^2 - 4(1)(-1)}}{2(1)}$$
$$= \frac{-1 \pm \sqrt{5}}{2}$$
$$\cos\theta = -1.6180 \quad \text{or} \qquad \cos\theta = 0.6180$$
$$\text{No solution} \qquad \text{and} \qquad \hat{\theta} = 51.8° \text{ and is in QI or QII}$$
$$\theta = 51.8° \text{ or } 308.2°$$

37.
$$2\sin^2\theta + 1 = 4\sin\theta$$
$$2\sin^2\theta - 4\sin\theta + 1 = 0 \quad \text{where } a = 2, \quad b = -4, \quad c = 1$$
$$\sin\theta = \frac{-(-4) \pm \sqrt{(-4)^2 - 4(2)(1)}}{2(2)}$$
$$= \frac{4 \pm \sqrt{8}}{4}$$
$$\sin\theta = 1.7071 \quad \text{or} \quad \sin\theta = 0.2929$$
$$\text{No solution} \qquad \hat{\theta} = 17.0° \text{ and } \theta \text{ is in QI or QII}$$
$$\theta = 17.0° \text{ or } 163.0°$$

39. From problem 1, $\hat{\theta} = 30°$ and θ is in QI or QII

$$\theta = 30° + 360°k \ \text{ or } \ \theta = 150° + 360°k$$

41. From problem 7, $\hat{t} = \dfrac{\pi}{3}$ and t is in QI or QII

$$t = \dfrac{\pi}{3} + 2k\pi \ \text{ or } \ t = \dfrac{2\pi}{3} + 2k\pi$$

43. From problem 11, we found that $t = \dfrac{3\pi}{2}$.

Therefore, $t = \dfrac{3\pi}{2} + 2k\pi$.

45. In problem 13 we found that $\theta = 48.6°$ or $131.4°$.

Therefore, $\theta = 48.6° + 360°k$ or $\theta = 131.4° + 360°k$

47. $\cos\left(2A - 50°\right) = \dfrac{\sqrt{3}}{2}$

$$2A - 50° = 30° + 360°k \qquad \text{or} \qquad 2A - 50° = 330° + 360°k$$
$$2A = 80° + 360°k \qquad\qquad\qquad 2A = 380° + 360°k$$
$$A = 40° + 180°k \qquad\qquad\qquad A = 190° + 180°k$$

49. $\sin\left(3A + 30°\right) = \dfrac{1}{2}$

$$3A + 30° = 30° + 360°k \qquad \text{or} \qquad 3A + 30° = 150° + 360°k$$
$$3A = 360°k \qquad\qquad\qquad 3A = 120° + 360°k$$
$$A = 120°k \qquad\qquad\qquad A = 40° + 120°k$$

51. $\cos\left(4A - 20°\right) = -\dfrac{1}{2}$

$$4A - 20° = 120° + 360°k \quad \text{or} \quad 4A - 20° = 240° + 360°k$$
$$4A = 140° + 360°k \qquad\qquad\quad 4A = 260° + 360°k$$
$$A = 35° + 90°k \qquad\qquad\quad A = 65° + 90°k$$

53. $\sin\left(5A+15°\right)=-\dfrac{1}{\sqrt{2}}$

$\begin{array}{ll} 5A+15°=225°+360°k & \text{or} \quad 5A+15°=315°+360°k \\ \quad\;\, 5A=210°+360°k & \qquad\qquad 5A=300°+360°k \\ \quad\;\;\; A=42°+72°k & \qquad\qquad\;\; A=60°+72°k \end{array}$

75. $h=-16t^2+vt\sin\theta$ where $v=1{,}500$ ft/sec and $\theta=30°$

$\quad=-16t^2+1500t\sin 30°$

$\quad=-16t^2+1500t\left(\dfrac{1}{2}\right)$

$\quad=-16t^2+750t$

77. $h=-16t^2+750t$ where $t=2\sec$

$\quad=-16\left(2^2\right)+750(2)$

$\quad=-64+1500$

$\quad=1{,}436\ \text{ft}$

79. $h=-16t^2+vt\sin\theta$ where $v=1{,}500$ ft/sec, $t=2$ sec, and $h=750$ ft

$750\;=-16\left(2^2\right)+1{,}500(2)\sin\theta$

$750=-64+3{,}000\sin\theta$

$814=3{,}000\sin\theta$

$\sin\theta=\dfrac{814}{3{,}000}$

$\theta=15.7°$

85. $\sin\left(\theta+45°\right)=\sin\theta\cos 45°+\cos\theta\sin 45°$

$\qquad\qquad\qquad=\sin\theta\left(\dfrac{1}{\sqrt{2}}\right)+\cos\theta\left(\dfrac{1}{\sqrt{2}}\right)$

$\qquad\qquad\qquad=\dfrac{1}{\sqrt{2}}\sin\theta+\dfrac{1}{\sqrt{2}}\cos\theta$

87. $\sin 75°=\sin\left(45°+30°\right)$

$\qquad\quad=\sin 45°\cos 30°+\cos 45°\sin 30°$

$\qquad\quad=\dfrac{\sqrt{2}}{2}\left(\dfrac{\sqrt{3}}{2}\right)+\dfrac{\sqrt{2}}{2}\left(\dfrac{1}{2}\right)$

$\qquad\quad=\dfrac{\sqrt{6}}{4}+\dfrac{\sqrt{2}}{4}=\dfrac{\sqrt{6}+\sqrt{2}}{4}$

89.

$$\frac{1-\tan^2 x}{1+\tan^2 x} = \frac{1-\dfrac{\sin^2 x}{\cos^2 x}}{1+\dfrac{\sin^2 x}{\cos^2 x}}$$ Ratio identity

$$= \frac{\cos^2 x - \sin^2 x}{\cos^2 x + \sin^2 x}$$ Multiply numerator and denominator by $\cos^2 x$

$$= \frac{\cos 2x}{1}$$ Double-angle identity and Pythagorean identity

$$= \cos 2x$$ Simplify

Problem Set 6.2

1.

$$\sqrt{3}\sec\theta = 2$$

$$\sec\theta = \frac{2}{\sqrt{3}}$$

$$\cos\theta = \frac{\sqrt{3}}{2}$$

$\hat{\theta} = 30°$ and θ is in QI or QIV

$\theta = 30°$ or $330°$

3.

$$\sqrt{2}\csc\theta + 5 = 3$$

$$\sqrt{2}\csc\theta = -2$$

$$\csc\theta = -\frac{2}{\sqrt{2}}$$

$$\sin\theta = -\frac{\sqrt{2}}{2}$$

$\hat{\theta} = 45°$ and θ is in QIII or QIV

$\theta = 225°$ or $315°$

5.

$$4\sin\theta - 2\csc\theta = 0$$

$$4\sin\theta - \frac{2}{\sin\theta} = 0 \qquad\qquad \csc\theta = \frac{1}{\sin\theta}$$

$$4\sin^2\theta - 2 = 0 \qquad\qquad \text{Multiply both sides by } \sin\theta$$

$$4\sin^2\theta = 2 \qquad\qquad \text{Add 2 to both sides}$$

$$\sin^2\theta = \frac{1}{2} \qquad\qquad \text{Divide both sides by 4}$$

$$\sin\theta = \pm\frac{1}{\sqrt{2}} \qquad\qquad \text{Take square root of both sides}$$

$\hat{\theta} = 45°$ and θ is in QI, QII, QIII, or QIV

$\theta = 45°, 135°, 225°, 315°$

7.

$$\sec\theta - 2\tan\theta = 0$$

$$\frac{1}{\cos\theta} - 2\left(\frac{\sin\theta}{\cos\theta}\right) = 0 \qquad \text{Reciprocal and ratio identities}$$

$$1 - 2\sin\theta = 0 \qquad \text{Multiply both sides by } \cos\theta \quad (\cos\theta \neq 0)$$

$$-2\sin\theta = -1 \qquad \text{Solve equation and check}$$

$$\sin\theta = \frac{1}{2}$$

$$\theta = 30° \text{ or } 150°$$

9.

$$\sin 2\theta - \cos\theta = 0$$

$$2\sin\theta\cos\theta - \cos\theta = 0 \qquad \text{Double-angle identity}$$

$$\cos\theta(2\sin\theta - 1) = 0 \qquad \text{Factor out } \cos\theta$$

$$\cos\theta = 0 \quad \text{or} \quad 2\sin\theta - 1 = 0 \qquad \text{Set each factor} = 0$$

$$\theta = 90°, 270° \quad 2\sin\theta = 1 \qquad \text{Solve each equation}$$

$$\sin\theta = \frac{1}{2}$$

$$\theta = 30° \text{ or } 150°$$

11.

$$2\cos\theta + 1 = \sec\theta$$

$$2\cos\theta + 1 = \frac{1}{\cos\theta} \qquad \text{Reciprocal identity}$$

$$2\cos^2\theta + \cos\theta = 1 \qquad \text{Multiply both sides by } \cos\theta \left(\cos\theta \neq 0\right)$$

$$2\cos^2\theta + \cos\theta - 1 = 0$$

$$(2\cos\theta - 1)(\cos + 1) = 0 \qquad \text{Factor}$$

$$2\cos\theta - 1 = 0 \quad \text{or} \quad \cos\theta + 1 = 0 \qquad \text{Set each factor} = 0$$

$$2\cos\theta = 1 \qquad\qquad \cos\theta = -1 \qquad \text{Solve each equation and check}$$

$$\cos\theta = \frac{1}{2} \qquad\qquad \theta = 180°$$

$$\theta = 60° \text{ or } 300°$$

13.

$$\cos 2x - 3\sin x - 2 = 0$$

$$1 - 2\sin^2 x - 3\sin x - 2 = 0 \qquad \text{Double-angle identity}$$

$$2\sin^2 x + 3\sin x + 1 = 0 \qquad \text{Multiply both sides by } -1 \text{ and simplify}$$

$$(2\sin x + 1)(\sin x + 1) = 0 \qquad \text{Factor}$$

$$2\sin x + 1 = 0 \quad \text{or} \quad \sin x + 1 = 0 \qquad \text{Set each factor} = 0$$

$$2\sin x = -1 \qquad\qquad \sin x = -1$$

$$\sin x = -\frac{1}{2} \qquad\qquad x = \frac{3\pi}{2} \qquad \text{Solve each equation}$$

$$x = \frac{7\pi}{6} \text{ or } \frac{11\pi}{6}$$

15.

$$\cos x - \cos 2x = 0$$

$$\cos x - \left(2\cos^2 x - 1\right) = 0 \qquad \text{Double-angle identity}$$

$$\cos x - 2\cos^2 x + 1 = 0 \qquad \text{Simplify}$$

$$2\cos^2 x - \cos x - 1 = 0 \qquad \text{Multiply both sides by } -1$$

$$\left(2\cos x + 1\right)\left(\cos x - 1\right) = 0 \qquad \text{Factor}$$

$$2\cos x + 1 = 0 \quad \text{or} \quad \cos x - 1 = 0 \qquad \text{Set each factor} = 0$$

$$2\cos x = -1 \qquad \cos x = 1 \qquad \text{Solve each equation}$$

$$\cos x = -\frac{1}{2} \qquad x = 0$$

$$\hat{x} = \frac{\pi}{3}$$

$$x = \frac{2\pi}{3} \quad \text{or} \quad \frac{4\pi}{3}$$

17.

$$2\cos^2 x + \sin x - 1 = 0$$

$$2\left(1 - \sin^2 x\right) + \sin x - 1 = 0 \qquad \text{Pythagorean identity}$$

$$2 - 2\sin^2 x + \sin x - 1 = 0 \qquad \text{Simplify}$$

$$2\sin^2 x - \sin x - 1 = 0 \qquad \text{Multiply both sides by } -1 \text{ and simplify}$$

$$\left(2\sin x + 1\right)\left(\sin x - 1\right) = 0 \qquad \text{Factor}$$

$$2\sin x + 1 = 0 \quad \text{or} \quad \sin x - 1 = 0 \qquad \text{Set each factor} = 0$$

$$2\sin x = -1 \qquad \sin x = 1 \qquad \text{Solve each equation}$$

$$\sin x = -\frac{1}{2} \qquad x = \frac{\pi}{2}$$

$$x = \frac{7\pi}{6} \quad \text{or} \quad \frac{11\pi}{6}$$

19.

$$4\sin^2 x + 4\cos x - 5 = 0$$

$$4\left(1 - \cos^2 x\right) + 4\cos x - 5 = 0 \qquad \text{Pythagorean identity}$$

$$4 - 4\cos^2 x + 4\cos x - 5 = 0 \qquad \text{Simplify}$$

$$-4\cos^2 x + 4\cos x - 1 = 0 \qquad \text{Simplify}$$

$$4\cos^2 x - 4\cos \theta + 1 = 0 \qquad \text{Multiply both sides by } -1$$

$$\left(2\cos x - 1\right)\left(2\cos x - 1\right) = 0 \qquad \text{Factor}$$

$$2\cos x - 1 = 0 \quad \text{or} \quad 2\cos x - 1 = 0 \qquad \text{Set each factor} = 0$$

$$2\cos x = 1 \qquad \text{Solve the equation}$$

$$\cos x = \frac{1}{2}$$

$$x = \frac{\pi}{3} \quad \text{or} \quad \frac{5\pi}{3}$$

21.

$$2\sin x + \cot x - \csc x = 0$$

$$2\sin x + \frac{\cos x}{\sin x} - \frac{1}{\sin x} = 0 \qquad \text{Ratio and reciprocal identities}$$

$$2\sin^2 x + \cos x - 1 = 0 \qquad \text{Multiply both sides by } \sin x \ (\sin x \neq 0)$$

$$2(1 - \cos^2 x) + \cos x - 1 = 0 \qquad \text{Pythagorean identity}$$

$$2 - 2\cos^2 x + \cos x - 1 = 0 \qquad \text{Simplify}$$

$$2\cos^2 x - \cos x - 1 = 0 \qquad \text{Multiply both sides by } -1 \text{ and simplify}$$

$$(2\cos x + 1)(\cos x - 1) = 0 \qquad \text{Factor}$$

$$2\cos x + 1 = 0 \quad \text{or} \quad \cos x - 1 = 0 \qquad \text{Set each factor} = 0$$

$$2\cos x = -1 \qquad\qquad \cos x = 1 \qquad \text{Solve each equation and check}$$

$$\cos x = -\frac{1}{2} \qquad\qquad x = 0 \text{ which is not possible because } \sin x \neq 0$$

$$x = \frac{2\pi}{3} \ \text{ or } \ \frac{4\pi}{3}$$

23.

$$\sin x + \cos x = \sqrt{2}$$

$$\sin^2 x + 2\sin x \cos x + \cos^2 x = 2 \qquad \text{Square both sides}$$

$$2\sin x \cos x + 1 = 2 \qquad \text{Pythagorean identity}$$

$$\sin 2x = 1 \qquad \text{Double-angle identity}$$

$$2x = \frac{\pi}{2} \ \text{ or } \ 2x = \frac{5\pi}{2} \qquad \text{Solve equation and check}$$

$$x = \frac{\pi}{4} \qquad\qquad x = \frac{5\pi}{4} \qquad \text{Possible solutions}$$

Check each possible solution:

$$\sin\frac{\pi}{4} + \cos\frac{\pi}{4} = \frac{\sqrt{2}}{2} + \frac{\sqrt{2}}{2}$$

$$= \frac{2\sqrt{2}}{2}$$

$$= \sqrt{2} \qquad \text{It checks}$$

$$\sin\frac{5\pi}{4} + \cos\frac{5\pi}{4} = \frac{\sqrt{2}}{2} - \frac{\sqrt{2}}{2}$$

$$= 0 \qquad \text{Does not check}$$

The answer is $\frac{\pi}{4}$ only.

25.

$$\sqrt{3}\sin\theta + \cos\theta = \sqrt{3}$$

$\cos\theta = \sqrt{3} - \sqrt{3}\sin\theta$	Subtract $\sqrt{3}\sin\theta$ from both sides
$\cos\theta = \sqrt{3}(1 - \sin\theta)$	Factor out $\sqrt{3}$
$\cos^2\theta = 3(1 - 2\sin\theta + \sin^2\theta)$	Square both sides
$1 - \sin^2\theta = 3 - 6\sin\theta + 3\sin^2\theta$	Pythagorean identity
$4\sin^2\theta - 6\sin\theta + 2 = 0$	Simplify
$2\sin^2\theta - 3\sin\theta + 1 = 0$	Divide both sides by 2
$(2\sin\theta - 1)(\sin\theta - 1) = 0$	Factor
$2\sin\theta - 1 = 0 \quad \text{or} \quad \sin\theta - 1 = 0$	Set each factor $= 0$
$2\sin\theta = 1 \qquad\qquad \sin\theta = 1$	Solve each equation and check
$\sin\theta = \dfrac{1}{2} \qquad\qquad \theta = 90°$	Possible solutions
$\theta = 30° \text{ or } 150°$	

Check each possible solution:

$$\sqrt{3}\sin 30° + \cos 30° = \sqrt{3}\left(\frac{1}{2}\right) + \frac{\sqrt{3}}{2}$$

$$\qquad\qquad\qquad = \sqrt{3} \quad \text{It checks}$$

$$\sqrt{3}\sin 90° + \cos 90° = \sqrt{3}(1) + 0$$

$$\qquad\qquad\qquad = \sqrt{3} \quad \text{It checks}$$

$$\sqrt{3}\sin 150° = \cos 150° = \sqrt{3}\left(\frac{1}{2}\right) + \left(-\frac{\sqrt{3}}{2}\right)$$

$$\qquad\qquad\qquad = 0 \quad \text{Does not check}$$

Answers: 30° or 90°

27.

$$\sqrt{3}\sin\theta - \cos\theta = 1$$

$\sqrt{3}\sin\theta = \cos\theta + 1$	Isolate $\sqrt{3}\sin\theta$
$3\sin^2\theta = \cos^2\theta + 2\cos\theta + 1$	Square both sides
$3(1 - \cos^2\theta) = \cos^2\theta + 2\cos\theta + 1$	Pythagorean identity
$3 - 3\cos^2\theta = \cos^2\theta + 2\cos\theta + 1$	Simplify
$4\cos^2\theta + 2\cos\theta - 2 = 0$	Put in standard form
$2(2\cos\theta - 1)(\cos\theta + 1) = 0$	Factor
$2\cos\theta - 1 = 0 \quad \text{or} \quad \cos\theta + 1 = 0$	Set each factor $= 0$
$2\cos\theta = 1 \qquad\qquad \cos\theta = -1$	Solve each equation and check
$\cos\theta = \dfrac{1}{2} \qquad\qquad \theta = 180°$	
$\theta = 60° \text{ or } 300°$	Possible solutions

The solution is 60° or 180°. (300° does not check.)

29.

$$\sin\frac{\theta}{2} - \cos\theta = 0$$

$$\sin\frac{\theta}{2} = \cos\theta \qquad \text{Add } \cos\theta \text{ to both sides}$$

$$\sin^2\frac{\theta}{2} = \cos^2\theta \qquad \text{Square both sides}$$

$$\frac{1-\cos\theta}{2} = \cos^2\theta \qquad \qquad \sin\frac{\theta}{2} = \pm\sqrt{\frac{1-\cos\theta}{2}}$$

$$1-\cos\theta = 2\cos^2\theta \qquad \text{Multiply both sides by 2}$$

$$2\cos^2\theta + \cos\theta - 1 = 0 \qquad \text{Rewrite in standard form}$$

$$(2\cos\theta - 1)(\cos\theta + 1) = 0 \qquad \text{Factor}$$

$$2\cos\theta - 1 = 0 \quad \text{or} \quad \cos\theta + 1 = 0 \qquad \text{Set each factor} = 0$$

$$2\cos\theta = 1 \qquad\qquad \cos\theta = -1 \qquad \text{Solve each equation and check}$$

$$\cos\theta = \frac{1}{2} \qquad\qquad \theta = 180° \qquad \text{Possible solutions}$$

$$\theta = 60° \text{ or } 300°$$

The solution is $60°$ or $300°$. ($180°$ does not check.)

31.

$$\cos\frac{\theta}{2} - \cos\theta = 1$$

$$\pm\sqrt{\frac{1+\cos\theta}{2}} - \cos\theta = 1 \qquad \text{Half-angle identity}$$

$$\pm\sqrt{\frac{1+\cos\theta}{2}} = \cos\theta + 1 \qquad \text{Add } \cos\theta \text{ to both sides}$$

$$\frac{1+\cos\theta}{2} = \cos^2\theta + 2\cos\theta + 1 \qquad \text{Square both sides}$$

$$1+\cos\theta = 2\cos^2\theta + 4\cos\theta + 2 \qquad \text{Multiply both sides by 2}$$

$$0 = 2\cos^2\theta + 3\cos\theta + 1 \qquad \text{Put in standard form}$$

$$0 = (2\cos\theta + 1)(\cos\theta + 1) \qquad \text{Factor}$$

$$2\cos\theta + 1 = 0 \quad \text{or} \quad \cos\theta + 1 = 0 \qquad \text{Set each factor} = 0$$

$$2\cos\theta = -1 \qquad\qquad \cos\theta = -1 \qquad \text{Solve each equation and check}$$

$$\cos\theta = -\frac{1}{2} \qquad\qquad \theta = 180° \qquad \text{Possible solutions}$$

$$\theta = 120° \text{ or } 240°$$

The solution is $120°$ or $180°$. ($240°$ does not check.)

33.

$$6\cos\theta + 7\tan\theta = \sec\theta$$

$$6\cos\theta + \frac{7\sin\theta}{\cos\theta} = \frac{1}{\cos\theta} \qquad \text{Ratio and reciprocal identities}$$

$$6\cos^2\theta + 7\sin\theta = 1 \qquad \text{Multiply both sides by } \cos\theta\,(\cos\theta \neq 0)$$

$$6\left(1 - \sin^2\theta\right) + 7\sin\theta = 1 \qquad \text{Pythagorean identity}$$

$$6 - 6\sin^2\theta + 7\sin^2\theta = 1 \qquad \text{Simplify}$$

$$-6\sin^2\theta + 7\sin\theta + 5 = 0 \qquad \text{Subtract 1 from both sides}$$

$$6\sin^2\theta - 7\sin\theta - 5 = 0 \qquad \text{Multiply both sides by } -1$$

$$\left(3\sin\theta - 5\right)\left(2\sin\theta + 1\right) = 0 \qquad \text{Factor}$$

$$3\sin\theta - 5 = 0 \quad \text{or} \quad 2\sin\theta + 1 = 0 \qquad \text{Set each factor} = 0$$

$$3\sin\theta = 5 \qquad\qquad 2\sin\theta = -1 \qquad \text{Solve each equation and check}$$

$$\sin\theta = \frac{5}{3} \qquad\qquad \sin\theta = -\frac{1}{2}$$

No solution $\qquad\qquad \theta = 210° \text{ or } 330°$ $\qquad$ Both check

35.

$$23\csc^2\theta - 22\cot\theta\csc\theta - 15 = 0$$

$$23\left(\frac{1}{\sin^2\theta}\right) - 22\left(\frac{\cos\theta}{\sin\theta}\right)\left(\frac{1}{\sin\theta}\right) - 15 = 0 \qquad \text{Reciprocal and ratio identities}$$

$$\frac{23}{\sin^2\theta} - \frac{22\cos\theta}{\sin^2\theta} - 15 = 0 \qquad \text{Simplify}$$

$$23 - 22\cos\theta - 15\sin^2\theta = 0 \qquad \text{Multiply both sides by } \sin^2\theta$$

$$23 - 22\cos\theta - 15\left(1 - \cos^2\theta\right) = 0 \qquad \text{Pythagorean identity}$$

$$23 - 22\cos\theta - 15 + 15\cos^2\theta = 0 \qquad \text{Simplify}$$

$$15\cos^2\theta - 22\cos^2\theta + 8 = 0 \qquad \text{Put in standard form}$$

$$\left(5\cos\theta - 4\right)\left(3\cos\theta - 2\right) = 0 \qquad \text{Factor}$$

$$5\cos\theta - 4 = 0 \quad \text{or} \quad 3\cos\theta - 2 = 0 \qquad \text{Set each factor} = 0$$

$$5\cos\theta = 4 \qquad\qquad 3\cos\theta = 2 \qquad \text{Solve each equations and check}$$

$$\cos\theta = \frac{4}{5} \qquad\qquad \cos\theta = \frac{2}{3}$$

$\theta = 36.9° \text{ or } 323.1° \quad \theta = 48.2° \text{ or } 311.8°$ $\qquad$ All check

37.

$$7\sin^2\theta - 9\cos 2\theta = 0$$

$$7\sin^2\theta - 9\left(1 - 2\sin^2\theta\right) = 0 \qquad \text{Double-angle identity}$$

$$7\sin^2\theta - 9 + 18\sin^2\theta = 0 \qquad \text{Simplify left side}$$

$$25\sin^2\theta = 9 \qquad \text{Add 9 to both sides}$$

$$\sin^2\theta = \frac{9}{25} \qquad \text{Divide both sides by 25}$$

This problem is continued o the next page

$$\sin\theta = \pm\frac{3}{5} \qquad \text{Take square root of both sides}$$

$$\hat{\theta} = 36.9° \text{ and } \theta = \text{QI, QII, QIII, or QIV}$$

$$\theta = 36.9°, 143.1°, 216.9° \text{ or } 323.1°$$

39. In problem 3, we get $225°$ or $315°$. All solutions would be $225° + 360°k$ or $315° + 360°k$.

41. In problem 23 we get $x = \frac{\pi}{4}$. All solutions would be $\frac{\pi}{4} + 2k\pi$.

43. In problem 31, we get $120°$ or $180°$. All solutions would be $120° + 360°k$ or $180° + 360°k$.

45.
$$r^4 \csc^2\theta - R^4 \csc\theta\cot\theta = r^4 \cdot \frac{1}{\sin^2\theta} - R^4 \cdot \frac{1}{\sin\theta} \cdot \frac{\cos\theta}{\sin\theta}$$

$$= \frac{r^4}{\sin^2\theta} - \frac{R^4\cos\theta}{\sin^2\theta}$$

$$= \frac{r^4 - R^4\cos\theta}{\sin^2\theta}$$

This expression is zero only when the numerator is zero. Therefore,
$$r^4 - R^4\cos\theta = 0$$

$$-R^4\cos\theta = -r^4$$

$$\cos\theta = \frac{r^4}{R^4}$$

Therefore, when $\cos\theta = \frac{r^4}{R^4}$, then $r^4\csc^2\theta - R^4\csc\theta\cot\theta = 0$

47.
$$2\sin^2\theta - 2\cos\theta - 1 = 0$$
$$2(1 - \cos^2\theta) - 2\cos\theta - 1 = 0 \qquad \text{Pythagorean identity}$$
$$2 - 2\cos^2\theta - 2\cos\theta - 1 = 0 \qquad \text{Simplify}$$
$$-2\cos^2\theta - 2\cos\theta + 1 = 0 \qquad \text{Put in standard form}$$
$$2\cos^2\theta + 2\cos\theta - 1 = 0 \qquad \text{Multiply both sides by } -1$$

This problem is continued on the next page

We apply the quadratic formula with $a = 2$, $b = 2$, and $c = -1$:

$$\cos\theta = \frac{-2 \pm \sqrt{2^2 - 4(2)(-1)}}{2(2)}$$

$$= \frac{-2 \pm \sqrt{12}}{4}$$

$$\cos\theta = \frac{-2 + 3.464}{4} \quad \text{or} \quad \cos\theta = \frac{-2 - 3.464}{4}$$

$$= 0.366 \qquad\qquad\qquad = -1.366$$

$$\hat{\theta} = 68.5° \text{ and } \theta \text{ in in QI, QIV} \qquad \text{No solution}$$

$$\theta = 68.5° \text{ or } 291.5°$$

49.

$$\cos^2\theta + \sin\theta = 0$$

$$1 - \sin^2\theta + \sin\theta = 0 \qquad\qquad \text{Pythagorean identity}$$

$$-\sin^2\theta + \sin\theta + 1 = 0 \qquad\qquad \text{Put in standard form}$$

$$\sin^2\theta - \sin\theta - 1 = 0 \qquad\qquad \text{Multiply both sides by } -1$$

We apply the quadratic formula with $a = 1$, $b = -1$, and $c = -1$:

$$\sin\theta = \frac{-(-1) \pm \sqrt{(-1)^2 - 4(1)(-1)}}{2(1)}$$

$$= \frac{1 \pm \sqrt{5}}{2}$$

$$\sin\theta = \frac{1 + 2.236}{2} \quad \text{or} \quad \sin\theta = \frac{1 - 2.236}{2}$$

$$= 1.618 \qquad\qquad\qquad = -0.618$$

$$\text{No solution} \qquad \hat{\theta} = 38.2° \text{ and } \theta \text{ in in QIII, QIV}$$

$$\theta = 218.2° \text{ or } 321.8°$$

51.

$$2\sin^2\theta = 3 - 4\cos\theta$$

$$2(1 - \cos^2\theta) = 3 - 4\cos\theta \qquad\qquad \text{Pythagorean identity}$$

$$2 - 2\cos^2\theta = 3 - 4\cos\theta \qquad\qquad \text{Simplify}$$

$$-2\cos^2\theta + 4\cos\theta - 1 = 0 \qquad\qquad \text{Put in standard form}$$

$$2\cos^2\theta - 4\cos\theta + 1 = 0 \qquad\qquad \text{Multiply both sides by } -1$$

We apply the quadratic formula with $a = 2$, $b = -4$, and $c = 1$:

$$\cos\theta = \frac{-(-4) \pm \sqrt{(-4)^2 - 4(2)(1)}}{2(2)}$$

$$= \frac{4 \pm \sqrt{8}}{4}$$

This problem is continued on the next page

$$\cos\theta = \frac{4+2.828}{4} \quad \text{or} \quad \cos\theta = \frac{4-2.828}{4}$$
$$= 1.707 \qquad\qquad = 0.293$$

$$\text{No solution} \qquad \hat{\theta} = 73.0° \text{ and } \theta \text{ is in QI and QIV}$$
$$\theta = 73.0° \text{ or } 287.0°$$

53. Graph $y_1 = \cos(x) + 3\sin(x) - 2$ on your graphing calculator. Set the window for x between 0 and 6.28 and y between -1 and 1. Use the zero or root finder to locate the zeros. The solution is 0.3630 or 2.1351.

55. Graph $y_1 = \left(\sin(x)\right)^2 - 3\sin(x) - 1$ on your graphing calculator. The window settings are the same as problem 53. Use the zero or root finder to locate the zeros. The solution is 3.4492 or 5.9756.

57. Graph $y_1 = \dfrac{1}{\cos(x)} + 2 - \dfrac{1}{\tan(x)}$ on your graphing calculator. The window settings are the same as problem 53. Use the zero or root finder to locate the zeros. The solution is 0.3166 or 1.9917.

59.
$$\cos A = \sqrt{1 - \sin^2 A} \qquad \text{with } A \text{ in QI} \qquad\qquad \sin\frac{A}{2} = \sqrt{\frac{1-\cos A}{2}}$$
$$= \sqrt{1 - \left(\frac{2}{3}\right)^2} \qquad\qquad\qquad\qquad\qquad = \sqrt{\frac{1-\sqrt{5}/3}{2}}$$
$$= \sqrt{1 - \frac{4}{9}} \qquad\qquad\qquad\qquad\qquad\qquad = \sqrt{\frac{3-\sqrt{5}}{6}}$$
$$= \sqrt{\frac{5}{9}}$$
$$= \frac{\sqrt{5}}{3}$$

61.
$$\csc\frac{A}{2} = \frac{1}{\sin\dfrac{A}{2}}$$
$$= \frac{1}{\sqrt{\dfrac{3-\sqrt{5}}{6}}} \qquad\qquad \text{(from problem 59)}$$
$$= \sqrt{\frac{6}{3-\sqrt{5}}}$$

63. $\tan\dfrac{A}{2} = \dfrac{\sin\dfrac{A}{2}}{\cos\dfrac{A}{2}}$

We know that $\cos A = \dfrac{\sqrt{5}}{3}$ and $\sin\dfrac{A}{2} = \sqrt{\dfrac{3-\sqrt{5}}{6}}$ from problem 59.

$$\cos\dfrac{A}{2} = \sqrt{\dfrac{1+\cos A}{2}}$$

Therefore, $\tan\dfrac{A}{2} = \dfrac{\sqrt{\dfrac{3-\sqrt{5}}{6}}}{\sqrt{\dfrac{3+\sqrt{5}}{6}}}$

$$= \sqrt{\dfrac{1+\sqrt{5}/3}{2}}$$

$$= \sqrt{\dfrac{3-\sqrt{5}}{3+\sqrt{5}}}$$

$$= \sqrt{\dfrac{3+\sqrt{5}}{6}}$$

$$= \dfrac{3-\sqrt{5}}{2} \quad \text{Rationalize the denominator}$$

67. $\sin 22.5° = \sin\dfrac{1}{2}(45°)$

$$= \sqrt{\dfrac{1-\cos 45°}{2}}$$

$$= \sqrt{\dfrac{1-\dfrac{\sqrt{2}}{2}}{2}}$$

$$= \sqrt{\dfrac{2-\sqrt{2}}{4}}$$

$$= \dfrac{\sqrt{2-\sqrt{2}}}{2}$$

Problem Set 6.3

1. $\sin 2\theta = \dfrac{\sqrt{3}}{2}$

$2\theta = 60° + 360°k$ or $2\theta = 120° + 360°k$

$\theta = 30° + 180°k$ $\theta = 60° + 180°k$

If we let $k = 0$ and 1, we get

$\theta = 30°$ $\theta = 60°$

$\theta = 210°$ $\theta = 240°$

3. $\tan 2\theta = -1$

$2\theta = 135° + 360°k$ or $2\theta = 315° + 360°k$

$\theta = 67.5° + 180°k$ $\theta = 157.5° + 180°k$

If we let $k = 0$ and 1, we get

$\theta = 67.5°$ $\theta = 157.5°$

$\theta = 247.5°$ $\theta = 337.5°$

5. $\cos 3\theta = -1$

$3\theta = 180° + 360°k$

$\theta = 60° + 120°k$

If we let $k = 0$, 1, and 2, we get

$\theta = 60°$ $\theta = 300°$
$\theta = 180°$

7. $\sin 2x = \dfrac{1}{\sqrt{2}}$

$2x = \dfrac{\pi}{4} + 2k\pi$ or $2x = \dfrac{3\pi}{4} + 2k\pi$

$x = \dfrac{\pi}{8} + k\pi$ or $x = \dfrac{3\pi}{8} + k\pi$

If we let $k = 0$ and 1, we get

$x = \dfrac{\pi}{8}$ $x = \dfrac{3\pi}{8}$

$x = \dfrac{9\pi}{8}$ $x = \dfrac{11\pi}{8}$

9. $\sec 3x = -1$

$\cos 3x = -1$ (Reciprocal identity)

$3x = \pi + 2k\pi$

$x = \dfrac{\pi}{3} + \dfrac{2k\pi}{3}$

If we let $k = 0$, 1, and 2, we get

$\theta = \dfrac{\pi}{3}$ $\theta = \dfrac{\pi}{3} + \dfrac{4\pi}{3} = \dfrac{5\pi}{3}$

$\theta = \dfrac{\pi}{3} + \dfrac{2\pi}{3} = \pi$

11. $\tan 2x = \sqrt{3}$

$2x = \dfrac{\pi}{3} + k\pi$

$x = \dfrac{\pi}{6} + \dfrac{k\pi}{2}$

If we let $k = 0$, 1, 2 and 3, we get

$x = \dfrac{\pi}{6}$ $x = \dfrac{7\pi}{6}$

$x = \dfrac{4\pi}{6} = \dfrac{2\pi}{3}$ $x = \dfrac{10\pi}{6} = \dfrac{5\pi}{3}$

13. $\sin 2\theta = \dfrac{1}{2}$

$2\theta = 30° + 360°k$ or $2\theta = 150° + 360°k$

$\theta = 15° + 180°k$ $\theta = 75° + 180°k$

15. $\cos 3\theta = 0$

$3\theta = 90° + 360°k$ or $3\theta = 270° + 360°k$

$\theta = 30° + 120°k$ $\theta = 90° + 120°k$

17. $\sin 10\theta = \dfrac{\sqrt{3}}{2}$

$$10\theta = 60° + 360° k \quad \text{or} \quad 10\theta = 120° + 360° k$$
$$\theta = 6° + 36° k \qquad\qquad \theta = 12° + 36° k$$

25. $2\sin 2x \cos x + \cos 2x \sin x = \dfrac{1}{2}$

$$\sin(2x + x) = \dfrac{1}{2} \quad \text{(Sum formula)}$$

$$3x = \dfrac{\pi}{6} + 2k\pi \quad \text{or} \quad 3x = \dfrac{5\pi}{6} + 2k\pi$$

$$x = \dfrac{\pi}{18} + \dfrac{2k\pi}{3} \qquad x = \dfrac{5\pi}{18} + \dfrac{2k\pi}{3}$$

$$x = \dfrac{\pi + 12k\pi}{18} \qquad x = \dfrac{5\pi + 12k\pi}{3}$$

If we let $k = 0$, 1, and 2, we get

$$x = \dfrac{\pi}{18} \qquad x = \dfrac{5\pi}{18}$$
$$x = \dfrac{13\pi}{18} \qquad x = \dfrac{17\pi}{18}$$
$$x = \dfrac{25\pi}{18} \qquad x = \dfrac{29\pi}{18}$$

27. $\cos 2x \cos x - \sin 2x \sin x = -\dfrac{\sqrt{3}}{2}$

$$\cos(2x + x) = -\dfrac{\sqrt{3}}{2} \quad \text{(Sum formula)}$$

$$\cos 3x = -\dfrac{\sqrt{3}}{2}$$

$$3x = \dfrac{5\pi}{6} + 2k\pi \quad \text{or} \quad 3x = \dfrac{7\pi}{6} + 2k\pi$$

$$x = \dfrac{5\pi}{18} + \dfrac{2k\pi}{3} \qquad x = \dfrac{7\pi}{18} + \dfrac{2k\pi}{3}$$

$$x = \dfrac{5\pi + 12k\pi}{18} \qquad x = \dfrac{7\pi + 12k\pi}{18}$$

If we let $k = 0$, 1, and 2, we get

$$x = \dfrac{5\pi}{18} \qquad x = \dfrac{7\pi}{18}$$
$$x = \dfrac{17\pi}{18} \qquad x = \dfrac{19\pi}{18}$$
$$x = \dfrac{29\pi}{18} \qquad x = \dfrac{31\pi}{18}$$

29. $\sin 3x \cos 2x + \cos 3x \sin 2x = 1$

$$\sin(3x + 2x) = 1$$

$$5x = \dfrac{\pi}{2} + 2k\pi$$

$$x = \dfrac{\pi}{10} + \dfrac{2k\pi}{5}$$

31.
$$\sin^2 4x = 1$$
$$\sin 4x = \pm 1$$

$$4x = \frac{\pi}{2} + 2k\pi \quad \text{or} \quad 4x = \frac{3\pi}{2} + 2k\pi$$

$$x = \frac{\pi}{8} + \frac{k\pi}{2} \qquad x = \frac{3\pi}{8} + \frac{k\pi}{2}$$

We could also write this as $4x = \dfrac{\pi}{2} + k\pi \quad \text{or} \quad x = \dfrac{\pi}{8} + \dfrac{k\pi}{4}$

33.
$$\cos^3 5x = -1$$
$$\cos x = -1$$
$$5x = \pi + 2k\pi$$
$$x = \frac{\pi}{5} + \frac{2k\pi}{5}$$

35.
$$2\sin^2 3\theta + \sin 3\theta - 1 = 0$$
$$(\sin 3\theta - 1)(\sin 3\theta + 1) = 0$$

$$2\sin 3\theta - 1 = 0 \qquad \text{or} \qquad \sin 3\theta + 1 = 0$$
$$2\sin 3\theta = 1 \qquad\qquad\qquad \sin 3\theta = -1$$
$$\sin 3\theta = \frac{1}{2} \qquad\qquad\qquad 3\theta = 270° + 360°k$$
$$3\theta = 30° + 360°k \quad \text{or} \quad 3\theta = 150° + 360°k \qquad \theta = 90° + 120°k$$
$$\theta = 10° + 120°k \qquad \theta = 50° + 120°k$$

37.
$$2\cos^2 2\theta + 3\cos 2\theta + 1 = 0$$
$$(2\cos 2\theta + 1)(\cos 2\theta + 1) = 0$$

$$2\cos 2\theta + 1 = 0 \qquad \text{or} \qquad \cos 2\theta + 1 = 0$$
$$2\cos 2\theta = -1 \qquad\qquad\qquad 2\theta = 180° + 360°k$$
$$\cos 2\theta = -\frac{1}{2} \qquad\qquad\qquad \theta = 90° + 180°k$$
$$2\theta = 120° + 360°k \quad \text{or} \quad 2\theta = 240° + 360°k$$
$$\theta = 60° + 180°k \qquad \theta = 120° + 180°k$$

39.
$$\tan^2 3\theta = 3$$
$$\tan 3\theta = \pm\sqrt{3}$$

$$3\theta = 60° + 180°k \qquad \text{or} \qquad 3\theta = 120° + 180°k$$
$$\theta = 20° + 60°k \qquad\qquad\qquad \theta = 40° + 60°k$$

41.

$$\cos\theta - \sin\theta = 1$$

$\cos^2\theta - 2\sin\theta\cos\theta + \sin^2\theta = 1$	Square both sides
$-2\sin\theta\cos\theta + 1 = 1$	Pythagorean identity
$\sin 2\theta = 0$	Double-angle identity

$2\theta = 0° + 360°k$ or $2\theta = 180° + 360°k$

$\theta = 180°k$ $\qquad\qquad$ $\theta = 90° + 180°k$

If we let $k = 0$ and 1, we get

$\qquad\theta = 0°\qquad\qquad\theta = 90°$

$\qquad\theta = 180°\qquad\quad\theta = 270°$

Since we squared both sides, we must check. Only $0°$ and $270°$ check.

43.

$$\sin\theta + \cos\theta = -1$$

$\sin^2 + 2\sin\theta\cos\theta + \cos^2\theta = 1$	Square both sides
$1 + 2\sin\theta\cos\theta = 1$	Pythagorean identity
$\sin 2\theta = 0$	Double-angle identity

$2\theta = 0° + 360°k$ or $2\theta = 180° + 360°k$

$\theta = 180°k$ $\qquad\qquad$ $\theta = 90° + 180°k$

If we let $k = 0$ and 1, we get

$\theta = 0°\qquad\qquad\theta = 90°$

$\theta = 180°\qquad\quad\theta = 270°$

Since we squared both sides, we must check. Only $180°$ and $270°$ check.

45. $\sin^2 2\theta - 4\sin 2\theta - 1 = 0$

We use the quadratic formula with $a = 1$, $b = -4$, $c = -1$,

$$\sin 2\theta = \frac{-(-4) \pm \sqrt{(-4)^2 - 4(1)(-1)}}{2(1)}$$

$$= \frac{4 \pm \sqrt{20}}{2}$$

$\sin 2\theta = \dfrac{4 + \sqrt{20}}{2}$ or $\sin 2\theta = \dfrac{4 - \sqrt{20}}{2}$

$\qquad\quad = 4.236$ $\qquad\qquad\qquad\qquad = -0.236$

$\qquad$ No solution $\qquad\qquad$ $2\theta = 193.7° + 360°k$ or $2\theta = 346.3° + 360°k$

$\qquad\qquad\qquad\qquad\qquad\qquad$ $\theta = 96.8° + 180°k$ $\qquad\qquad$ $\theta = 173.2° + 180°k$

$\qquad\qquad\qquad\qquad\qquad\qquad$ If we let $k = 0$ and 1, we get:

$\qquad\qquad\qquad\qquad\qquad\qquad$ $\theta = 96.8°$ $\qquad\qquad\qquad$ $\theta = 173.2°$

$\qquad\qquad\qquad\qquad\qquad\qquad$ $\theta = 276.8°$ $\qquad\qquad\qquad$ $\theta = 353.2°$

47.

$$4\cos^2 3\theta - 8\cos 3\theta + 1 = 0 \qquad\qquad a = 4, b = -8, c = 1$$

$$\cos 3\theta = \frac{-(-8) \pm \sqrt{(-8)^2 - 4(4)(1)}}{2(4)}$$

$$= \frac{8 \pm \sqrt{48}}{8}$$

$$\cos 3\theta = \frac{8 \pm 6.9282}{8}$$

$$\cos 3\theta = 1.8660 \qquad \text{or} \qquad \cos 3\theta = 0.1340$$

No solution

$$3\theta = 82.3° + 360°k \quad 3\theta = 277.7° + 360°k$$

$$\theta = 27.4° + 120°k \quad \theta = 92.6° + 120°k$$

If we let $k = 0$, 1 and 2, we get

$\theta = 27.4°$	$\theta = 92.6°$
$\theta = 147.4°$	$\theta = 212.6°$
$\theta = 267.4°$	$\theta = 332.6°$

49.

$$2\cos^2 4\theta + 2\sin 4\theta = 1$$

$$2(1 - \sin^2 4\theta) + 2\sin 4\theta - 1 = 0 \qquad\qquad \text{Pythagorean identity}$$

$$2 - 2\sin^2 4\theta + 2\sin 4\theta - 1 = 0$$

$$-2\sin^2 4\theta + 2\sin 4\theta + 1 = 0$$

$$2\sin^2 4\theta - 2\sin 4\theta - 1 = 0$$

We apply the quadratic formula with $a = 2$, $b = -2$, and $c = -1$:

$$\sin 4\theta = \frac{-(-2) \pm \sqrt{(-2)^2 - 4(2)(-1)}}{2(2)}$$

$$= \frac{2 \pm \sqrt{12}}{4}$$

$$\sin 4\theta = \frac{2 + \sqrt{12}}{4} \qquad \text{or} \qquad \sin 4\theta = \frac{2 - \sqrt{12}}{4}$$

$$= 1.366 \qquad\qquad\qquad = -0.366$$

No solution

$$4\theta = 201.5° + 360°k \quad \text{or} \quad 4\theta = 338.5° + 360°k$$

$$\theta = 50.4° + 90°k \qquad\qquad \theta = 84.6° + 90°k$$

If we let $k = 0$, 1, 2, and 3, we get

$\theta = 50.4°$	$\theta = 84.6°$
$\theta = 140.4°$	$\theta = 174.6°$
$\theta = 230.4°$	$\theta = 264.6°$
$\theta = 320.4°$	$\theta = 354.6°$

51. We want to find t when $h = 100$.

$$100 = 139 - 125\cos\frac{\pi}{10}t$$

$$125\cos\frac{\pi}{10}t = 39$$

$$\cos\frac{\pi}{10}t = 0.312$$

$$\frac{\pi}{10}t = 1.253 + 2\pi k \quad \text{or} \quad \frac{\pi}{10}t = (2\pi - 1.253) + 2\pi k$$

$$t = 4.0 + 20k \qquad \qquad \frac{\pi}{10}t = 5.030 + 2\pi k$$

$$t = 16.0 + 20k \text{ (where } k = 0, 1, 2, 3, \ldots, 9)$$

It will be at 100 ft after 4.0 min, 16.0 min, 24.0 min. 28.0 min, and so on.

53. $l = 2r\sin\frac{180°}{n}$ where $l = r$

$$r = 2r\sin\frac{180°}{n}$$

$$\sin\frac{180°}{n} = \frac{r}{2r}$$

$$\sin\frac{180°}{n} = \frac{1}{2}$$

$$\frac{180°}{n} = 30° \qquad \text{or} \qquad \frac{180°}{n} = 150°$$

$$180° = 30°n \qquad \qquad 180° = 150°n$$

$$n = 6 \qquad \qquad n = 1.2$$

$$\qquad \qquad \qquad \text{Not possible}$$

The polygon has 6 sides.

55. $d = 10\tan\pi t$ where $d = 10$

$$10 = 10\tan\pi t$$

$$\tan\pi t = 1$$

$$\pi t = \frac{\pi}{4} + k\pi$$

$$t = \frac{1}{4} + k$$

t is $\frac{1}{4}$ second and every second after that.

57.

$$\sin 2\pi t = \frac{1}{2}$$

$$2\pi t = \frac{\pi}{6} + 2k\pi$$

$$t = \frac{1}{12} + k \qquad \text{Let } k = 0, \text{ then } t = \frac{1}{12}.$$

59.

$$\frac{\sin x}{1 + \cos x} = \frac{\sin x}{1 + \cos x} \cdot \frac{1 - \cos x}{1 - \cos x} \qquad \text{Multiply by a fraction equal to one}$$

$$= \frac{\sin x (1 - \cos x)}{1 - \cos^2 x} \qquad \text{Multiply}$$

$$= \frac{\sin x (1 - \cos x)}{\sin^2 x} \qquad \text{Pythagorean identity}$$

$$= \frac{1 - \cos x}{\sin x} \qquad \text{Reduce}$$

61.

$$\frac{1}{1 + \cos t} + \frac{1}{1 - \cos t} = \frac{1}{1 + \cos t} \cdot \frac{1 - \cos t}{1 - \cos t} + \frac{1}{1 - \cos t} \cdot \frac{1 + \cos t}{1 + \cos t}$$

$$= \frac{1 - \cos t}{1 - \cos^2 t} + \frac{1 + \cos t}{1 - \cos^2 t}$$

$$= \frac{1 - \cos t + 1 + \cos t}{1 - \cos^2 t}$$

$$= \frac{2}{\sin^2 t}$$

$$= 2 \csc^2 t$$

63.

$$\tan \frac{A}{2} = \frac{\sin \frac{A}{2}}{\cos \frac{A}{2}} \qquad \text{Ratio identity}$$

$$= \frac{\sqrt{\frac{1 - \cos A}{2}}}{\sqrt{\frac{1 + \cos A}{2}}} \qquad \text{Half-angle identities}$$

$$= \frac{\sqrt{1 - \cos A}}{\sqrt{1 + \cos A}} \qquad \text{Divide}$$

This problem is continued on the next page

$$= \frac{\sqrt{1-\cos A}}{\sqrt{1+\cos A}} \cdot \frac{\sqrt{1+\cos A}}{\sqrt{1+\cos A}} \quad \text{Multiply by a fraction equal to one}$$

$$= \frac{\sqrt{1-\cos^2 A}}{1+\cos A} \quad \text{Multiply}$$

$$= \frac{\sqrt{\sin^2 A}}{1+\cos A} \quad \text{Pythagorean identity}$$

$$= \frac{\sin A}{1+\cos A} \quad \text{Simplify}$$

65. If $\sin A = \frac{1}{3}$ and A is in QII, then

$$\cos A = -\sqrt{1-\sin^2 A}$$

$$= \sqrt{1-\frac{1}{9}} = -\sqrt{\frac{8}{9}} = -\frac{2\sqrt{2}}{3}$$

$$\sin 2A = 2\sin A \cos A$$

$$= 2\left(\frac{1}{3}\right)\left(-\frac{2\sqrt{2}}{3}\right)$$

$$= -\frac{4\sqrt{2}}{9}$$

67. If $90° \le A \le 180°$, then $45° \le \frac{A}{2} \le 90°$

Also, $\sin A = \frac{1}{3}$ and $\cos A = -\frac{2\sqrt{2}}{3}$ (from problem 65)

Therefore, $\cos \frac{A}{2} = \sqrt{\frac{1+\cos A}{2}}$

$$= \sqrt{\frac{1+\left(\frac{-2\sqrt{2}}{3}\right)}{2}}$$

$$= \sqrt{\frac{3-2\sqrt{2}}{6}}$$

69. If $\sin B = \frac{3}{5}$ with B in QI, then

$$\cos B = \sqrt{1-\sin^2 B}$$

$$= \sqrt{1-\frac{9}{25}} = \sqrt{\frac{16}{25}} = \frac{4}{5}$$

This problem is continued on the next page

From problem 65, we have $\cos A = -\dfrac{2\sqrt{2}}{3}$.

Substituting these above, we get:

$\sin(A+B) = \sin A \cos B + \cos A \sin B$

$$= \frac{1}{3}\left(\frac{4}{5}\right) + \left(-\frac{2\sqrt{2}}{3}\right)\left(\frac{3}{5}\right)$$

$$= \frac{4}{15} - \frac{6\sqrt{2}}{15} = \frac{4-6\sqrt{2}}{15}$$

71.
$$\csc(A+B) = \frac{1}{\sin(A+B)}$$

$$= \frac{1}{\dfrac{4-6\sqrt{2}}{15}}$$

$$= \frac{15}{4-6\sqrt{2}}$$

Problem Set 6.4

1. $\sin t = x$ and $\cos t = y$

$\sin^2 t + \cos^2 t = 1$	Pythagorean identity
$x^2 + y^2 = 1$	Substitute known values

The graph is a circle with its center at the origin and $r = 1$.

3. $3\cos t = x$ and $3\sin t = y$

$$\cos t = \frac{x}{3} \qquad \sin t = \frac{y}{3}$$

$\cos^2 t + \sin^2 t = 1$	Pythagorean identity
$\left(\dfrac{x}{3}\right)^2 + \left(\dfrac{y}{3}\right)^2 = 1$	Substitute known values
$\dfrac{x^2}{9} + \dfrac{y^2}{9} = 1$	Simplify
$x^2 + y^2 = 9$	

The graph is a circle with its center at the origin and $r = 3$.

5. $2\sin t = x$ and $4\cos t = y$

$\sin t = \dfrac{x}{2}$ $\cos t = \dfrac{y}{4}$

$\sin^2 t + \cos^2 t = 1$

$\left(\dfrac{x}{2}\right)^2 + \left(\dfrac{y}{4}\right)^2 = 1$

$\dfrac{x^2}{4} + \dfrac{y^2}{16} = 1$

The graph is an ellipse with center at the origin. The intercepts on the major axis are (0, 4) and (0, –4). The intercepts on the minor axis are (2, 0) and (–2,0).

7. $2 + \sin t = x$ and $3 + \cos t = y$

$\sin t = x - 2$ $\cos t = y - 3$

$\sin^2 t + \cos^2 t = 1$

$(x-2)^2 + (y-3)^2 = 1$

The graph is a circle with center at (2, 3) and $r = 1$.

9. $\sin t - 2 = x$ and $\cos t - 3 = y$

$\sin t = x + 2$ $\cos t = y + 3$

$\sin^2 t + \cos^2 t = 1$

$(x+2)^2 + (y+3)^2 = 1$

The graph is a circle with center at $(-2, -3)$ and $r = 1$.

11. $3 + 2\sin t = x$ and $1 + 2\cos t = y$

$2\sin t = x - 3$ $2\cos t = y - 1$

$\sin t = \dfrac{x-3}{2}$ $\cos t = \dfrac{y-1}{2}$

$\sin^2 t + \cos^2 t = 1$

$\left(\dfrac{x-3}{2}\right)^2 + \left(\dfrac{y-1}{2}\right)^2 = 1$

$\dfrac{(x-3)^2}{4} + \dfrac{(y-1)^2}{4} = 1$

$(x-3)^2 + (y-1)^2 = 4$

The graph is a circle with center at (3, 1) and $r = 2$.

13.

$$3\cos t - 3 = x \quad \text{and} \quad 3\sin t + 1 = y$$
$$3\cos t = x + 3 \qquad\qquad 3\sin t = y - 1$$
$$\cos t = \frac{x+3}{3} \qquad\qquad \sin t = \frac{y-1}{3}$$
$$\cos^2 t + \sin^2 t = 1$$
$$\left(\frac{x+3}{3}\right)^2 + \left(\frac{y-1}{3}\right)^2 = 1$$
$$\frac{(x+3)^2}{9} + \frac{(y-1)^2}{9} = 1$$
$$(x+3)^2 + (y-1)^2 = 9$$

The graph is a circle with center at $(-3, 1)$ and $r = 3$.

15.

$$\sec t = x \qquad \text{and} \qquad \tan t = y$$

$$\sec^2 t = \tan^2 t + 1 \qquad\qquad \text{Pythagorean identity}$$
$$x^2 = y^2 + 1 \qquad\qquad\qquad \text{Substitute known values}$$
$$x^2 - y^2 = 1 \qquad\qquad\qquad \text{Simplify}$$

17.

$$3\sec t = x \qquad \text{and} \qquad 3\tan t = y$$
$$\sec t = \frac{x}{3} \qquad\qquad\qquad \tan t = \frac{y}{3}$$
$$\sec^2 t = \tan^2 t + 1$$
$$\left(\frac{x}{3}\right)^2 = \left(\frac{y}{3}\right)^2 + 1$$
$$\frac{x^2}{9} = \frac{y^2}{9} + 1$$
$$\frac{x^2}{9} - \frac{y^2}{9} = 1$$

19.

$$2 + 3\tan t = x \quad \text{and} \quad 4 + 3\sec t = y$$
$$3\tan t = x - 2 \qquad\qquad 3\sec t = y - 4$$
$$\tan t = \frac{x-2}{3} \qquad\qquad \sec t = \frac{y-4}{3}$$
$$\sec^2 t = \tan^2 t + 1$$
$$\left(\frac{y-4}{3}\right)^2 = \left(\frac{x-2}{3}\right)^2 + 1$$
$$\frac{(y-4)^2}{9} = \frac{(x-2)^2}{9} + 1$$
$$\frac{(y-4)^2}{9} - \frac{(x-2)^2}{9} = 1$$

21. $\cos 2t = x$ and $\sin t = y$

$\cos 2t = 1 - 2\sin^2 t$ Double-angle identity

$\quad x = 1 - 2y^2$ Substitute known values

23. $\sin t = x$ and $\sin t = y$

$\sin t = \sin t$ Reflective property of equality

$\quad x = y$

25. $3\sin t = x$ and $2\sin t = y$

$\sin t = \dfrac{x}{3}$ $\quad\quad\quad$ $\sin t = \dfrac{y}{2}$

$\sin t = \sin t$ Reflective property or equality

$\dfrac{x}{3} = \dfrac{y}{2}$ Substitute known values

$2x = 3y$ Multiply both sides by 6

27.   From the diagram at the right,

$\cos\theta = \dfrac{x}{98.5}$ $\quad\quad$ $\sin\theta = \dfrac{h}{98.5}$

$x = 98.5\cos\theta$ $\quad\quad$ $h = 98.5\sin\theta$

$\quad\quad\quad\quad\quad\quad\quad\quad\quad\quad y = 110.5 + h$

$\quad\quad\quad\quad\quad\quad\quad\quad\quad\quad = 110.5 + 98.5\sin\theta$

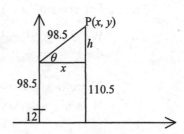

Next, we write θ in terms of t using angular velocity:

$\omega = \dfrac{1\text{ rev}}{15\text{ min}} \cdot \dfrac{2\pi\text{ rad}}{1\text{ rev}} = \dfrac{2\pi}{15}\dfrac{\text{rad}}{\text{min}}$

$\omega = \dfrac{\theta}{t}$ (from Section 3.5) $\quad\quad$ We want the starting point to be at $(0, 0)$. So we must

$\dfrac{2\pi}{15} = \dfrac{\theta}{t}$ $\quad\quad\quad\quad\quad$ subtract $\dfrac{\pi}{2}$ from $\dfrac{2\pi}{15}t$.

$\theta = \dfrac{2\pi}{15}t$

Therefore: $x = 98.5\cos\left(\dfrac{2\pi}{15}t - \dfrac{\pi}{2}\right)$

$\quad\quad\quad\quad\quad y = 110.5 + 98.5\sin\left(\dfrac{2\pi}{15}t - \dfrac{\pi}{2}\right)$

Check on your graphing calculator.

29.   Let $\alpha = \cos^{-1}\dfrac{1}{2}$. Then $\cos\alpha = \dfrac{1}{2}$ and $0 \le \alpha \le \pi$.

Therefore, $\alpha = \dfrac{\pi}{3}$. We want to find $\sin\alpha = \sin\dfrac{\pi}{3} = \dfrac{\sqrt{3}}{2}$

31. Let $\alpha = \tan^{-1}\dfrac{1}{3}$ and $\beta = \sin^{-1}\dfrac{1}{4}$.

Then $\tan\alpha = \dfrac{1}{3}$ and $\sin\beta = \dfrac{1}{4}$

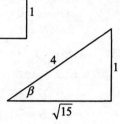

Next, we draw 2 triangles and label the sides.
We then find the missing sides:

hypotenuse $= \sqrt{3^2 + 1^2}$ adjacent side $= \sqrt{4^2 - 1^2}$

$\qquad\qquad = \sqrt{10}$ $\qquad\qquad = \sqrt{15}$

Now, we can find $\sin(\alpha + \beta) = \sin\alpha\cos\beta + \cos\alpha\sin\beta$

$$= \frac{1}{\sqrt{10}} \cdot \frac{\sqrt{15}}{4} + \frac{3}{\sqrt{10}} \cdot \frac{1}{4}$$

$$= \frac{\sqrt{15}}{4\sqrt{10}} + \frac{3}{4\sqrt{10}}$$

$$= \frac{3 + \sqrt{15}}{4\sqrt{10}}$$

33. Let $\alpha = \sin^{-1}x$. Then $\sin\alpha = \dfrac{x}{1}$ and $-\dfrac{\pi}{2} \le \alpha \le \dfrac{\pi}{2}$.

We draw a triangle and we find the adjacent side using the Pythagorean Theorem.

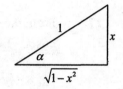

Therefore, $\cos\alpha = \dfrac{\sqrt{1-x^2}}{1} = \sqrt{1-x^2}$.

35. Let $\alpha = \tan^{-1}x$. Then $\tan\alpha = \dfrac{x}{1}$ and $-\dfrac{\pi}{2} < \alpha < \dfrac{\pi}{2}$.

We draw a triangle and we find the hypotenuse using the Pythagorean Theorem.

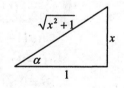

Therefore, $\cos 2\alpha = \cos^2\alpha - \sin^2\alpha$

$$= \left(\frac{1}{\sqrt{x^2+1}}\right)^2 - \left(\frac{x}{\sqrt{x^2+1}}\right)^2$$

$$= \frac{1}{x^2+1} - \frac{x^2}{x^2+1} = \frac{1-x^2}{x^2+1}$$

37. $8\sin 3x\cos 2x = 8 \cdot \dfrac{1}{2}\big[\sin(3x+2x) + \sin(3x-2x)\big]$

$\qquad\qquad\quad = 4(\sin 5x + \sin x)$

Chapter 6 Test

1. $2\sin - 1 = 0$

        $2\sin\theta = 1$ Add 1 to both sides

        $\sin\theta = \dfrac{1}{2}$ Divide both sides by 2

        $\hat{\theta} = 30°$ and θ is in QI or QII

        $\theta = 30°$ or $150°$

2. $\sqrt{3}\tan\theta + 1 = 0$

        $\sqrt{3}\tan\theta = -1$ Subtract 1 from both sides

        $\tan\theta = -\dfrac{1}{\sqrt{3}}$ Divide both sides by $\sqrt{3}$

        $\hat{\theta} = 30°$ and θ is in QII or QIV

        $\theta = 150°$ or $330°$

3. $\cos\theta - 2\sin\theta\cos\theta = 0$

        $\cos\theta(1 - 2\sin\theta) = 0$ Factor

    $\cos\theta = 0$ or $1 - 2\sin\theta = 0$ Set each factor = 0

    $\theta = 90°$ or $270°$ $-2\sin\theta = -1$ Solve each equation

                           $\sin\theta = \dfrac{1}{2}$

                          $\theta = 30°$ or $150°$

4. $\tan\theta - 2\cos\theta\tan\theta = 0$

        $\tan\theta(1 - 2\cos\theta) = 0$ Factor

    $\tan\theta = 0$ or $1 - 2\cos\theta = 0$ Set each factor = 0

     $\theta = 0°$ or $180°$ $-2\cos\theta = -1$ Solve each equation

                           $\cos\theta = \dfrac{1}{2}$

                          $\theta = 60°$ or $300°$

5. $4\cos\theta - 2\sec\theta = 0$

        $4\cos\theta - \dfrac{2}{\cos\theta} = 0$ Reciprocal identity

        $4\cos^2\theta - 2 = 0$ Multiply both sides by $\cos\theta$

        $4\cos^2\theta = 2$ Add 2 to both sides

    This problem is continued on the next page

$$\cos^2\theta = \frac{1}{2}$$ Divide both sides by 4

$$\cos\theta = \pm\frac{1}{\sqrt{2}}$$ Take square root of both sides

$$\theta = 45°, 135°, 225°, 315°$$ $$\hat{\theta} = 45°$$

6.
$$2\sin\theta - \csc\theta = 1$$

$$2\sin\theta - \frac{1}{\sin\theta} = 1$$ Reciprocal identity

$$2\sin^2\theta - 1 = \sin\theta$$ Multiply both sides by $\sin\theta$

$$2\sin^2\theta - \sin\theta - 1 = 0$$ Rewrite in standard form

$$(2\sin\theta + 1)(\sin\theta - 1) = 0$$ Factor

$$2\sin\theta + 1 = 0 \quad \text{or} \quad \sin\theta - 1 = 0$$ Set each factor = 0

$$2\sin\theta = -1 \qquad \sin\theta = 1$$ Solve each equation

$$\sin\theta = -\frac{1}{2} \qquad \theta = 90°$$

$$\theta = 210° \text{ or } 330°$$

7.
$$\sin\frac{\theta}{2} + \cos\theta = 0$$

$$\pm\sqrt{\frac{1-\cos\theta}{2}} + \cos\theta = 0$$ Half-angle identity

$$\pm\sqrt{\frac{1-\cos\theta}{2}} = -\cos\theta$$ Subtract $\cos\theta$ from both sides

$$\frac{1-\cos\theta}{2} = \cos^2\theta$$ Square both sides

$$1 - \cos\theta = 2\cos^2\theta$$ Multiply both sides by 2

$$2\cos^2\theta + \cos\theta - 1 = 0$$ Rewrite in standard form

$$(2\cos\theta - 1)(\cos\theta + 1) = 0$$ Factor

$$2\cos\theta - 1 = 0 \quad \text{or} \quad \cos\theta + 1 = 0$$ Set each factor = 0

$$2\cos\theta = 1 \qquad \cos\theta = -1$$ Solve each equation and check

$$\cos\theta = \frac{1}{2} \qquad \theta = 180°$$ Possible solutions

$$\theta = 60° \text{ or } 300°$$

We check each possible solution and the only one that checks is $180°$.

8.

$$\cos\frac{\theta}{2} - \cos\theta = 0$$

$$\pm\sqrt{\frac{1+\cos\theta}{2}} - \cos\theta = 0 \qquad \text{Half-angle identity}$$

$$\pm\sqrt{\frac{1+\cos\theta}{2}} = \cos\theta \qquad \text{Add } \cos\theta \text{ to both sides}$$

$$\frac{1+\cos\theta}{2} = \cos^2\theta \qquad \text{Square both sides}$$

$$1 + \cos\theta = 2\cos^2\theta \qquad \text{Multiply both sides by 2}$$

$$2\cos^2\theta - \cos\theta - 1 = 0 \qquad \text{Rewrite in standard form}$$

$$(2\cos\theta + 1)(\cos\theta - 1) = 0 \qquad \text{Factor}$$

$$2\cos+1 = 0 \quad \text{or} \quad \cos\theta - 1 = 0 \qquad \text{Set each factor} = 0$$

$$2\cos\theta = -1 \qquad\qquad \cos\theta = 1 \qquad \text{Solve each equation and check}$$

$$\cos\theta = -\frac{1}{2} \qquad\qquad \theta = 0° \qquad \text{Possible solutions}$$

$$\theta = 120° \text{ or } 240°$$

We check all the possible solutions and 0° and 240° check.

9.

$$4\cos 2\theta + 2\sin\theta = 1$$

$$4\left(1 - 2\sin^2\theta\right) + 2\sin\theta = 1 \qquad \text{Double-angle identity}$$

$$4 - 8\sin^2\theta + 2\sin\theta = 1 \qquad \text{Simplify}$$

$$-8\sin^2\theta + 2\sin\theta + 3 = 0 \qquad \text{Subtract 1 from both sides}$$

$$8\sin^2\theta - 2\sin\theta - 3 = 0 \qquad \text{Multiply both sides by } -1$$

$$(4\sin\theta - 3)(2\sin\theta + 1) = 0 \qquad \text{Factor}$$

$$4\sin\theta - 3 = 0 \quad \text{or} \quad 2\sin\theta + 1 = 0 \qquad \text{Set each factor} = 0$$

$$4\sin\theta = 3 \qquad\qquad 2\sin\theta = -1 \qquad \text{Solve each equation}$$

$$\sin\theta = \frac{3}{4} \qquad\qquad \sin\theta = -\frac{1}{2}$$

$$\theta = 48.6° \text{ or } 131.4° \quad \theta = 210° \text{ or } 330°$$

10.

$$\sin\left(3\theta - 45°\right) = -\frac{\sqrt{3}}{2} \qquad \text{(The reference angle is } 60°)$$

$$3\theta - 45° = 240° + 360°k \quad \text{or} \quad 3\theta - 45° = 300° + 360°k$$

$$3\theta = 285° + 360°k \qquad\qquad 3\theta = 345° + 360°k$$

$$\theta = 95° + 120°k \qquad\qquad \theta = 115° + 120°k$$

Let $k = 0$, 1, and 2, then

$$\theta = 95° \qquad\qquad \theta = 115°$$

$$\theta = 215° \qquad\qquad \theta = 235°$$

$$\theta = 335° \qquad\qquad \theta = 355°$$

11.

$$\sin\theta + \cos\theta = 1$$

$\sin\theta = 1 - \cos\theta$	Subtract $\cos\theta$ from both sides
$\sin^2\theta = 1 - 2\cos + 2\cos^2\theta$	Square both sides
$1 - \cos^2\theta = 1 - 2\cos\theta + \cos^2\theta$	Pythagorean identity
$2\cos^2\theta - 2\cos\theta = 0$	Rewrite in standard form
$2\cos\theta(\cos\theta - 1) = 0$	Factor
$\cos\theta = 0$ or $\cos\theta - 1 = 0$	Set each factor $= 0$
$\theta = 90°$ or $270°$ $\cos\theta = 1$	Solve each equation and check
$\theta = 0°$	

We check the possible solutions and $0°$ and $90°$ check.

12.

$$\sin\theta - \cos\theta = 1$$

$\sin\theta = 1 + \cos\theta$	Add $\cos\theta$ to both sides
$\sin^2\theta = 1 + 2\cos\theta + \cos^2\theta$	Square both sides
$1 - \cos^2\theta = 1 + 2\cos\theta + \cos^2\theta$	Pythagorean identity
$2\cos^2\theta + 2\cos\theta = 0$	Rewrite in standard form
$2\cos\theta(\cos\theta + 1) = 0$	Factor
$\cos\theta = 0$ or $\cos\theta + 1 = 0$	Set each factor $= 0$
$\theta = 90°$ or $270°$ $\cos\theta = -1$	Solve each equation and check
$\theta = 180°$	

We check the possible solutions and $90°$ and $180°$ check.

13. $\cos 3\theta = -\dfrac{1}{2}$ (Reference angle is $60°$)

$3\theta = 120° + 360°k$ or $3\theta = 240° + 360°k$

$\theta = 40° + 120°k$ $\theta = 80° + 120°k$

We let $k = 0$, 1, and 2:

$\theta = 40°$ $\theta = 80°$

$\theta = 160°$ $\theta = 200°$

$\theta = 280°$ $\theta = 320°$

14. $\tan 2\theta = 1$ (Reference angle is $45°$)

$2\theta = 45° + 360°k$ or $2\theta = 225° + 360°k$

$\theta = 22.5° + 180°k$ $\theta = 112.5° + 180°k$

We let $k = 0$ and 1:

$\theta = 22.5°$ $\theta = 112.5°$

$\theta = 202.5°$ $\theta = 292.5°$

15.
$$\cos 2x - 3\cos x = -2$$

$2\cos^2 x - 1 - 3\cos x = -2$	Double-angle identity
$2\cos^2 x - 3\cos + 1 = 0$	Add 2 to both sides
$(2\cos x - 1)(\cos x - 1) = 0$	Factor
$2\cos x - 1 = 0 \quad \text{or} \quad \cos x - 1 = 0$	Set each factor = 0
$2\cos x = 1 \qquad\qquad \cos x = 1$	Solve each equation

$$x = 0 + 2k\pi$$

$$\cos x = \frac{1}{2} \qquad\qquad x = 2k\pi$$

$$x = \frac{\pi}{3} + 2k\pi \text{ or}$$

$$x = \frac{5\pi}{3} + 2k\pi$$

16.
$$\sqrt{3}\sin x - \cos x = 0$$

$\sqrt{3}\sin x = \cos x$	Add $\cos x$ to both sides
$\sqrt{3}\dfrac{\sin x}{\cos x} = 1$	Divide both sides by $\cos x$ $\quad(\cos x \neq 0)$
$\sqrt{3}\tan x = 1$	Ratio identity
$\tan x = \dfrac{1}{\sqrt{3}}$	Divide both sides by $\sqrt{3}$.
$x = \dfrac{\pi}{6} + k\pi$	(This answer can also be written as $\dfrac{\pi}{6} + 2k\pi$ or $\dfrac{7\pi}{6} + 2k\pi$)

17.
$$\sin 2x \cos x + \cos 2x \sin x = -1$$

$\sin(2x + x) = -1$	Sum identity
$\sin 3x = -1$	Simplify

$$3x = \frac{3\pi}{2} + 2k\pi$$

$$x = \frac{\pi}{2} + \frac{2k\pi}{3}$$

18.
$$\sin^3 4x = 1$$

$\sin 4x = 1$	Take cube root of both sides

$$4x = \frac{\pi}{2} + 2k\pi$$

$$x = \frac{\pi}{8} + \frac{k\pi}{2}$$

19.

$$5\sin^2\theta - 3\sin\theta = 2$$

$$5\sin^2\theta - 3\sin\theta - 2 = 0 \qquad \text{Subtract 2 from both sides}$$

$$(5\sin\theta + 2)(\sin\theta - 1) = 0 \qquad \text{Factor}$$

$$5\sin\theta + 2 = 0 \quad \text{or} \quad \sin\theta - 1 = 0 \qquad \text{Set each factor} = 0$$

$$5\sin\theta = -2 \qquad \qquad \sin\theta = 1 \qquad \text{Solve each equation}$$

$$\sin\theta = -\frac{2}{5} \qquad \qquad \theta = 90°$$

$$\sin\theta = -0.4$$

$$\hat{\theta} = 23.6°$$

$$\theta = 203.6° \text{ or } 336.4°$$

20.

$$4\cos^2\theta - 4\cos\theta = 2$$

$$4\cos^2\theta - 4\cos\theta - 2 = 0 \qquad \text{Subtract 2 from both sides}$$

$$2\cos^2\theta - 2\cos\theta - 1 = 0 \qquad \text{Divide both sides by 2}$$

$$\cos\theta = \frac{-(-2) \pm \sqrt{(-2)^2 - 4(2)(-1)}}{2(2)} \qquad a = 2, b = -2, c = -1$$

$$= \frac{2 \pm \sqrt{12}}{4}$$

$$\cos\theta = \frac{2 \pm 3.4641}{4}$$

$$\cos\theta = 1.3660 \qquad \text{or} \qquad \cos\theta = -0.3660$$

$$\text{No solution} \qquad \qquad \theta = 111.5° \text{ or } 248.5°$$

21. Graph $y_1 = 3\sin(x) - 2$ on your graphing calculator. Use the zero or root finder to locate the zeros of the function. The solution is 0.7297 or 2.4419.

22. Graph $y_1 = \cos(x) - 4\sin(x) + 3$ on your graphing calculator. Use the zero or root finder to locate the zeros of the function. The solution is 1.0598 or 2.5717.

23. Graph $y_1 = (\sin(x))^2 + 3\sin(x) - 1$ on your graphing calculator. Use the zero or root finder to locate the zeros of the function. The solution is 0.3076 or 2.8340.

24. Graph $y_1 = \sin(2x) - \frac{3}{5}$ on your graphing calculator. Use the zero or root finder to locate the zeros of the function. The solution is 0.3218, 1.2490, 3.4633, or 4.3906.

25. We let $h = 150$ ft and solve for t:

$$150 = 139 - 125\cos\frac{\pi}{10}t$$

$$11 = -125\cos\frac{\pi}{10}t$$

$$\cos\frac{\pi}{10}t = -\frac{11}{125} \qquad\qquad \text{(Reference angle is 1.48)}$$

$$\frac{\pi}{10}t = \pi - 1.48 \qquad \text{or} \qquad \frac{\pi}{10}t = \pi + 1.48$$

$$t = \frac{10}{\pi}(\pi - 1.48) \qquad\qquad t = \frac{10}{\pi}(\pi + 1.48)$$

$$t = 5.3 \text{ min} \qquad\qquad\qquad t = 14.7 \text{ min}$$

26. $\quad 3\cos t = x \qquad$ and $\qquad 3\sin t = y$

$$\cos t = \frac{x}{3} \qquad\qquad \sin t = \frac{y}{3}$$

$\cos^2 t + \sin^2 t = 1 \qquad$ Pythagorean identity

$\left(\dfrac{x}{3}\right)^2 + \left(\dfrac{y}{3}\right)^2 = 1 \qquad$ Substitute known values

$\dfrac{x^2}{9} + \dfrac{y^2}{9} = 1 \qquad$ Simplify

$$x^2 + y^2 = 9$$

The graph is a circle with a center at $(0, 0)$ and $r = 3$.

27. $\quad \sec t = x \qquad$ and $\qquad \tan t = y$

$$\sec^2 t = \tan^2 t + 1$$

$$x^2 = y^2 + 1$$

$$x^2 - y^2 = 1$$

The graph is a hyperbola with center at $(0, 0)$, vertices at $(\pm 1, 0)$, and asymptotes of $y = x$ and $y = -x$.

28. $\quad 3 + 2\sin t = x \qquad\qquad 1 + 2\cos t = y$

$$2\sin t = x - 3 \qquad\qquad 2\cos t = y - 1$$

$$\sin t = \frac{x-3}{2} \qquad\qquad \cos t = \frac{y-1}{2}$$

$$\sin^2 t + \cos^2 t = 1$$

$$\left(\frac{x-3}{2}\right)^2 + \left(\frac{y-1}{2}\right)^2 = 1$$

This problem is continued on the next page.

$$\frac{(x-3)^2}{4}+\frac{(y-1)^2}{4}=1$$
$$(x-3)^2+(y-1)^2=4$$

The graph is a circle with center at (3, 1) and $r=2$.

29.

$$3\cos t - 3 = x \qquad \text{and} \qquad 3\sin t + 1 = y$$
$$3\cos t = x + 3 \qquad\qquad 3\sin t = y - 1$$
$$\cos t = \frac{x+3}{3} \qquad\qquad \sin t = \frac{y-1}{3}$$
$$\cos^2 t + \sin^2 t = 1$$
$$\left(\frac{x+3}{3}\right)^2 + \left(\frac{y-1}{3}\right)^2 = 1$$
$$\frac{(x+3)^2}{9} + \frac{(y-1)^2}{9} = 1$$

$$(x+3)^2 + (y-1)^2 = 9$$

The graph is a circle with center at $(-3, 1)$ and $r=3$.

30. From the diagram at the right,

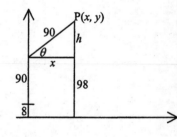

$$\cos\theta = \frac{x}{90} \qquad\qquad \sin\theta = \frac{h}{90}$$
$$x = 90\cos\theta \qquad\qquad h = 90\sin\theta$$
$$\qquad\qquad\qquad\qquad y = 98 + 90\sin\theta$$

Next, we write θ in terms of t using angular velocity:

$$\omega = \frac{1\,\text{rev}}{3\,\text{min}} \cdot \frac{2\pi\,\text{rad}}{1\,\text{rev}} = \frac{2\pi}{3}\frac{\text{rad}}{\text{min}}$$

$$\omega = \frac{\theta}{t} \qquad\qquad \text{We want the starting point to be at (0, 0). So we must}$$

$$\frac{2\pi}{3} = \frac{\theta}{t} \qquad\qquad \text{subtract } \frac{\pi}{2} \text{ from } \frac{2\pi}{3}t.$$

$$\theta = \frac{2\pi}{3}t$$

Therefore:
$$x = 90\cos\left(\frac{2\pi}{3}t - \frac{\pi}{2}\right)$$
$$y = 98 + 90\sin\left(\frac{2\pi}{3}t - \frac{\pi}{2}\right)$$

Check on your graphing calculator.

CHAPTER 7   Triangles

Problem Set 7.1

1.
$$\frac{b}{\sin B} = \frac{a}{\sin A}$$ Law of Sines

$$\frac{b}{\sin 60^\circ} = \frac{12}{\sin 40^\circ}$$ Substitute known values

$$b = \frac{12 \sin 60^\circ}{\sin 40^\circ}$$ Multiply both sides by sin 60°

$$= 16 \text{ cm}$$ Round to 2 significant digits

3.
$$\frac{b}{\sin B} = \frac{c}{\sin C}$$ Law of Sines

$$\frac{b}{\sin 120^\circ} = \frac{28}{\sin 20^\circ}$$ Substitute known values

$$b = \frac{28 \sin 120^\circ}{\sin 20^\circ}$$ Multiply both sides by sin 120°

$$= 71 \text{ inches}$$ Round to 2 significant digits

5.
$$\frac{c}{\sin C} = \frac{a}{\sin A}$$ Law of Sines

$$\frac{c}{\sin 100^\circ} = \frac{24}{\sin 10^\circ}$$ Substitute known values

$$c = \frac{24 \sin 100^\circ}{\sin 10^\circ}$$ Multiply both sides by sin 100°

$$= 140 \text{ yards}$$ Round to 2 significant digits

7.
$$C = 180^\circ - (A + B)$$
$$= 180^\circ - (50^\circ + 60^\circ)$$
$$= 180^\circ - 110^\circ = 70^\circ$$

$$\frac{c}{\sin C} = \frac{a}{\sin A}$$ Law of Sines

$$\frac{c}{\sin 70^\circ} = \frac{36}{\sin 50^\circ}$$ Substitute known values

$$c = \frac{36 \sin 70^\circ}{\sin 50^\circ}$$ Multiply both sides by sin 70°

$$= 44 \text{ km}$$ Round to 2 significant digits

9. $C = 180° - (A + B)$
$= 180° - (52° + 48°)$
$= 180° - 100° = 80°$

$\dfrac{a}{\sin A} = \dfrac{c}{\sin C}$ Law of Sines

$\dfrac{a}{\sin 52°} = \dfrac{14}{\sin 80°}$ Substitute known values

$a = \dfrac{14 \sin 52°}{\sin 80°}$ Multiply both sides by sin 52°

$= 11$ cm   Round to 2 significant digits

11. $C = 180° - (A + B)$
$= 180° - (42.5° + 71.4°)$
$= 180° - 113.9° = 66.1°$

$\dfrac{c}{\sin C} = \dfrac{a}{\sin A}$ $\dfrac{b}{\sin B} = \dfrac{a}{\sin A}$

$\dfrac{c}{\sin 66.1°} = \dfrac{215}{\sin 42.5°}$ $\dfrac{b}{\sin 71.4°} = \dfrac{215}{\sin 42.5°}$

$c = \dfrac{215 \sin 66.1°}{\sin 42.5°}$ $b = \dfrac{215 \sin 71.4°}{\sin 42.5°}$

$= 291$ inches $= 302$ inches

13. $C = 180° - (A + B)$
$= 180° - (46° + 95°)$
$= 180° - 141° = 39°$

$\dfrac{a}{\sin A} = \dfrac{c}{\sin C}$ $\dfrac{b}{\sin B} = \dfrac{c}{\sin C}$

$\dfrac{a}{\sin 46°} = \dfrac{6.8}{\sin 39°}$ $\dfrac{b}{\sin 95°} = \dfrac{6.8}{\sin 39°}$

$a = \dfrac{6.8 \sin 46°}{\sin 39°}$ $b = \dfrac{6.8 \sin 95°}{\sin 39°}$

$= 7.8$ meters $= 11$ meters

15. $B = 180° - (A + C)$
$= 180° - (43.5° + 120.5°)$
$= 180° - 164° = 16°$

This problem is continued on the next page

$$\frac{b}{\sin B} = \frac{a}{\sin A} \qquad\qquad \frac{c}{\sin C} = \frac{a}{\sin A}$$

$$\frac{b}{\sin 16^\circ} = \frac{3.48}{\sin 43.5^\circ} \qquad\qquad \frac{c}{\sin 120.5^\circ} = \frac{3.48}{\sin 43.5^\circ}$$

$$b = \frac{3.48\sin 16^\circ}{\sin 43.5^\circ} = 1.39 \text{ ft} \qquad\qquad c = \frac{3.48\sin 120.5^\circ}{\sin 43.5^\circ} = 4.36 \text{ ft}$$

17. $A = 180^\circ - (B + C)$

$ = 180^\circ - (13.4^\circ + 24.8^\circ)$

$ = 180^\circ - 38.2^\circ = 141.8^\circ$

$$\frac{b}{\sin B} = \frac{a}{\sin A} \qquad\qquad \frac{c}{\sin C} = \frac{b}{\sin B}$$

$$\frac{b}{\sin 13.4^\circ} = \frac{315}{\sin 141.8^\circ} \qquad\qquad \frac{c}{\sin 24.8^\circ} = \frac{315}{\sin 141.8^\circ}$$

$$b = \frac{315\sin 13.4^\circ}{\sin 141.8^\circ} \qquad\qquad c = \frac{315\sin 24.8^\circ}{\sin 141.8^\circ}$$

$$= 118 \text{ cm} \qquad\qquad\qquad = 214 \text{ cm}$$

19.
$$\frac{\sin B}{b} = \frac{\sin A}{a}$$

$$\frac{\sin B}{20} = \frac{\sin 30^\circ}{2}$$

$$\sin B = \frac{20\sin 30^\circ}{2} = 5$$

This is impossible because the sine function must be between -1 and 1.

21. $s = r\theta \quad (\theta \text{ is } \angle C)$ \qquad\qquad Arc length formula

$11 = 12 \cdot \theta$ \qquad\qquad\qquad Substitute known values

$\theta = \dfrac{11}{12}$ \qquad\qquad\qquad\qquad Divide both sides by 12

$\angle C = \dfrac{11}{12}$ radians. Converting this to degrees, we get:

$\angle C = \left(\dfrac{11}{12} \cdot \dfrac{180}{\pi} \right)^\circ = 53^\circ$ \qquad Also: $D = 180^\circ - (C + A)$

$\phantom{\angle C = \left(\dfrac{11}{12} \cdot \dfrac{180}{\pi} \right)^\circ = 53^\circ \qquad} = 180^\circ - (53^\circ + 31^\circ)$

$\phantom{\angle C = \left(\dfrac{11}{12} \cdot \dfrac{180}{\pi} \right)^\circ = 53^\circ \qquad} = 180^\circ - 84^\circ = 96^\circ$

This problem is continued on the next page

Using the Law of Sines, we find x:

$$\frac{x+r}{\sin D} = \frac{r}{\sin A}$$

$$\frac{x+12}{\sin 96°} = \frac{12}{\sin 31°}$$

$$x+12 = \frac{12\sin 96°}{\sin 31°}$$

$$x+12 = 23$$

$$x = 11$$

23.

$$s = r\theta \quad (\theta \text{ is } \angle C) \qquad \text{Arc length formula}$$

$$18 = 15 \cdot \theta \qquad\qquad\qquad \text{Substitute known values}$$

$$\theta = \frac{18}{15} = \frac{6}{5} \qquad\qquad\quad \text{Divide both sides by 12}$$

$\angle C = \dfrac{6}{5}$ radians. Converting this to degrees, we get:

$$\angle C = \left(\frac{6}{5} \cdot \frac{180}{\pi}\right)^° = 69° \qquad \text{Also: } D = 180° - (C + A)$$

$$= 180° - (69° + 45°)$$

$$= 180° - 114° = 66°$$

Using the Law of Sines, we find y:

$$\frac{y}{\sin C} = \frac{r}{\sin A} \qquad\qquad y = \frac{15\sin 69°}{\sin 45°}$$

$$\frac{y}{\sin 69°} = \frac{15}{\sin 45°} \qquad\qquad y = 20$$

25. We find the missing angles first:

$$\angle ABD = 180° - 64° = 116°$$

$$\angle ADB = 180° - (46° + 116°) = 18°$$

Now we find BD using the Law of Sines:

$$\frac{BD}{\sin A} = \frac{AB}{\sin ADB}$$

$$\frac{BD}{\sin 46°} = \frac{100}{\sin 18°}$$

$$BD = \frac{100\sin 46°}{\sin 18°} = 233$$

Then we find h, using the sine ratio:

$$\sin 64° = \frac{h}{233}$$

$$h = 233\sin 64° = 209 \text{ feet}$$

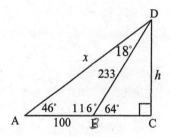

27. First, we'll find C:

$$C = 180° - (36° + 35°)$$

$$= 180° - 71° = 109°$$

Next, we'll find the distance that the man travels from A to B using the distance formula:

$$d = r \cdot t$$

$$= 5\frac{\text{ft}}{\text{sec}} \cdot 90 \text{ sec} = 450 \text{ ft}$$

Now we use the Law of Sines to find x:

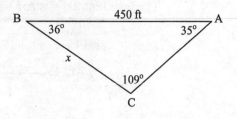

$$\frac{x}{\sin 35°} = \frac{450}{\sin 109°}$$

$$x = \frac{450 \sin 35°}{\sin 109°} = 273 \text{ ft}$$

29. We find the missing angles first:

$$\angle ADB = 90° - 35° = 55°$$

$$\angle ADC = 180° - 55° = 125°$$

$$\angle C = 180° - (125° - 0.5°) = 54.5°$$

Now we find AD using the Law of Sines:

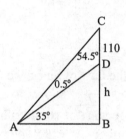

$$\frac{AD}{\sin 54.5°} = \frac{110}{\sin 0.5°}$$

$$AD = \frac{110 \sin 54.5}{\sin 0.5} = 10,262$$

Then we find h, using the sine ratio:

$$\sin 35° = \frac{h}{10,262}$$

$$h = 10,262 \sin 35° = 5,900 \text{ ft}$$

31. Using triangle ABC on the ground, $C = 180° - (105° + 44°) = 31°$.

Now we can use the Law of Sines to find b:

$$\frac{b}{\sin 44°} = \frac{25}{\sin 31°}$$

$$b = \frac{25 \sin 44°}{\sin 31°} = 34$$

Then we can find h using the tangent ratio:

$$\tan 51° = \frac{h}{34}$$

$$h = 34 \tan 51° = 42 \text{ ft}$$

33. We find the missing angle first: $\angle C = 180° - (53° + 31°) = 96°$

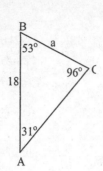

Then we find the missing sides using the Law of Sines:

$$\frac{18}{\sin 96°} = \frac{a}{\sin 31°} \qquad\qquad \frac{18}{\sin 96°} = \frac{b}{\sin 53°}$$

$$a = \frac{18 \sin 31°}{\sin 96°} = 9.3 \text{ miles} \qquad b = \frac{18 \sin 53°}{\sin 96°} = 14 \text{ miles}$$

35. We can redraw the two tension vectors **AC** and **BC** and the vector **W** due to gravity. We know that the magnitude of **W** is 125 pounds.

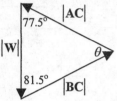

First, we find the missing angle:

$$\theta = 180° - (77.5° + 81.5°) = 21°$$

Then, we find $|\mathbf{AC}|$ and $|\mathbf{BC}|$ using the Law of Sines:

$$\frac{|\mathbf{AC}|}{\sin 81.5°} = \frac{125}{\sin 21°} \qquad\qquad \frac{|\mathbf{BC}|}{\sin 77.5°} = \frac{125}{\sin 21°}$$

$$|\mathbf{AC}| = \frac{125 \sin 81.5°}{\sin 21°} = 345 \text{ lbs} \qquad |\mathbf{BC}| = \frac{125 \sin 77.5°}{\sin 21°} = 341 \text{ lbs}$$

37. We can redraw the two tension vectors **AC** and **BC** and the vector **W** due to gravity. We know that the magnitude of **W** is 1850 pounds.

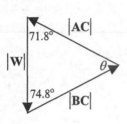

First, we find the missing angle:

$$\theta = 180° - (71.8° + 74.8°) = 33.4°$$

Then, we find $|\mathbf{AC}|$ and $|\mathbf{BC}|$ using the Law of Sines:

$$\frac{|\mathbf{AC}|}{\sin 71.8°} = \frac{1850}{\sin 33.4°} \qquad\qquad \frac{|\mathbf{BC}|}{\sin 74.8°} = \frac{1850}{\sin 33.4°}$$

$$|\mathbf{AC}| = \frac{1850 \sin 71.8°}{\sin 33.4°} = 3190 \text{ lbs} \qquad |\mathbf{BC}| = \frac{1850 \sin 74.8°}{\sin 33.4°} = 3240 \text{ lbs}$$

39.
$$2\sin\theta - \sqrt{2} = 0$$
$$2\sin\theta = \sqrt{2}$$
$$\sin\theta = \frac{\sqrt{2}}{2}$$
$$\hat{\theta} = 45°$$
$$\theta = 45° \text{ or } 135°$$

41.
$$\sin\theta\cos\theta - 2\cos\theta = 0$$
$$\cos\theta(\sin\theta - 2) = 0$$
$$\sin\theta = 2 \text{ or } \cos\theta = 0$$

No solution $\theta = 90° \text{ or } 270°$

43.

$$2\sin^2\theta - 3\sin\theta + 1 = 0$$
$$(2\sin\theta - 1)(\sin\theta - 1) = 0$$
$$2\sin\theta - 1 = 0 \quad \text{or} \quad \sin\theta - 1 = 0$$
$$\sin\theta = \frac{1}{2} \qquad \sin\theta = 1$$
$$\theta = 30° \text{ or } 150° \quad \theta = 90°$$

45.

$$\cos^2\theta - 4\cos\theta + 2 = 0 \quad \text{where } a = 1, b = -4, \text{ and } c = 2$$
$$\cos\theta = \frac{-(-4) \pm \sqrt{(-4)^2 - 4(1)(2)}}{2(1)}$$
$$= \frac{4 \pm \sqrt{8}}{2}$$
$$\cos\theta = 3.4142 \quad \text{or} \quad \cos\theta = 0.5858$$
$$\text{No solution} \qquad \theta = 54.1° \text{ or } 305.9°$$

47.

$$(\sin x + 1)(2\sin x - 1) = 0$$
$$\sin x + 1 = 0 \quad \text{or} \qquad\qquad 2\sin x - 1 = 0$$
$$\sin x = -1 \qquad\qquad\qquad \sin x = \frac{1}{2}$$
$$x = \frac{3\pi}{2} + 2k\pi \qquad\qquad x = \frac{\pi}{6} + 2k\pi \text{ or } x = \frac{5\pi}{6} + 2k\pi$$

49.

$$\sin\theta = 0.7380$$
$$\hat{\theta} = 47.6° \text{ and } \theta \text{ is in QI or QII}$$
$$\theta = 47.6° \text{ or } 132.4°$$

51.

$$\sin\theta = 0.9668$$
$$\hat{\theta} = \sin^{-1}(0.9668) = 75.2°$$
$$\theta = 75.2° \text{ or } 104.8°$$

Problem Set 7.2

1.
$$\sin B = \frac{b \sin A}{a}$$
$$= \frac{40 \sin 30°}{10} = 2$$
Since $\sin B$ can never be greater than 1, no triangle exists.

3.
$$\sin B = \frac{b \sin A}{a}$$
$$= \frac{20 \sin 120°}{30}$$
$$= 0.5774$$
$$B = 35°$$
Only one triangle is possible because there can only be one obtuse angle in a triangle.

5.
$$\sin B = \frac{b \sin A}{a}$$
$$= \frac{18 \sin 60°}{16}$$
$$= 0.9743$$
$$B = 77° \text{ or } B' = 180° - 77° = 103°$$
Since $a < b$, there are 2 possible triangles.

7.
$$\sin B = \frac{b \sin A}{a}$$
$$= \frac{54 \sin 38°}{41}$$
$$= 0.8109$$
$$B = 54° \text{ or } B^1 = 180° - 54° = 126°$$

Since $a < b$, there are 2 possible triangles.
$$C = 180° - (38° + 54°) = 88° \qquad C' = 180° - (38° + 126°) = 16°$$
$$c = \frac{a \sin C}{\sin A} \qquad\qquad c' = \frac{a \sin C'}{\sin A}$$
$$= \frac{41 \sin 88°}{\sin 38°} = 67 \text{ ft} \qquad = \frac{41 \sin 16°}{\sin 38°} = 18 \text{ ft}$$

9.
$$\sin B = \frac{b \sin A}{a}$$
$$= \frac{22.3 \sin 112.2°}{43.8}$$
$$= 0.4714$$
$$B = 28.1°$$
There is only one triangle possible, because there can only be one obtuse angle in a triangle.

$$C = 180° - (112.2° + 28.1°)$$
$$= 180° - 140.3° = 39.7°$$
$$c = \frac{a \sin C}{\sin A}$$
$$= \frac{43.8 \sin 39.7°}{\sin 112.2°}$$
$$= 30.2 \text{ cm}$$

11.

$$\sin B = \frac{b \sin C}{c}$$

$$= \frac{425 \sin 27°50'}{347}$$

$$= 0.5719$$

$B = 34°50'$ or $B' = 180° - 34°50' = 145°10'$ There are two triangles possible because c < b.

$A = 180° - (34°50' + 27°50')$ $A' = 180° - (145°10' + 27°50')$

$\quad = 180° - 62°40' = 117°20'$ $\quad = 180° - 173° = 7°$

$a = \dfrac{c \sin A}{\sin C}$ $a' = \dfrac{c \sin A'}{\sin C}$

$\quad = \dfrac{347 \sin 117°20'}{\sin 27°50'}$ $\quad = \dfrac{347 \sin 7°}{\sin 27°50'}$

$\quad = 660 \text{ m}$ $\quad = 90.6 \text{ m}$

13.

$$\sin C = \frac{c \sin B}{b}$$

$$= \frac{1.12 \sin 45°10'}{1.79}$$

$$= 0.4437$$

$C = 26.3°$ or $C' = 180° - 26.3°$

$C = 26°20' \quad C' = 153.7°$

$\qquad$ This is impossible because
$\qquad 45.2° + 153.7° = 198.9°$.

$A = 180° - (45°10' + 26°20') = 108°30'$

$a = \dfrac{b \sin A}{\sin B}$

$\quad = \dfrac{1.79 \sin 108°30'}{\sin 45°10'}$

$\quad = 2.39 \text{ in}$

15.

$$\sin A = \frac{a \sin B}{b}$$

$$= \frac{0.92 \sin 118°}{0.68}$$

$$= 1.1946$$

Since sin A can never be greater than 1, no triangle exists.

17.

$$\sin B = \frac{b \sin A}{a}$$

$$= \frac{2.9 \sin 142°}{1.4}$$

$$= 1.2753$$

Since sin B can never be greater than 1, no triangle exists.

19.

$$\sin B = \frac{b \sin C}{c}$$

$$= \frac{36.8 \sin 26.8°}{36.8}$$

$$= 0.4509$$

Only one triangle is possible because
$b = c$. This is an isosceles triangle.

$$B = 26.8°$$

$$A = 180° - (26.8° + 26.8°) = 126.4°$$

$$a = \frac{c \sin A}{\sin C}$$

$$= \frac{36.8 \sin 126.4°}{\sin 26.8°} = 65.7 \text{ km}$$

21.

$$\sin C = \frac{c \sin A}{a}$$

$$= \frac{50 \sin 58°}{44}$$

$$= 0.9637$$

$$C = 75° \quad \text{or} \quad C' = 105°$$

$$B = 180° - (58° + 75°)$$

$$= 180° - 133° = 47°$$

$$b = \frac{44 \sin 47°}{\sin 58°}$$

$$= 38 \text{ feet}$$

$$B' = 180° - (58° + 105°)$$

$$= 180° - 163° = 17°$$

$$b' = \frac{44 \sin 17°}{\sin 58°}$$

$$= 15 \text{ feet}$$

27.

$$\sin B = \frac{340 \sin 8°}{55}$$

$$= 0.8603$$

$$B = 59° \text{ or } B' = 121°$$

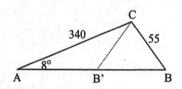

There are two triangles possible.
We want to find c and c':

$$C = 180° - (59° + 8°)$$

$$= 180° - 67° = 113°$$

$$c = \frac{55 \sin 113°}{\sin 8°}$$

$$= 360 \text{ mph}$$

$$C' = 180° - (121° + 8°)$$

$$= 180° - 129° = 51°$$

$$c' = \frac{55 \sin 51°}{\sin 8°}$$

$$= 310 \text{ mph}$$

29.

$$\sin C = \frac{12 \sin 30°}{6}$$

$$= 1$$

$$C = 90°$$

This is a 30°-60°-90° right triangle.
Therefore, $x = 6\sqrt{3}$ or 10 mph.

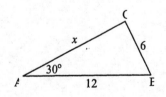

31.
$$\sin B = \frac{30\sin 48^\circ}{32}$$
$$= 0.5772$$
$$B = 44^\circ \quad or \quad B' = 136^\circ$$
$$\text{This is impossible}$$
$$C = 180^\circ - \left(48^\circ + 44^\circ\right) = 88^\circ$$

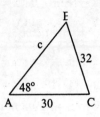

33.

$$4\sin\theta - \csc\theta = 0 \qquad\qquad (2\sin\theta - 1)(2\sin\theta + 1) = 0$$

$$4\sin\theta - \frac{1}{\sin\theta} = 0 \qquad\qquad 2\sin\theta - 1 = 0 \text{ or } 2\sin\theta + 1 = 0$$

$$4\sin^2\theta - 1 = 0, \quad \sin\theta \neq 0 \qquad \sin\theta = \frac{1}{2} \qquad\qquad \sin\theta = -\frac{1}{2}$$

$$\theta = 30^\circ \text{ or } 150^\circ \qquad \theta = 210^\circ \text{ or } 330^\circ$$

35.

$$2\cos\theta - \sin 2\theta = 0$$
$$2\cos\theta - 2\sin\theta\cos\theta = 0$$
$$2\cos\theta(1 - \sin\theta) = 0$$
$$2\cos\theta = 0 \quad or \quad 1 - \sin\theta = 0$$
$$\cos\theta = 0 \qquad\qquad \sin\theta = 1$$
$$\theta = 90^\circ \text{ or } 270^\circ \qquad \theta = 90^\circ$$

37.

$$18\sec^2\theta - 17\tan\theta\sec\theta - 12 = 0 \qquad\qquad 12\sin^2\theta - 17\sin\theta + 6 = 0$$

$$18\left(\frac{1}{\cos^2\theta}\right) - 17\left(\frac{\sin\theta}{\cos\theta}\right)\left(\frac{1}{\cos\theta}\right) - 12 = 0 \qquad (4\sin\theta - 3)(3\sin\theta - 2) = 0$$

$$18 - 17\sin\theta - 12\cos^2\theta = 0, \quad \cos\theta \neq 0 \qquad (4\sin\theta - 3)(3\sin\theta - 2) = 0$$

$$18 - 17\sin\theta - 12\left(1 - \sin^2\theta\right) = 0 \qquad 4\sin\theta - 3 = 0 \quad or \quad 3\sin\theta - 2 = 0$$

$$18 - 17\sin\theta - 12 + 12\sin^2\theta = 0 \qquad \sin\theta = \frac{3}{4} \qquad\qquad \sin\theta = \frac{2}{3}$$

$$12\sin^2\theta - 17\sin\theta + 6 = 0 \qquad \theta = 48.6^\circ \text{ or } 131.4^\circ \qquad \theta = 41.8^\circ \text{ or } 138.2^\circ$$

39.
$$2\cos x - \sec x + \tan x = 0$$
$$2\cos x - \frac{1}{\cos x} + \frac{\sin x}{\cos x} = 0$$
$$2\cos^2 x - 1 + \sin x = 0, \quad \cos x \neq 0$$
$$2(1 - \sin^2 x) - 1 + \sin x = 0$$
$$2 - 2\sin^2 x - 1 + \sin x = 0$$
$$2\sin^2 x - \sin x - 1 = 0$$
$$(2\sin x + 1)(\sin x - 1) = 0$$

$$\sin x - 1 = 0 \qquad \text{or} \qquad 2\sin x + 1 = 0$$

$$\sin x = 1 \qquad\qquad \sin x = -\frac{1}{2}$$

$$x = \frac{\pi}{2} + 2k\pi \qquad\qquad x = \frac{7\pi}{6} + 2k\pi \text{ or } \frac{11\pi}{6} + 2k\pi$$

This answer doesn't check.

41.
$$\sin x + \cos x = 0$$
$$\sin x = -\cos x$$
$$\frac{\sin x}{\cos x} = -1, \quad \cos x \neq 0$$
$$\tan x = -1$$

$$\hat{x} = \frac{\pi}{4} \quad \text{and } x \text{ is in QII or QIV}$$

$$x = \frac{3\pi}{4} + 2k\pi \text{ or } x = \frac{7\pi}{4} + 2k\pi$$

Problem Set 7.3

1.
$$c^2 = a^2 + b^2 - 2ab\cos C$$
$$= (120)^2 + (66)^2 - 2(120)(66)\cos 60°$$
$$= 10,836$$
$$c = 100 \text{ inches (rounded to 2 significant digits)}$$

3.
$$\cos C = \frac{c^2 - a^2 - b^2}{-2ab}$$
$$= \frac{26^2 - 22^2 - 24^2}{-2(22)(24)}$$
$$= 0.3636$$
$$C = 69° \text{ (The largest angle is opposite the longest side)}$$

5.
$$a^2 = b^2 + c^2 - 2bc\cos A$$
$$= (4.2)^2 + (6.8)^2 - 2(4.2)(6.8)\cos 116°$$
$$= 88.92$$
$$a = 9.4 \text{ meters}$$

7.
$$\cos A = \frac{a^2 - b^2 - c^2}{-2bc}$$
$$= \frac{38^2 - 10^2 - 31^2}{-2(10)(31)}$$
$$= -0.6177$$
$$A = 128°$$

9.

$$b^2 = a^2 + c^2 - 2ac\cos B$$

$$= (410)^2 + (340)^2 - 2(410)(340)\cos 151.5°$$

$$= 528,714.211$$

$$b = 727 \text{ meters}$$

$$\sin A = \frac{a\sin B}{b}$$

$$= \frac{410\sin 151.5°}{727}$$

$$= 0.2691$$

$$A = 15.6°$$

$$C = 180° - (15.6° + 151.5°) = 12.9°$$

11.

$$\cos C = \frac{c^2 - a^2 - b^2}{-2ab}$$

$$= \frac{0.75^2 - 0.48^2 - 0.63^2}{-2(0.48)(0.63)}$$

$$= 0.1071$$

$$C = 84°$$

$$\sin B = \frac{b\sin C}{c}$$

$$= \frac{0.63\sin 84°}{0.75}$$

$$= 0.8354$$

$$B = 57°$$

$$A = 180° - (57° + 84°) = 39°$$

13.

$$a^2 = b^2 + c^2 - 2bc\cos A$$

$$= (0.923)^2 + (0.387)^2 - 2(0.923)(0.387)\cos 43°20'$$

$$= 0.4821$$

$$a = 0.694 \text{ kilometers}$$

$$\sin C = \frac{c\sin A}{a}$$

$$= \frac{0.387\sin 43°20'}{0.694} = 0.3827$$

$$C = 22°30'$$

$$B = 180° - (22°30' + 43°20') = 114°10'$$

Note: *Answers may differ depending on which angle you solve for first.*

15.

$$\cos C = \frac{c^2 - a^2 - b^2}{-2ab}$$

$$= \frac{5.22^2 - 4.38^2 - 3.79^2}{-2(4.38)(3.79)}$$

$$= 0.1898$$

$$C = 79.1°$$

$$\sin B = \frac{b\sin C}{c}$$

$$= \frac{3.79\sin 79.1°}{5.22}$$

$$= 0.7130$$

$$B = 45.5°$$

$$A = 180° - (79.1° + 45.5°) = 55.4°$$

Note: *Answers may differ depending on which angle you solve for first.*

17.

$$a^2 = b^2 + c^2 - 2bc\cos A$$

$$a^2 = b^2 + c^2 - 2bc\cos 90°$$

$$a^2 = b^2 + c^2 - 2bc(0)$$

$$a^2 = b^2 + c^2$$

19. The diagonals of a parallelogram bisect each other.
The angle opposite side x is $180° - 120° = 60°$.
$$x^2 = 17^2 + 28^2 - 2(17)(28)\cos 60°$$
$$= 597$$
$$x = 24 \text{ inches}$$

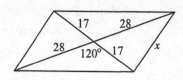

21.
$$\begin{aligned} d_1 &= r_1 t_1 & d_2 &= r_2 t_2 \\ &= 130(1.5) & &= 150(1.5) \\ &= 195 \text{ miles} & &= 225 \text{ miles} \end{aligned}$$

$$a^2 = b^2 + c^2 - 2bc\cos A$$
$$= (225)^2 + (195)^2 - 2(225)(195)\cos 36°$$
$$= 17658.76$$
$$a = 130 \text{ miles (rounded to 2 significant digits)}$$

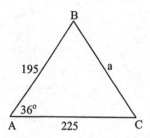

23. Distance of first plane is 246 (2) or 492 miles.
Distance of second plane is 357 (2) or 714 miles.
Angle between the two planes is $175° - 135°$ or $40°$.
We will use the Law of Cosines to find x:
$$x^2 = (492)^2 + (714)^2 - 2(492)(714)\cos 40°$$
$$= 213655.56$$
$$x = 462 \text{ miles (rounded to 3 significant digits)}$$

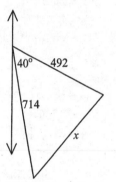

25.
$$|\mathbf{V} + \mathbf{W}|^2 = |\mathbf{V}|^2 + |\mathbf{W}|^2 - 2|\mathbf{V}||\mathbf{W}|\cos\theta$$
$$= (35)^2 + (160)^2 - 2(35)(160)\cos 165°$$
$$= 37,643$$
$$|\mathbf{V} + \mathbf{W}| = 190 \text{ mph (to 2 significant digits)}$$

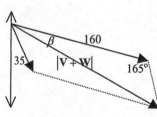

$$\sin\beta = \frac{35\sin 165°}{190}$$
$$= 0.0477$$
$$\beta = 3° \text{ (to the nearest degree)}$$
The true course is $150° + 3° = 153°$. The ground speed of the plane is 190 mph.

27. The angle between the plane vector and its airspeed is $34.0° - 30.0°$ or $4.0°$.
We will use the Law of Cosines to find the speed of the wind:

$$|\mathbf{w}|^2 = (195)^2 + (207)^2 - 2(195)(207)\cos 4.0°$$
$$= 340.65$$
$$|\mathbf{w}| = 18.5 \text{ mph}$$

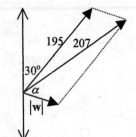

Next, we will use the Law of Sines to find α:

$$\sin\alpha = \frac{195\sin 4.0°}{18.5}$$
$$= 0.7353$$
$$\alpha = 47.3°$$

The wind is 18.5 mph at $34.0° + 47.3°$ or $81.3°$ from due north.

29.
$$x^2 = (47.0)^2 + (47.5)^2 - 2(47.0)(47.5)\cos 78.0°$$
$$= 3537$$
$$x = 59.5 \text{ cm}$$

The length of the down tube is 59.5 cm.

31.
$$x^2 = (52.3)^2 + (48.0)^2 - 2(52.3)(48.0)\cos 75.0°$$
$$= 3740$$
$$x = 61.2 \text{ cm}$$

$$\sin\theta = \frac{52.3\sin 75.0°}{61.2}$$
$$= 0.8281$$
$$\theta = 55.6°$$

The length of the down tube is 61.2 cm and the angle between the seat tube and the down tube is $55.6°$.

33. $\sin 3x = \dfrac{1}{2}$

$3x = \dfrac{\pi}{6} + 2k\pi$ or $3x = \dfrac{5\pi}{6} + 2k\pi$

$x = \dfrac{\pi}{18} + \dfrac{2k\pi}{3}$ $\qquad$ $x = \dfrac{5\pi}{18} + \dfrac{2k\pi}{3}$

35. $\tan^2 3x = 1$

$\tan 3x = 1$ or $\tan 3x = -1$

$3x = \dfrac{\pi}{4} + k\pi$ $\qquad$ $3x = \dfrac{3\pi}{4} + k\pi$

$x = \dfrac{\pi}{12} + \dfrac{k\pi}{3}$ $\qquad$ $x = \dfrac{\pi}{4} + \dfrac{k\pi}{3}$

37.
$$2\cos^2 3\theta - 9\cos 3\theta + 4 = 0$$
$$(2\cos 3\theta - 1)(\cos 3\theta - 4) = 0$$
$$\cos 3\theta - 4 = 0 \quad \text{or} \quad 2\cos 3\theta - 1 = 0$$
$$\cos 3\theta = 4 \qquad\qquad \cos 3\theta = \frac{1}{2}$$

No Solution
$$3\theta = 60° + 360° k \quad \text{or} \quad 3\theta = 300° + 360° k$$
$$\theta = 20° + 120° k \qquad \theta = 100° + 120° k$$

39.
$$\sin 4\theta \cos 2\theta + \cos 4\theta \sin 2\theta = -1$$
$$\sin(4\theta + 2\theta) = -1$$
$$\sin 6\theta = -1$$
$$6\theta = 270° + 360° k$$
$$\theta = 45° + 60° k$$

41.
$$\sin\theta + \cos\theta = 1$$
$$(\sin\theta + \cos\theta)^2 = 1^2$$
$$\sin^2\theta + 2\sin\theta\cos\theta + \cos^2\theta = 1$$
$$1 + \sin 2\theta = 1$$
$$\sin 2\theta = 0$$
$$2\theta = 0° \text{ or } 180°$$
$$\theta = 0° \text{ or } 90°$$

Both answers check.

43. We can use the Pythagorean Theorem to find the ground speed:
$$|\mathbf{g}| = \sqrt{176^2 + 45.5^2}$$
$$= \sqrt{33,046.25}$$
$$= 182 \text{ mph}$$
We can use the tangent ratio to find the course of the plane:
$$\tan\theta = \frac{45.5}{176}$$
$$\theta = \tan^{-1}(0.2585)$$
$$= 14.5°$$
The true course of the plane is 40.0° + 14.4° or 54.5° from due north and the ground speed is 182 mph.

45. We can use the Pythagorean Theorem to find the ground speed:
$$x = \sqrt{195^2 + 32.5^2}$$
$$= \sqrt{39,081.25}$$
$$= 198 \text{ mph}$$
We can use the tangent ratio to find the course of the plane:
$$\tan\theta = \frac{32.5}{195}$$
$$\theta = \tan^{-1}(0.1667)$$
$$= 9.5° \qquad\qquad \alpha = 30.0° - 9.5° \text{ or } 20.5°$$

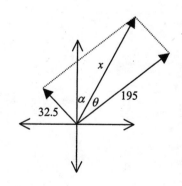

The true course of the plane is N 20.5° E and the ground speed is 198 mph..

Problem Set 7.4

1. $S = \dfrac{1}{2} ab \sin C$ Formula for area of a triangle

$\quad = \dfrac{1}{2}(50)(70)\sin 60°$ Substitute known values

$\quad = 1{,}520 \ \text{cm}^2$ Round to 3 significant digits

3. $S = \dfrac{1}{2} ac \sin B$ Formula for area of a triangle

$\quad = \dfrac{1}{2}(41.5)(34.5)\sin 151.5°$ Substitute known values

$\quad = 342 \ \text{m}^2$ Round to 3 significant digits

5. $S = \dfrac{1}{2} bc \sin A$ Formula for area of a triangle

$\quad = \dfrac{1}{2}(0.923)(0.387)\sin 43°20'$ Substitute known values

$\quad = 0.123 \ \text{km}^2$ Round to 3 significant digits

7. $C = 180° - (46° + 95°) = 39°$ $S = \dfrac{c^2 \sin A \sin B}{2 \sin C}$ Formula for area of a triangle

$\quad = \dfrac{(6.8)^2 \sin 46° \sin 95°}{2 \sin 39°}$ Substitute known values

$\quad = 26.3 \, \text{m}^2$ Round to 3 significant digits

9. $C = 180° - (42.5° + 71.4°) = 66.1°$ $S = \dfrac{a^2 \sin B \sin C}{2 \sin A}$ Formula for area of a triangle

$\quad = \dfrac{(210)^2 \sin 71.4° \sin 66.1°}{2 \sin 42.5°}$ Substitute known values

$\quad = 28{,}300 \, \text{in}^2$ Round to 3 significant digits

11. $B = 180° - (43°30' + 120°30') = 16°$ $S = \dfrac{a^2 \sin B \sin C}{2 \sin A}$ Formula for area of a triangle

$\quad = \dfrac{(3.48)^2 \sin 16° \sin 120.5°}{2 \sin 43.5°}$ Substitute known values

$\quad = 2.09 \ \text{ft}^2$ Round to 3 significant digits

13.

$$s = \frac{1}{2}(a+b+c)$$ Formula for half the perimeter

$$= \frac{1}{2}(44+66+88) = 99$$ Substitute known values and simplify

$$S = \sqrt{s(s-a)(s-b)(s-c)}$$ Formula for area of a triangle

$$= \sqrt{99(99-44)(99-66)(99-88)}$$ Substitute known values

$$= \sqrt{1,976,535}$$ Simplify

$$= 1,410 \text{ in}^2$$ Round to 3 significant digits

15.

$$s = \frac{1}{2}(a+b+c)$$ Formula for half the perimeter

$$= \frac{1}{2}(4.8+6.3+7.5) = 9.3$$ Substitute known values and simplify

$$S = \sqrt{s(s-a)(s-b)(s-c)}$$ Formula for area of a triangle

$$= \sqrt{9.3(9.3-4.8)(9.3-6.3)(9.3-7.5)}$$ Substitute known values

$$= \sqrt{225.99}$$ Simplify

$$= 15.0 \text{ yd}^2$$ Round to 3 significant digits

17.

$$s = \frac{1}{2}(a+b+c)$$ Formula for half the perimeter

$$= \frac{1}{2}(4.38+3.79+5.22) = 6.695$$ Substitute known values and simplify

$$S = \sqrt{s(s-a)(s-b)(s-c)}$$ Formula for area of a triangle

$$= \sqrt{6.695(6.695-4.38)(6.695-3.79)(6.695-5.22)}$$ Substitute known values

$$= \sqrt{66.41}$$ Simplify

$$= 8.15 \text{ ft}^2$$ Round to 3 significant digits

19. Area of the parallelogram is twice the area of the triangle:

$$\text{Area of parallelogram} = 2\left[\frac{1}{2}(12)(15)\sin 120°\right]$$

$$= 156 \text{ in}^2$$

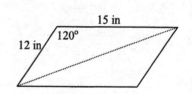

21. $C = 180° - (30° + 50°) = 100°$

$$S = \frac{c^2 \sin A \sin B}{2 \sin C}$$ Formula for area of a triangle

$$c^2 = \frac{2S \sin C}{\sin A \sin B}$$ Solve for c^2

$$= \frac{2(40) \sin 100°}{\sin 30° \sin 50°}$$ Substitute known values

$$= 205.69$$ Simplify

$$c = 14.3 \text{ cm}$$ Round to 3 significant digits

23. $\cos^2 t + \sin^2 t = 1$

$x^2 + y^2 = 1$

The graph is a circle with center at $(0,0)$ and radius of 1.

25. $3 + 2\sin t = x$ $1 + 2\cos t = y$ $\sin^2 t + \cos^2 t = 1$

$2\sin t = x - 3$ $2\cos t = y - 1$ $\left(\dfrac{x-3}{2}\right)^2 + \left(\dfrac{y-1}{2}\right)^2 = 1$

$\sin t = \dfrac{x-3}{2}$ $\cos t = \dfrac{y-1}{2}$ $\dfrac{(x-3)^2}{4} + \dfrac{(y-1)^2}{4} = 1$

$(x-3)^2 + (y-1)^2 = 4$

The graph is a circle with center at $(3,1)$ and radius of 2.

27. $2\tan t = x$ $3\sec t = y$ $\sec^2 t = \tan^2 t + 1$

$\tan t = \dfrac{x}{2}$ $\sec t = \dfrac{y}{3}$ $\left(\dfrac{y}{3}\right)^2 = \left(\dfrac{x}{2}\right)^2 + 1$

$\dfrac{y^2}{9} - \dfrac{x^2}{4} = 1$

29. $\cos 2t = y$ and $\sin t = x$

$\cos 2t = 1 - 2\sin^2 t$

$y = 1 - 2x^2$

Problem Set 7.5

17.   Let $\mathbf{v} = \langle -5, 6 \rangle$

$|\mathbf{v}| = \sqrt{(-5)^2 + 6^2}$

$= \sqrt{25 + 36}$

$= \sqrt{61}$

19.   Let $\mathbf{v} = \langle 2, 0 \rangle$

$|\mathbf{v}| = \sqrt{2^2 + 0^2}$

$= \sqrt{4}$

$= 2$

21. Let $\mathbf{v} = \langle -2, -5 \rangle$

$$|\mathbf{v}| = \sqrt{(-2)^2 + (-5)^2}$$
$$= \sqrt{4 + 25}$$
$$= \sqrt{29}$$

23.
$$|\mathbf{v}| = \sqrt{a^2 + b^2}$$
$$= \sqrt{3^2 + 4^2}$$
$$= \sqrt{9 + 16}$$
$$= \sqrt{25} = 5$$

25.
$$|\mathbf{u}| = \sqrt{a^2 + b^2}$$
$$= \sqrt{5^2 + (-12)^2}$$
$$= \sqrt{25 + 144}$$
$$= \sqrt{169} = 13$$

27.
$$|\mathbf{w}| = \sqrt{a^2 + b^2}$$
$$= \sqrt{1^2 + 2^2}$$
$$= \sqrt{1 + 4}$$
$$= \sqrt{5}$$

29.
$$\mathbf{u} + \mathbf{v} = \langle 4, 4 \rangle + \langle 4, -4 \rangle$$
$$= \langle 4 + 4, 4 + (-4) \rangle$$
$$= \langle 8, 0 \rangle$$

$$\mathbf{u} - \mathbf{v} = \langle 4, 4 \rangle - \langle 4, -4 \rangle$$
$$= \langle 4 - 4, 4 - (-4) \rangle$$
$$= \langle 0, 8 \rangle$$

$$2\mathbf{u} - 3\mathbf{v} = 2\langle 4, 4 \rangle - 3\langle 4, -4 \rangle$$
$$= \langle 8, 8 \rangle - \langle 12, -12 \rangle$$
$$= \langle -4, 20 \rangle$$

31.
$$\mathbf{u} + \mathbf{v} = \langle 2, 0 \rangle + \langle 0, -7 \rangle$$
$$= \langle 2 + 0, 0 + (-7) \rangle$$
$$= \langle 2, -7 \rangle$$

$$\mathbf{u} - \mathbf{v} = \langle 2, 0 \rangle - \langle 0, -7 \rangle$$
$$= \langle 2 - 0, 0 - (-7) \rangle$$
$$= \langle 2, 7 \rangle$$

$$2\mathbf{u} - 3\mathbf{v} = 2\langle 2, 0 \rangle - 3\langle 0, -7 \rangle$$
$$= \langle 4, 0 \rangle - \langle 0, -21 \rangle$$
$$= \langle 4, 21 \rangle$$

33.
$$\mathbf{u} + \mathbf{v} = \langle 4, 1 \rangle + \langle -5, 2 \rangle$$
$$= \langle 4 + (-5), 1 + 2 \rangle$$
$$= \langle -1, 3 \rangle$$

$$\mathbf{u} - \mathbf{v} = \langle 4, 1 \rangle - \langle -5, 2 \rangle$$
$$= \langle 4 - (-5), 1 - 2 \rangle$$
$$= \langle 9, -1 \rangle$$

$$2\mathbf{u} - 3\mathbf{v} = 2\langle 4, 1 \rangle - 3\langle -5, 2 \rangle$$
$$= \langle 8, 2 \rangle - \langle -15, 6 \rangle$$
$$= \langle 23, -4 \rangle$$

35.
$$\mathbf{u} + \mathbf{v} = (\mathbf{i} + \mathbf{j}) + (\mathbf{i} - \mathbf{j})$$
$$= 2\mathbf{i}$$

$$\mathbf{u} - \mathbf{v} = (\mathbf{i} + \mathbf{j}) - (\mathbf{i} - \mathbf{j})$$
$$= \mathbf{i} + \mathbf{j} - \mathbf{i} + \mathbf{j}$$
$$= 2\mathbf{j}$$

$$3\mathbf{u} + 2\mathbf{v} = 3(\mathbf{i} + \mathbf{j}) + 2(\mathbf{i} - \mathbf{j})$$
$$= 3\mathbf{i} + 3\mathbf{j} + 2\mathbf{i} - 2\mathbf{j}$$
$$= 5\mathbf{i} + \mathbf{j}$$

37.
$$\mathbf{u} + \mathbf{v} = (6\mathbf{i}) + (-8\mathbf{j})$$
$$= 6\mathbf{i} - 8\mathbf{j}$$

$$\mathbf{u} - \mathbf{v} = (6\mathbf{i}) - (-8\mathbf{j})$$
$$= 6\mathbf{i} + 8\mathbf{j}$$

$$3\mathbf{u} + 2\mathbf{v} = 3(6\mathbf{i}) + 2(-8\mathbf{j})$$
$$= 18\mathbf{i} - 16\mathbf{j}$$

39.
$$\mathbf{u} + \mathbf{v} = (2\mathbf{i} + 5\mathbf{j}) + (5\mathbf{i} + 2\mathbf{j})$$
$$= 7\mathbf{i} + 7\mathbf{j}$$

$$\mathbf{u} - \mathbf{v} = (2\mathbf{i} + 5\mathbf{j}) - (5\mathbf{i} + 2\mathbf{j})$$
$$= 2\mathbf{i} + 5\mathbf{j} - 5\mathbf{i} - 2\mathbf{j}$$
$$= -3\mathbf{i} + 3\mathbf{j}$$

$$3\mathbf{u} + 2\mathbf{v} = 3(2\mathbf{i} + 5\mathbf{j}) + 2(5\mathbf{i} + 2\mathbf{j})$$
$$= 6\mathbf{i} + 15\mathbf{j} + 10\mathbf{i} + 4\mathbf{j}$$
$$= 16\mathbf{i} + 19\mathbf{j}$$

41. $\mathbf{v} = \langle 18\cos 40°, 18\sin 40° \rangle$
$= \langle 14, 12 \rangle$
$= 14\mathbf{i} + 12\mathbf{j}$

43. $\mathbf{w} = \langle 8\cos 230°, 8\sin 230° \rangle$
$= \langle -5.1, -6.1 \rangle$
$= -5.1\mathbf{i} - 6.1\mathbf{j}$

45. $\tan\theta = \dfrac{3}{3} = 1$ $\qquad |\mathbf{u}| = \sqrt{3^2 + 3^2}$
$\theta = 45°$ $\qquad\qquad = \sqrt{9+9}$
$\qquad\qquad\qquad = \sqrt{18} = 3\sqrt{2}$

47. $\tan\theta = \dfrac{-\sqrt{3}}{-1} = \sqrt{3}$ $\qquad |\mathbf{w}| = \sqrt{(-1)^2 + \left(-\sqrt{3}\right)^2}$
$\hat{\theta} = 60°$ $\qquad\qquad\qquad = \sqrt{1+3}$
$\theta = 240°$ in QIII $\qquad\qquad = \sqrt{4} = 2$

49. We are given that $|\mathbf{W}| = 10$ pounds.
$\mathbf{F}$ is a horizontal vector. $\qquad \mathbf{N}$ is a vertical vector.
$|\mathbf{F}| = |\mathbf{W}|\cos 75°$ $\qquad\qquad |\mathbf{N}| = |\mathbf{W}|\sin 75°$
$\quad = 10\cos 75°$ $\qquad\qquad\qquad = 10\sin 75°$
$\quad = 2.6$ pounds $\qquad\qquad\qquad = 9.7$ pounds

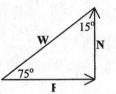

51. We redraw the vectors and find the missing angles:
$\theta = 90° - 25.5° = 64.5°$
$\mathbf{H} = -|\mathbf{H}|\mathbf{i}$, $\mathbf{W} = -95.5\mathbf{j}$, and $\mathbf{T} = |\mathbf{T}|\cos 64.5°\,\mathbf{i} + |\mathbf{T}|\sin 64.5°\,\mathbf{j}$
Also, $\mathbf{H} + \mathbf{W} + \mathbf{T} = 0$

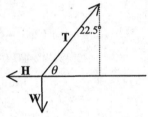

Collecting all the $\mathbf{i}$ components and all the $\mathbf{j}$ components together,
we have: $\left(-|\mathbf{H}| + |\mathbf{T}|\cos 64.5°\right)\mathbf{i} + \left(-95.5 + |\mathbf{T}|\sin 64.5°\right)\mathbf{j} = 0$

Therefore, $-|\mathbf{H}| + |\mathbf{T}|\cos 64.5° = 0$ and $-95.5 + |\mathbf{T}|\sin 64.5° = 0$
Solving the second equation for $|\mathbf{T}|$, we get: $|\mathbf{T}|\sin 64.5° = 95.5$
$$|\mathbf{T}| = \frac{95.5}{\sin 64.5°}$$
$$= 105.807$$

We substitute this into the first equation and solve for $|\mathbf{H}|$:
$$-|\mathbf{H}| + 105.807\cos 64.5° = 0$$
$$|\mathbf{H}| = 105.807\cos 64.5°$$
$$= 46.5 \text{ pounds}$$

53. We are given that $|\mathbf{W}| = 22$ pounds.

$$\mathbf{T_1} = |\mathbf{T_1}|\cos 30° \mathbf{i} + |\mathbf{T_1}|\sin 30° \mathbf{j} \qquad \mathbf{T_2} = |\mathbf{T_2}|\cos 135° \mathbf{i} + |\mathbf{T_2}|\sin 135° \mathbf{j}$$

$$= |\mathbf{T_1}|\left(\frac{\sqrt{3}}{2}\right)\mathbf{i} + |\mathbf{T_1}|\left(\frac{1}{2}\right)\mathbf{j} \qquad = |\mathbf{T_2}|\left(-\frac{\sqrt{2}}{2}\right)\mathbf{i} + |\mathbf{T_2}|\left(\frac{\sqrt{2}}{2}\right)\mathbf{j}$$

$$\mathbf{W} = 22\cos 270° \mathbf{i} + 22\sin 270° \mathbf{j}$$
$$= 0\mathbf{i} - 22\mathbf{j}$$

Since $\mathbf{T_1} + \mathbf{T_2} = \mathbf{W}$, we add the corresponding $\mathbf{i}$ and $\mathbf{j}$ components:

(1) $\quad |\mathbf{T_1}|\left(\frac{\sqrt{3}}{2}\right) + |\mathbf{T_2}|\left(-\frac{\sqrt{2}}{2}\right) = 0 \quad$ and $\quad$ (2) $\quad |\mathbf{T_1}|\left(\frac{1}{2}\right) + |\mathbf{T_2}|\left(\frac{\sqrt{2}}{2}\right) = -22$

Solving this system using the elimination method, we get:

$$\frac{\sqrt{3}+1}{2}|\mathbf{T_1}| = |-22| \quad \text{or} \quad |\mathbf{T_1}| = \left|\frac{-22}{\frac{\sqrt{3}+1}{2}}\right| = 16 \text{ pounds}$$

Substituting the value for $|\mathbf{T_1}|$ in equation (1) we get:

$$16\left(\frac{\sqrt{3}}{2}\right) + |\mathbf{T_2}|\left(-\frac{\sqrt{2}}{2}\right) = 0 \quad \text{or} \quad |\mathbf{T_2}| = \frac{8\sqrt{3}}{\frac{\sqrt{2}}{2}} = 20 \text{ pounds}$$

Problem Set 7.6

1. $\quad \langle 6,6 \rangle \cdot \langle 3,5 \rangle = 6(3) + 6(5)$
$\qquad\qquad = 18 + 30 = 48$

3. $\quad \langle -23,4 \rangle \cdot \langle 15,-6 \rangle = -23(15) + 4(-6)$
$\qquad\qquad\qquad\qquad = -345 - 24 = -369$

5. $\quad \mathbf{u} \cdot \mathbf{v} = 1(1) + 1(-1)$
$\qquad\quad = 1 - 1 = 0$

7. $\quad \mathbf{u} \cdot \mathbf{v} = 6(0) + 0(-8)$
$\qquad\quad = 0 + 0 = 0$

9. $\quad \mathbf{u} \cdot \mathbf{v} = 2(5) + 5(2)$
$\qquad\quad = 10 + 10 = 20$

11. $\quad$ First, we must find $\mathbf{u} \cdot \mathbf{v}, |\mathbf{u}|,$ and $|\mathbf{v}|$:

$\qquad \mathbf{u} \cdot \mathbf{v} = \langle 13,0 \rangle \cdot \langle 0,-6 \rangle \qquad |\mathbf{u}| = 13 \text{ and } |\mathbf{v}| = 6$

$\qquad\qquad = 13(0) + 0(-6) = 0$

$\qquad\qquad \cos\theta = \frac{\mathbf{u} \cdot \mathbf{v}}{|\mathbf{u}||\mathbf{v}|} = \frac{0}{13(6)} = 0$

$\qquad\qquad \theta = 90.0°$

13. First, we must find $\mathbf{u} \cdot \mathbf{v}, |\mathbf{u}|,$ and $|\mathbf{v}|$:

$\mathbf{u} \cdot \mathbf{v} = \langle 4, 5 \rangle \cdot \langle 7, -4 \rangle$ $|\mathbf{u}| = \sqrt{4^2 + 5^2}$ and $|\mathbf{v}| = \sqrt{7^2 + (-4)^2}$

$\quad\quad = 4(7) + 5(-4) = 8$ $= \sqrt{16 + 25} = \sqrt{41}$ $= \sqrt{49 + 16} = \sqrt{64} = 8$

$\cos\theta = \dfrac{\mathbf{u} \cdot \mathbf{v}}{|\mathbf{u}||\mathbf{v}|} = \dfrac{8}{8\sqrt{41}} = 0.1562$

$\theta = 81.1°$

15. First, we must find $\mathbf{u} \cdot \mathbf{v}, |\mathbf{u}|,$ and $|\mathbf{v}|$:

$\mathbf{u} \cdot \mathbf{v} = \langle 13, -8 \rangle \cdot \langle 2, 11 \rangle$ $|\mathbf{u}| = \sqrt{13^2 + (-8)^2}$ and $|\mathbf{v}| = \sqrt{2^2 + 11^2}$

$\quad\quad = 13(2) + (-8)(11) = -62$ $= \sqrt{169 + 64} = \sqrt{233}$ $= \sqrt{4 + 121} = \sqrt{125}$

$\cos\theta = \dfrac{\mathbf{u} \cdot \mathbf{v}}{|\mathbf{u}||\mathbf{v}|} = \dfrac{-62}{\sqrt{233}\sqrt{125}} = -0.3633$

$\theta = 111.3°$

17. To show that $\mathbf{i}$ and $\mathbf{j}$ are perpendicular, we must show that their dot product equals zero.

$\mathbf{i} = \langle 1, 0 \rangle$ and $\mathbf{j} = \langle 0, 1 \rangle$

$\mathbf{i} \cdot \mathbf{j} = 1(0) + 0(1) = 0$

Therefore, they are perpendicular.

19. To show that $-\mathbf{i}$ and $\mathbf{j}$ are perpendicular, we must show that their dot product equals zero.

$-\mathbf{i} = \langle -1, 0 \rangle$ and $\mathbf{j} = \langle 0, 1 \rangle$

$-\mathbf{i} \cdot \mathbf{j} = -1(0) + 0(1) = 0$

Therefore, they are perpendicular.

21. $\mathbf{V} = \langle a, b \rangle$ and $\mathbf{W} = \langle -b, a \rangle$

$\mathbf{V} \cdot \mathbf{W} = a(-b) + b(a)$

$\quad\quad = -ab + ab = 0$

Therefore, they are perpendicular.

23. Work $= \mathbf{F} \cdot \mathbf{d}$

$\quad = \langle 22, 9 \rangle \cdot \langle 30, 4 \rangle$

$\quad = 22(30) + 9(4) = 696$ ft-lb

25. Work $= \mathbf{F} \cdot \mathbf{d}$

$\quad = \langle -67, 39 \rangle \cdot \langle -96, -28 \rangle$

$\quad = -67(-96) + 39(-28) = 5,340$ ft-lb

27. Work $= \mathbf{F} \cdot \mathbf{d}$

$\quad = \langle 85, 0 \rangle \cdot \langle 6, 0 \rangle$

$\quad = 85(6) + 0(0) = 510$ ft-lb

29. Work $= \mathbf{F} \cdot \mathbf{d}$

$\quad = \langle 0, 39 \rangle \cdot \langle 72, 0 \rangle$

$\quad = 0(72) + 39(0) = 0$ ft-lb

31. Let $\mathbf{U} = a\mathbf{i} + b\mathbf{j}$ and $\mathbf{V} = c\mathbf{i} + d\mathbf{j}$

Then $\mathbf{U} - \mathbf{V} = (a-c)\mathbf{i} + (b-d)\mathbf{j}$

$$|\mathbf{U} - \mathbf{V}|^2 = (a-c)^2 + (b-d)^2$$
$$= a^2 - 2ac + c^2 + b^2 - 2bd + d^2$$
$$(1) \qquad = a^2 + b^2 + c^2 + d^2 - 2(ac + bd)$$

Using the Law of Cosines, we also have:

$(2)\ |\mathbf{U} - \mathbf{V}|^2 = |\mathbf{U}|^2 + |\mathbf{V}|^2 - 2|\mathbf{U}||\mathbf{V}|\cos\theta$

We know that $|\mathbf{U}|^2 = a^2 + b^2$, $\ |\mathbf{V}|^2 = c^2 + d^2$, and $\mathbf{U} \cdot \mathbf{V} = ac + bd$

Substituting these values into (2), we get:

$(3)\ |\mathbf{U} - \mathbf{V}|^2 = a^2 + b^2 + c^2 + d^2 - 2|\mathbf{U}||\mathbf{V}|\cos\theta$

Setting (1) equal to (3) and simplifying, we get:

$$a^2 + b^2 + c^2 + d^2 - 2(ac + bd) = a^2 + b^2 + c^2 + d^2 - 2|\mathbf{U}||\mathbf{V}|\cos\theta$$
$$-2(ac + bd) = -2|\mathbf{U}||\mathbf{V}|\cos\theta$$
$$ac + bd = |\mathbf{U}||\mathbf{V}|\cos\theta$$
$$\mathbf{U} \cdot \mathbf{V} = |\mathbf{U}||\mathbf{V}|\cos\theta$$

33. $\mathbf{d} = \langle 75, 0 \rangle$

$\mathbf{F} = \langle 40\cos 20°, 40\sin 20° \rangle = \langle 37.59, 13.68 \rangle$

Work $= \mathbf{F} \cdot \mathbf{d}$

$\qquad = \langle 75, 0 \rangle \cdot \langle 37.59, 13.68 \rangle$

$\qquad = 75(37.59) + 0(13.68) = 2800$ ft-lb (rounded to 2 significant digits)

35. $\mathbf{d} = \langle 100, 0 \rangle$

$\mathbf{F} = \langle 85\cos 15°, 85\sin 15° \rangle = \langle 82.10, 22.00 \rangle$

Work $= \mathbf{F} \cdot \mathbf{d}$

$\qquad = \langle 100, 0 \rangle \cdot \langle 82.10, 22.00 \rangle$

$\qquad = 100(82.10) + 0(22.00) = 8200$ ft-lb (rounded to 2 significant digits)

Chapter 7 Test

1. $b = \dfrac{a \sin B}{\sin A}$

$= \dfrac{3.8 \sin 70°}{\sin 32°}$

$= 6.7 \text{ in}$

2. $b = \dfrac{c \sin B}{\sin C}$

$= \dfrac{2.9 \sin 118°}{\sin 37°}$

$= 4.3 \text{ in}$

3. $C = 180° - (A + B)$

$= 180° - (38.2° + 63.4°) = 78.4°$

$a = \dfrac{c \sin A}{\sin C}$

$= \dfrac{42.0 \sin 38.2°}{\sin 78.4°}$

$= 26.5 \text{ cm}$

$b = \dfrac{c \sin B}{\sin C}$

$= \dfrac{42.0 \sin 63.4°}{\sin 78.4°}$

$= 38.3 \text{ cm}$

4. $B = 180° - (A + C)$

$= 180° - (24.7° + 106.1°) = 49.2°$

$a = \dfrac{b \sin A}{\sin B}$

$= \dfrac{34.0 \sin 24.7°}{\sin 49.2°}$

$= 18.8 \text{ cm}$

$c = \dfrac{b \sin C}{\sin B}$

$= \dfrac{34.0 \sin 106.1°}{\sin 49.2°}$

$= 43.2 \text{ cm}$

5. $\sin B = \dfrac{b \sin A}{a}$

$= \dfrac{42 \sin 60°}{12} = 3.0311$

Since $\sin B$ is never greater than 1, no triangle exists.

6. $\sin B = \dfrac{b \sin A}{a}$

$= \dfrac{21 \sin 42°}{29} = 0.4845$

$B = 29°$ or $B' = 180° - 29° = 151°$

This is impossible because the sum of angles is greater than $180°$.

Therefore, the only possibility is $29°$ which means that only one triangle exists.

7. $\sin B = \dfrac{b \sin A}{a}$

$= \dfrac{7.9 \sin 51°}{6.5} = 0.9445$

$B = 71°$ or $B' = 180° - 71°$

$= 109°$

$C = 180° - (51° + 71°) = 58°$

$c = \dfrac{a \sin C}{\sin A}$

$= \dfrac{6.5 \sin 58°}{\sin 51°}$

$= 7.1 \text{ ft}$

$C' = 180° - (51° + 109°) = 20°$

$c' = \dfrac{a \sin C'}{\sin A}$

$= \dfrac{6.5 \sin 20°}{\sin 51°}$

$= 2.9 \text{ ft}$

8. $\sin B = \dfrac{b \sin A}{a}$ $\qquad$ $C = 180° - (26° + 59°) = 95°$ $\qquad$ $C' = 180° - (26° + 121°) = 33°$

$\qquad = \dfrac{9.4 \sin 26°}{4.8} = 0.8585$ $\qquad$ $c = \dfrac{a \sin C}{\sin A}$ $\qquad\qquad$ $c' = \dfrac{a \sin C'}{\sin A}$

$\qquad B = 59°$ or $B' = 180° - 59°$ $\qquad = \dfrac{4.8 \sin 95°}{\sin 26°}$ $\qquad\qquad = \dfrac{4.8 \sin 33°}{\sin 26°}$

$\qquad\qquad\qquad\quad = 121°$ $\qquad\qquad = 11\,\text{ft}$ $\qquad\qquad\qquad = 6.0\,\text{ft}$

9. $c^2 = a^2 + b^2 - 2ab \cos C$ $\qquad\qquad$ **10.** $c^2 = a^2 + b^2 - 2ab \cos C$

$\qquad\quad = 10^2 + 12^2 - 2(10)(12) \cos 60°$ $\qquad\qquad\quad = 10^2 + 12^2 - 2(10)(12) \cos 120°$

$\qquad\quad = 124$ $\qquad\qquad\qquad\qquad\qquad\qquad\quad = 364$

$\qquad c = 11\,\text{cm}$ $\qquad\qquad\qquad\qquad\qquad\quad c = 19\,\text{cm}$

11. $\cos C = \dfrac{a^2 + b^2 - c^2}{2ab}$ $\qquad\qquad$ **12.** $\cos B = \dfrac{a^2 + c^2 - b^2}{2ac}$

$\qquad\qquad = \dfrac{5^2 + 7^2 - 9^2}{2(5)(7)}$ $\qquad\qquad\qquad\qquad = \dfrac{10^2 + 11^2 - 12^2}{2(10)(11)}$

$\qquad\qquad = -0.1$ $\qquad\qquad\qquad\qquad\qquad\quad = 0.35$

$\qquad C = 95.7°$ $\qquad\qquad\qquad\qquad\qquad B = 69.5°$

13. $c^2 = a^2 + b^2 - 2ab \cos C$ $\qquad$ $\sin B = \dfrac{b \sin C}{c}$ $\qquad$ $A = 180° - (18° + 119°)$

$\qquad\quad = 6.4^2 + 2.8^2 - 2(6.4)(2.8) \cos 119°$ $\qquad = \dfrac{2.8 \sin 119°}{8.1}$ $\qquad = 180° - 137° = 43°$

$\qquad\quad = 66.18$ $\qquad\qquad\qquad\qquad\qquad = 0.3023$

$\qquad c = 8.1\,\text{cm}$ $\qquad\qquad\qquad\qquad\quad B = 18°$

14. $a^2 = b^2 + c^2 - 2bc \cos A$ $\qquad$ $\sin B = \dfrac{b \sin A}{a}$ $\qquad$ $C = 180° - (35° + 34°)$

$\qquad\quad = 3.7^2 + 6.2^2 - 2(3.7)(6.2) \cos 35°$ $\qquad = \dfrac{3.7 \sin 35°}{3.8}$ $\qquad = 180° - 69° = 111°$

$\qquad\quad = 14.547$ $\qquad\qquad\qquad\qquad\qquad = 0.5585$

$\qquad a = 3.8\,\text{m}$ $\qquad\qquad\qquad\qquad\quad B = 34°$

15. $S = \dfrac{c^2 \sin A \sin B}{2 \sin C}$ $\qquad\qquad$ **16.** $S = \dfrac{1}{2} ab \sin C$

$\qquad = \dfrac{42.0^2 \sin 38.2° \sin 63.4°}{2 \sin 78.4°}$ $\qquad\qquad\quad = \dfrac{1}{2}(18.8)(34.0) \sin 106.1°$

$\qquad = 498\,\text{cm}^2$ $\qquad\qquad\qquad\qquad\qquad = 307\,\text{cm}^2$

17.
$$S = \frac{1}{2}ab\sin C$$
$$= \frac{1}{2}(10)(12)\sin 60°$$
$$= 52\,\text{cm}^2$$

18.
$$S = \frac{1}{2}ab\sin C$$
$$= \frac{1}{2}(10)(12)\sin 120°$$
$$= 52\,\text{cm}^2$$

19.
$$s = \frac{1}{2}(a+b+c)$$
$$= \frac{1}{2}(5+7+9) = 10.5$$

$$S = \sqrt{s(s-a)(s-b)(s-c)}$$
$$= \sqrt{10.5(10.5-5)(10.5-7)(10.5-9)}$$
$$= \sqrt{303.1875} = 17\,\text{km}^2$$

20.
$$s = \frac{1}{2}(a+b+c)$$
$$= \frac{1}{2}(10+12+11) = 16.5$$

$$S = \sqrt{s(s-a)(s-b)(s-c)}$$
$$= \sqrt{16.5(16.5-10)(16.5-12)(16.5-11)}$$
$$= \sqrt{2,654.4375} = 52\,\text{km}^2$$

21.
$$\cos\theta = \frac{38^2 + 48^2 - 38^2}{2(38)(48)}$$
$$= 0.6316$$
$$\theta = 51°$$

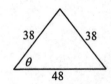

22.
$$\tan 48° = \frac{h}{53}$$
$$h = 53\tan 48° = 59\,\text{ft}$$

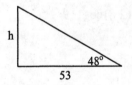

23.   First, we find the missing angles of triangle ABD:
$$\angle ABD = 180° - 64° = 116°$$
$$\angle ADB = 180° - (43° + 116°) = 21°$$
Next, we find x using the Law of Sines:
$$x = \frac{240\sin 43°}{\sin 21°}$$
$$= 457$$
Then, we find h using the sine relationship:
$$\sin 64° = \frac{h}{457}$$
$$h = 457\sin 64°$$
$$= 410\,\text{ft (rounded to 2 significant digits)}$$

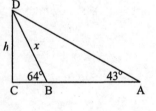

24. From Geometry, we know that the diagonals of a parallelogram bisect each other. Since we want to find the shorter side, we must find θ:

$$\theta = 180° - 134.5° = 45.5°$$

Now we can find x using the Law of Cosines:

$$x^2 = 13.4^2 + 19.7^2 - 2(13.4)(19.7)\cos 45.5°$$
$$= 197.598$$
$$x = 14.1\,\text{m}$$

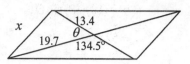

25.
$$CD = r + 4.55$$
$$= 3960 + 4.55 = 3964.55$$
$$\sin D = \frac{3960 \sin 90.8°}{3964.55}$$
$$= 0.9988$$
$$D = 87.14°$$
$$C = 180° - (90.8° + 87.14°) = 2.06°$$

$$s = 3960(2.06°)\left(\frac{\pi}{180}\right)$$
$$= 142 \text{ mi}$$

26. First, we find angle B, using alternate interior angles of parallel lines:

$$B = 55° + 44° = 99°$$

Next, we find b using the Law of Cosines:

$$b^2 = 2.2^2 + 3.3^2 - 2(2.2)(3.3)\cos 99°$$
$$= 18.001$$
$$b = 4.2 \text{ mi}$$

To find his bearing, we must find θ:

$$\sin \theta = \frac{2.2 \sin 99°}{4.2}$$
$$= 0.5174$$
$$\theta = 31° \qquad \text{The bearing is S 75° W.}$$

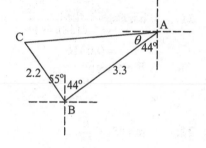

27.
$$C = 180° - (A + B)$$
$$= 180° - (47° + 37°) = 96°$$
$$c^2 = a^2 + b^2 - 2ab\cos C$$
$$= 56^2 + 65^2 - 2(56)(65)\cos 96°$$
$$= 8122$$
$$c = 90 \text{ ft}$$

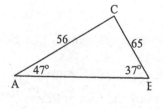

28. First, we find angle A: $\quad A = 95.5° - 90° = 5.5°$

Next, we find angle B using the Law of Sines:

$$\sin B = \frac{b \sin A}{a}$$
$$= \frac{345 \sin 5.5°}{55}$$

This problem is continued on the next page

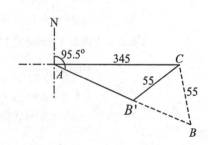

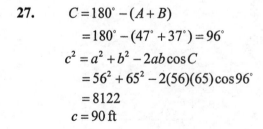

$$= 0.6012$$
$$B = 37° \quad \text{or} \quad B' = 180° - 37° = 143°$$

Next, we find angles C and C':
$$C = 180° - (5.5° + 37°) = 137.5° \qquad C' = 180° - (5.5° + 143°) = 31.5°$$

Last, we find c and c' using the Law of Sines:

$$c = \frac{a \sin C}{\sin A} \qquad\qquad c' = \frac{a \sin C'}{\sin A}$$
$$= \frac{55 \sin 137.5°}{\sin 5.5°} = 388 \text{ mph} \qquad = \frac{55 \sin 31.5°}{\sin 5.5°} = 300 \text{ mph}$$

29. First, we find the angles of triangle ABC:

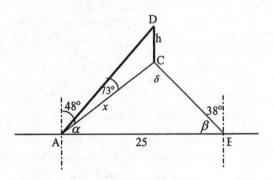

$$\alpha = 90° - 48° = 42°$$

$$\beta = 90° - 38° = 52°$$

$$\delta = 180° - \left(42° + 52°\right) = 86°$$

Next, we find x using the Law of Sines:
$$x = \frac{25 \sin 52°}{\sin 86°} = 19.7$$

Then, we find h using the tangent ratio:
$$h = 19.7 \tan 73° = 65 \text{ ft}$$

Therefore, the tree is 65 feet tall.

30. From Geometry:

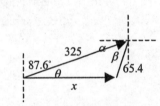

$$\alpha = 90° - 87.6° = 2.4$$

$$\beta = 270° - \left(262.6° + 2.4°\right) = 5.0°$$

We find x using the Law of Cosines:
$$x^2 = 65.4^2 + 325^2 - 2(65.4)(325)\cos 5.0°$$
$$= 67553.92$$
$$x = 260 \text{ mph}$$

Last, we find θ using the Law of Sines:

$$\sin \theta = \frac{65.4 \sin 5.0°}{260}$$
$$= 0.02192$$
$$\theta = 1.3° \qquad \theta + 87.6° = 88.9°$$

The true course of the plane is $88.9°$ from due north and the ground speed is 260 mph.

31.
$$|\mathbf{U}| = \sqrt{a^2 + b^2}$$
$$= \sqrt{5^2 + 12^2}$$
$$= \sqrt{169} = 13$$

32.
$$3\mathbf{U} + 5\mathbf{V} = 3(5\mathbf{i} + 12\mathbf{j}) + 5(-4\mathbf{i} + \mathbf{j})$$
$$= 15\mathbf{i} + 36\mathbf{j} - 20\mathbf{i} + 5\mathbf{j}$$
$$= -5\mathbf{i} + 41\mathbf{j}$$

33.
$$3\mathbf{U} - 5\mathbf{V} = 3(5\mathbf{i} + 12\mathbf{j}) - 5(-4\mathbf{i} + \mathbf{j})$$
$$= 15\mathbf{i} + 36\mathbf{j} + 20\mathbf{i} - 5\mathbf{j}$$
$$= 35\mathbf{i} + 31\mathbf{j}$$

34.
$$2\mathbf{V} - \mathbf{W} = 2(-4\mathbf{i} + \mathbf{j}) - (\mathbf{i} - 4\mathbf{j})$$
$$= -8\mathbf{i} + 2\mathbf{j} - \mathbf{i} + 4\mathbf{j}$$
$$= -9\mathbf{i} + 6\mathbf{j}$$
$$|2\mathbf{V} - \mathbf{W}| = \sqrt{(-9)^2 + 6^2} = \sqrt{117}$$

35.
$$\mathbf{V} \cdot \mathbf{W} = (-4\mathbf{i} + \mathbf{j}) \cdot (\mathbf{i} - 4\mathbf{j})$$
$$= -4(1) + 1(-4)$$
$$= -8$$

36.
$$|\mathbf{U}| = 13 \quad \text{(from problem 31)}$$

$$|\mathbf{V}| = \sqrt{(-4)^2 + 1^2} = \sqrt{17}$$

$$\mathbf{U} \cdot \mathbf{V} = (5\mathbf{i} + 12\mathbf{j}) \cdot (-4\mathbf{i} + \mathbf{j})$$
$$= 5(-4) + 12(1) = -8$$

$$\cos\theta = \frac{\mathbf{U} \cdot \mathbf{V}}{|\mathbf{U}||\mathbf{V}|}$$
$$= \frac{-8}{13\sqrt{17}}$$
$$= -0.1493$$
$$\theta = 98.6°$$

37. For vector **V**, the horizontal component is a and the vertical component is b. The vector goes from $(0, 0)$ to (a, b). We can find the slope between these two points:
$$m = \frac{b - 0}{a - 0} = \frac{b}{a}$$

38. We will find their dot product:
$$\mathbf{v} \cdot \mathbf{w} = \langle 3, 6 \rangle \cdot \langle -8, 4 \rangle$$
$$= 3(-8) + 6(4)$$
$$= -24 + 24 = 0$$
Since their dot product is 0, the vectors are perpendicular.

39. We will find their dot product:
$$\mathbf{u} \cdot \mathbf{v} = \langle 5, 12 \rangle \cdot \langle 4, b \rangle$$
$$= 5(4) + 12(b) = 20 + 12b$$
If the vectors are perpendicular, their dot product is 0:
$$20 + 12b = 0$$
$$12b = -20$$
$$b = -\frac{20}{12} = -\frac{5}{3}$$

40.
$$\text{Work} = \mathbf{F} \cdot \mathbf{d}$$
$$= \langle 33, -4 \rangle \cdot \langle 56, 10 \rangle$$
$$= 33(56) + (-4)(10)$$
$$= 1848 - 40 = 1808 \text{ units}$$

CHAPTER 8 Complex Numbers and Polar Coordinates

Problem Set 8.1

1. $\sqrt{-16} = i\sqrt{16}$
$= 4i$

3. $\sqrt{-121} = i\sqrt{121}$
$= 11i$

5. $\sqrt{-18} = i\sqrt{18}$
$= 3i\sqrt{2}$

7. $\sqrt{-8} = i\sqrt{8}$
$= 2i\sqrt{2}$

9. $\sqrt{-4}\sqrt{-9} = i\sqrt{4}\,i\sqrt{9}$
$= 2 \cdot 3 \cdot i^2$
$= 6(-1) = -6$

11. $\sqrt{-1}\sqrt{-9} = i \cdot i\sqrt{9}$
$= 3i^2$
$= 3(-1) = -3$

13. $4 = 6x$ and $7 = -14y$

$x = \dfrac{4}{6} = \dfrac{2}{3}$ $y = -\dfrac{7}{14} = -\dfrac{1}{2}$

15. $5x + 2 = 4$ and $-7 = 2y + 1$

$5x = 2$ $-8 = 2y$

$x = \dfrac{2}{5}$ $y = -4$

17. $x^2 - x - 6 = 0$ and $y^2 = 9$

$(x-3)(x+2) = 0$ $y = \pm 3$

$x - 3 = 0$ or $x + 2 = 0$
$x = 3$ $x = -2$

19. $\cos x = \sin x$ and $\sin y = 1$

$1 = \dfrac{\sin x}{\cos x}$ $y = \dfrac{\pi}{2}$

$\tan x = 1$

$x = \dfrac{\pi}{4}, \dfrac{5\pi}{4}$

21. $\sin^2 x - 2\sin x + 1 = 0$ and $\tan y = 1$

$(\sin x - 1)(\sin x - 1) = 0$ $y = \dfrac{\pi}{4}, \dfrac{5\pi}{4}$

$\sin x - 1 = 0$
$\sin x = 1$

$x = \dfrac{\pi}{2}$

23. $7 + 2i + 3 - 4i = 10 - 2i$

25. $(6 + 7i) - (4 + i) = 6 + 7i - 4 - i$
$= (6 - 4) + (7 - 1)i$
$= 2 + 6i$

27. $(7-3i)-(4+10i)=7-3i-4-10i$
$$=(7-4)+(-3-10)i$$
$$=3-13i$$

29. $(3\cos x+4i\sin y)+(2\cos x-7i\sin y)$
$$=(3\cos x+2\cos x)+(4\sin y-7\sin y)i$$
$$=5\cos x-3i\sin y$$

31. $[(3+2i)-(6+i)]+(5+i)=3+2i-6-i+5+i$
$$=(3-6+5)+(2i-i+i)$$
$$=2+2i$$

33. $(7-4i)-[(-2+i)-(3+7i)]=7-4i-(-2+i-3-7i)$
$$=7-4i+2-i+3+7i$$
$$=12+2i$$

35. $i^{12}=\left(i^4\right)^3=1^3=1$

37. $i^{14}=\left(i^4\right)^3\cdot i^2$
$$=1(-1)=-1$$

39. $i^{32}=\left(i^4\right)^8$
$$=1^8=1$$

41. $i^{33}=\left(i^4\right)^8\cdot i$
$$=1^8\cdot i=i$$

43. $-6i(3-8i)=-18i+48i^2$
$$=-18i+48(-1)=-48-18i$$

45. $(2-4i)(3+i)=6+2i-12i-4i^2$
$$=6-10i-4(-1)$$
$$=6-10i+4=10-10i$$

47. $(3+2i)^2=9+12i+4i^2$
$$=9+12i+4(-1)$$
$$=9+12i-4=5+12i$$

49. $(5+4i)(5-4i)=25-16i^2$
$$=25-16(-1)$$
$$=25+16=41$$

51. $(7+2i)(7-2i)=49-4i^2$
$$=49-4(-1)$$
$$=49+4=53$$

53. $2i(3+i)(2+4i)=2i(6+14i+4i^2)$
$$=12i+28i^2+8i^3$$
$$=12i+28(-1)+8(-i)=-28+4i$$

55. $3i(1+i)^2=3i(1+2i+i^2)$
$$=3i+6i^2-3i$$
$$=6(-1)=-6$$

57. $\dfrac{2i}{3+i}\cdot\dfrac{3-i}{3-i}=\dfrac{6-2i^2}{9-i^2}$
$$=\dfrac{6i-2(-1)}{9-(-1)}$$
$$=\dfrac{2+6i}{10}=\dfrac{1}{5}+\dfrac{3}{5}i$$

59. $\dfrac{2+3i}{2-3i}\cdot\dfrac{2+3i}{2+3i}=\dfrac{4+12i+9i^2}{4-9i^2}$
$$=\dfrac{4+12i-9}{4-(-9)}$$
$$=\dfrac{-5+12i}{13}=-\dfrac{5}{13}+\dfrac{12}{13}i$$

61.
$$\frac{5-2i}{i} \cdot \frac{i}{i} = \frac{5i-2i^2}{i^2}$$
$$= \frac{5i-2(-1)}{-1}$$
$$= \frac{2+5i}{-1} = -2-5i$$

63.
$$\frac{2+i}{5-6i} \cdot \frac{5+6i}{5+6i} = \frac{10+17i+6i^2}{25-36i^2}$$
$$= \frac{10+17i-6}{25-(-36)}$$
$$= \frac{4+17i}{61} = \frac{4}{61} + \frac{17}{61}i$$

65.
$$z_1 z_2 = (2+3i)(2-3i)$$
$$= 4 - 9i^2$$
$$= 4 - (-9) = 13$$

67.
$$z_1 z_3 = (2+3i)(4+5i)$$
$$= 8 + 22i + 15i^2$$
$$= 8 + 22i - 15 = -7 + 22i$$

69.
$$2z_1 + 3z_2 = 2(2+3i) + 3(2-3i)$$
$$= 4 + 6i + 6 - 9i$$
$$= 10 - 3i$$

71.
$$z_3(z_1 + z_2) = (4+5i)\left[(2+3i) + (2-3i)\right]$$
$$= (4+5i)(4)$$
$$= 16 + 20i$$

73.
$$(x+3i)(x-3i) = x^2 - 9i^2$$
$$= x^2 + 9$$

75.
$$x^2 - 4x + 13 = (2+3i)^2 - 4(2+3i) + 13$$
$$= 4 + 12i + 9i^2 - 8 - 12i + 13$$
$$= 4 + 12i - 9 - 8 - 12i + 13$$
$$= 0$$

77.
$$x^2 - 2ax + (a^2 + b^2)$$
$$= (a+bi)^2 - 2a(a+bi) + a^2 + b^2$$
$$= a^2 + 2abi + b^2i^2 - 2a^2 - 2abi + a^2 + b^2$$
$$= b^2(-1) + b^2 = 0$$

79. If $x + y = 8$, then $y = 8 - x$. Substituting this into the second equation, we get:
$$x(8 - x) = 20$$
$$8x - x^2 = 20$$
$$x^2 - 8x + 20 = 0 \quad \text{where } a = 1, \ b = -8, \text{ and } c = 20$$
$$x = \frac{-(-8) \pm \sqrt{(-8)^2 - 4(1)(20)}}{2(1)}$$
$$= \frac{8 \pm \sqrt{-16}}{2}$$
$$= \frac{8 \pm 4i}{2} = 4 \pm 2i$$

Now, we find the corresponding y-values:

If $x = 4 + 2i$, then $y = 8 - (4 + 2i) = 4 - 2i$

If $x = 4 - 2i$, then $y = 8 - (4 - 2i) = 4 + 2i$

81. If $2x + y = 4$, then $y = 4 - 2x$. Substituting this into the second equation, we get:
$$x(4 - 2x) = 8$$
$$4x - 2x^2 = 8$$
This problem is continued on the next page.

$$2x^2 - 4x + 8 = 0 \quad \text{or} \quad x^2 - 2x + 4 = 0 \quad \text{where } a = 1, b = -2, \text{ and } c = 4$$

$$x = \frac{-(-2) \pm \sqrt{(-2)^2 - 4(1)(4)}}{2(1)}$$

Now, we find the corresponding y-values:

$$= \frac{2 \pm \sqrt{-12}}{2}$$

If $x = 1 + i\sqrt{3}$, then $y = 4 - 2(1 + i\sqrt{3}) = 2 - 2i\sqrt{3}$.

$$= \frac{2 \pm 2i\sqrt{3}}{2} = 1 \pm i\sqrt{3}$$

If $x = 1 - i\sqrt{3}$, then $y = 4 - 2(1 - i\sqrt{3}) = 2 + 2i\sqrt{3}$.

83.
$$(a + bi)(a - bi) = a^2 - b^2 i^2$$
$$= a^2 - b^2(-1)$$
$$= a^2 + b^2$$

This is a real number. Therefore, the product of a complex number and its conjugate is a real number.

85.
$$(a + bi) + (c + di) = (a + c) + (b + d)i$$
$$= (c + a) + (d + b)i$$
$$= (c + di) + (a + bi)$$

Therefore, addition of complex numbers is commutative.

87.
$$r = \sqrt{3^2 + (-4)^2} \qquad \sin\theta = -\frac{4}{5}$$
$$= \sqrt{25} = 5 \qquad \cos\theta = \frac{3}{5}$$

89.
$$r = \sqrt{a^2 + b^2} \qquad \sin\theta = \frac{b}{\sqrt{a^2 + b^2}}$$
$$\cos\theta = \frac{a}{\sqrt{a^2 + b^2}}$$

91.
$$\hat{\theta} = 45° \text{ and is in QII}$$
$$\theta = 180° - 45° = 135°$$

93.
$$a^2 = b^2 + c^2 - 2bc\cos A$$
$$= 243^2 + 157^2 - 2(243)(157)\cos 73.1°$$
$$= 61{,}517$$
$$a = 248 \text{ cm}$$

$$\sin C = \frac{c\sin A}{a}$$
$$= \frac{157\sin 73.1°}{248} = 0.6057$$
$$C = 37.3°$$
$$B = 180° - (73.1° + 37.3°) = 69.6°$$

95.
$$\cos C = \frac{a^2 + b^2 - c^2}{2ab}$$
$$= \frac{42.1^2 + 56.8^2 - 63.4^2}{2(42.1)(56.8)} = 0.2047$$
$$C = 78.2°$$

$$\sin B = \frac{b\sin C}{c}$$
$$= \frac{56.8\sin 78.2°}{63.4} = 0.8770$$
$$B = 61.3°$$
$$A = 180° - (78.2° + 61.3°) = 40.5°$$

Problem Set 8.2

1. $|3+4i| = \sqrt{3^2 + 4^2} = \sqrt{25} = 5$

3. $|1+i| = \sqrt{1^2 + 1^2} = \sqrt{2}$

5. $|0-5i| = \sqrt{0^2 + (-5)^2} = \sqrt{25} = 5$

7. $|2+0i| = \sqrt{2^2 + 0^2} = \sqrt{4} = 2$

9. $|-4-3i| = \sqrt{(-4)^2 + (-3)^2} = \sqrt{25} = 5$

11. Opposite of $2 - i = -(2-i) = -2+i$
    Conjugate of $2 - i = 2 + i$

13. Opposite of $4i = -4i$
    Conjugate of $4i = -4i$

15. Opposite of $-3 = 3$
    Conjugate of $-3 = -3$

17. Opposite of $-5 - 2i = -(-5-2i) = 5+2i$

    Conjugate of $-5 - 2i = -5+2i$

19. $2(\cos 30° + i\sin 30°) = 2\left[\dfrac{\sqrt{3}}{2} + i\left(\dfrac{1}{2}\right)\right]$

    $= \sqrt{3} + i$

21. $4(\cos 120° + i\sin 120°) = 4\left[-\dfrac{1}{2} + i\left(\dfrac{\sqrt{3}}{2}\right)\right]$

    $= -2 + 2i\sqrt{3}$

23. $\cos 210° + i\sin 210° = -\dfrac{\sqrt{3}}{2} + i\left(-\dfrac{1}{2}\right)$

    $= -\dfrac{\sqrt{3}}{2} - \dfrac{1}{2}i$

25. $\cos 315° + i\sin 315° = \dfrac{\sqrt{2}}{2} + i\left(-\dfrac{\sqrt{2}}{2}\right)$

    $= \dfrac{\sqrt{2}}{2} - \dfrac{\sqrt{2}}{2}i$

27. $10(\cos 12° + i\sin 12°) = 10\left[0.978 + i(0.208)\right]$

    $= 9.78 + 2.08i$

29. $100(\cos 143° + i\sin 143°) = 100\left[-0.7986 + i(0.6018)\right]$
    $= -79.86 + 60.18i$

31. $\cos 205° + i\sin 205° = -0.91 + i(-0.42)$
    $= -0.91 - 0.42i$

33. $10(\cos 342° + i\sin 342°) = 10\left[0.951 + i(-0.309)\right]$
    $= 9.51 - 3.09i$

35. We have $x = -1$ and $y = 1$; therefore $r = \sqrt{(-1)^2 + 1^2} = \sqrt{2}$.

We also know that $\tan \theta = \dfrac{1}{-1}$ and θ is in QII. Therefore, $\theta = 135°$.

In trigonometric form, $z = r(\cos \theta + i \sin \theta) = \sqrt{2}(\cos 135° + i \sin 135°)$.

37. We have $x = 1$ and $y = -1$; therefore $r = \sqrt{1^2 + (-1)^2} = \sqrt{2}$.

We also know that $\tan \theta = \dfrac{-1}{1}$ and θ is in QIV. Therefore, $\theta = 315°$.

In trigonometric form, $z = r(\cos \theta + i \sin \theta) = \sqrt{2}(\cos 315° + i \sin 315°)$.

39. We have $x = 3$ and $y = 3$; therefore $r = \sqrt{3^2 + 3^2} = \sqrt{18} = 3\sqrt{2}$.

We also know that $\tan \theta = \dfrac{3}{3} = 1$ and θ is in QI. Therefore, $\theta = 45°$.

In trigonometric form, $z = r(\cos \theta + i \sin \theta) = 3\sqrt{2}(\cos 45° + i \sin 45°)$.

41. We have $x = 0$ and $y = 8$; therefore $r = \sqrt{0^2 + 8^2} = \sqrt{64} = 8$.

We also know that $\theta = 90°$ from its graph.

In trigonometric form, $z = r(\cos \theta + i \sin \theta) = 8(\cos 90° + i \sin 90°)$.

43. We have $x = -9$ and $y = 0$; therefore $r = \sqrt{(-9)^2 + 0^2} = \sqrt{81} = 9$.

We also know that $\theta = 180°$ from its graph.

In trigonometric form, $z = r(\cos \theta + i \sin \theta) = 9(\cos 180° + i \sin 180°)$.

45. We have $x = -2$ and $y = 2\sqrt{3}$; therefore $r = \sqrt{(-2)^2 + \left(2\sqrt{3}\right)^2} = \sqrt{16} = 4$.

We also know that $\tan \theta = \dfrac{2\sqrt{3}}{-2} = -\sqrt{3}$ and θ is in QII. Therefore, $\theta = 120°$.

In trigonometric form, $z = r(\cos \theta + i \sin \theta) = 4(\cos 120° + i \sin 120°)$.

47. We have $x = 3$ and $y = 4$; therefore $r = \sqrt{3^2 + 4^2} = \sqrt{25} = 5$.

We also know that $\tan \theta = \dfrac{4}{3}$ and θ is in QI. Therefore, $\theta = 53.13°$.

In trigonometric form, $z = r(\cos \theta + i \sin \theta) = 5(\cos 53.13° + i \sin 53.13°)$.

49. We have $x = 20$ and $y = 21$; therefore $r = \sqrt{20^2 + 21^2} = \sqrt{841} = 29$.

We also know that $\tan\theta = \dfrac{21}{20}$ and θ is in QI. Therefore, $\theta = 46.40°$.

In trigonometric form, $z = r(\cos\theta + i\sin\theta) = 29(\cos 46.40° + i\sin 46.40°)$.

51. We have $x = 7$ and $y = -24$; therefore $r = \sqrt{7^2 + (-24)^2} = \sqrt{625} = 25$.

We also know that $\tan\theta = -\dfrac{24}{7}$ and θ is in QIV. Therefore, $\hat{\theta} = 73.74°$ and $\theta = 286.26°$.

In trigonometric form, $z = r(\cos\theta + i\sin\theta) = 5(\cos 286.26° + i\sin 286.26°)$.

53. We have $x = 11$ and $y = 2$; therefore $r = \sqrt{11^2 + 2^2} = \sqrt{125} = 5\sqrt{5}$.

We also know that $\tan\theta = \dfrac{2}{11}$ and θ is in QI. Therefore, $\theta = 10.30°$.

In trigonometric form, $z = r(\cos\theta + i\sin\theta) = 5\sqrt{5}(\cos 10.30° + i\sin 10.30°)$.

63. $2i = 2(\cos 90° + i\sin 90°)$ and $3i = 3(\cos 90° + i\sin 90°)$

$2i \cdot 3i = \left[2(\cos 90° + i\sin 90°)\right]\left[3(\cos 90° + i\sin 90°)\right]$

$\qquad = 2 \cdot 3\left(\cos^2 90° + 2i\cos 90° \sin 90° + i^2 \sin^2 90°\right)$

$\qquad = 6\left[0 + 2i(0)(1) + (-1)(1)^2\right]$

$\qquad = 6(-1) = -6$

65. $2(\cos 30° + i\sin 30°) = 2\left(\dfrac{\sqrt{3}}{2} + i\cdot\dfrac{1}{2}\right) = \sqrt{3} + i$

$2\left[\cos(-30°) + i\sin(-30°)\right] = 2\left[\dfrac{\sqrt{3}}{2} + i\left(-\dfrac{1}{2}\right)\right] = \sqrt{3} - i$

67. $|z| = \sqrt{\cos^2\theta + \sin^2\theta} = \sqrt{1} = 1$

69. $\cos 75° = \cos(30° + 45°)$

$\qquad = \cos 30° \cos 45° - \sin 30° \sin 45°$

$\qquad = \dfrac{\sqrt{3}}{2}\cdot\dfrac{\sqrt{2}}{2} - \dfrac{1}{2}\cdot\dfrac{\sqrt{2}}{2}$

$\qquad = \dfrac{\sqrt{6} - \sqrt{2}}{4}$

71. $\sin A = \dfrac{3}{5}$ $\qquad\qquad\qquad\qquad$ $\sin B = \dfrac{5}{13}$

$\cos A = \sqrt{1-\left(\dfrac{3}{5}\right)^2}$ $\qquad\qquad$ $\cos B = \sqrt{1-\left(\dfrac{5}{13}\right)^2}$

$\qquad = \sqrt{1-\dfrac{9}{25}} = \sqrt{\dfrac{16}{25}} = \dfrac{4}{5}$ $\qquad\qquad = \sqrt{1-\dfrac{25}{169}} = \sqrt{\dfrac{144}{169}} = \dfrac{12}{13}$

$\sin(A+B) = \sin A \cos B + \cos A \sin B$

$\qquad = \dfrac{3}{5}\cdot\dfrac{12}{13} + \dfrac{4}{5}\cdot\dfrac{5}{13} = \dfrac{36}{65} + \dfrac{20}{65} = \dfrac{56}{65}$

73. $\sin 30° \cos 90° + \cos 30° \sin 90° = \sin\left(30° + 90°\right)$

$\qquad\qquad\qquad\qquad\qquad\qquad = \sin 120° = \dfrac{\sqrt{3}}{2}$

75. $\cos 18° \cos 32° - \sin 18° \sin 32° = \cos(18° + 32°)$

$\qquad\qquad\qquad\qquad\qquad\qquad = \cos 50°$

77. $\sin B = \dfrac{b \sin A}{a}$ $\qquad\qquad$ **79.** $\sin B = \dfrac{b \sin A}{a}$

$\qquad = \dfrac{567 \sin 45.6°}{234} = 1.7312$ $\qquad\qquad\qquad = \dfrac{567 \sin 45.6°}{456} = 0.8884$

No triangle exists because sin B is $\qquad\qquad$ $B = 62.7°$ or $117.3°$

greater than 1 (which is impossible). $\qquad\qquad$ Two triangles exist because $a < b$.

Problem Set 8.3

1. $3(\cos 20° + i \sin 20°) \cdot 4(\cos 30° + i \sin 30°) = 3 \cdot 4 \left[\cos(20° + 30°) + i \sin(20° + 30°)\right]$

$\qquad\qquad\qquad\qquad\qquad\qquad\qquad = 12(\cos 50° + i \sin 50°)$

3. $7(\cos 110° + i \sin 110°) \cdot 8(\cos 47° + i \sin 47°) = 7 \cdot 8 \left[\cos(110° + 47°) + i \sin(110° + 47°)\right]$

$\qquad\qquad\qquad\qquad\qquad\qquad\qquad\qquad = 56(\cos 157° + i \sin 157°)$

5. $2(\cos 135° + i \sin 135°) \cdot 2(\cos 45° + i \sin 45°) = 2 \cdot 2 \left[\cos(135° + 45°) + i \sin(135° + 45°)\right]$

$\qquad\qquad\qquad\qquad\qquad\qquad\qquad = 4(\cos 180° + i \sin 180°) = 4\text{cis}180°$

7.

$$z_1 z_2 = (1+i)(-1+i)$$
$$= -1 + i - i + i^2$$
$$= -1 - 1 = -2$$

$z_1 = 1 + i$ where $x = 1$, $y = 1$, and $r = \sqrt{1^2 + 1^2} = \sqrt{2}$

Also $\tan\theta = 1$ and θ is in QI. Therefore, $\theta = 45°$.

$z_1 = \sqrt{2}\,\text{cis}45°$ in trigonometric form

$z_2 = -1 + i$ where $x = -1$, $y = 1$, and $r = \sqrt{(-1)^2 + 1^2} = \sqrt{2}$

Also $\tan\theta = -1$ and θ is in QII. Therefore, $\theta = 135°$.

$z_2 = \sqrt{2}\,\text{cis}(135°)$ in trigonometric form

$$z_1 z_2 = \sqrt{2}\,\text{cis}(45°) \cdot \sqrt{2}\,\text{cis}(135°)$$
$$= \sqrt{2} \cdot \sqrt{2}\,\text{cis}(45° + 135°)$$
$$= 2\,\text{cis}(180°) = 2(-1 + i \cdot 0) = -2$$

9.

$$z_1 z_2 = (1 + i\sqrt{3})(-\sqrt{3} + i)$$
$$= -\sqrt{3} - 2i + i^2\sqrt{3}$$
$$= -\sqrt{3} - 2i - \sqrt{3} = -2\sqrt{3} - 2i$$

$z_1 = 1 + i\sqrt{3}$ where $x = 1$, $y = \sqrt{3}$, and $r = \sqrt{1^2 + \left(\sqrt{3}\right)^2} = 2$

Also $\tan\theta = \sqrt{3}$ and θ is in QI. Therefore, $\theta = 60°$.

$z_1 = 2\,\text{cis}(60°)$ in trigonometric form

$z_2 = -\sqrt{3} + i$ where $x = -\sqrt{3}$, $y = 1$, and

$$r = \sqrt{\left(-\sqrt{3}\right)^2 + 1^2} = 2$$

Also $\tan\theta = \dfrac{-1}{\sqrt{3}}$ and θ is in QII. Therefore, $\theta = 150°$.

$z_2 = 2\,\text{cis}(150°)$ in trigonometric form

$$z_1 z_2 = 2\,\text{cis}(60°) \cdot 2\,\text{cis}(150°)$$
$$= 2 \cdot 2\,\text{cis}(60° + 150°)$$
$$= 4\,\text{cis}(210°) = 4\left(-\frac{\sqrt{3}}{2} - i \cdot \frac{1}{2}\right) = -2\sqrt{3} - 2i$$

11.

$$z_1 z_2 = (3i)(-4i)$$
$$= -12i^2$$
$$= -12(-1) = 12$$

$z_1 = 3i$ where $x = 0$, $y = 3$, and $r = 3$, and $\theta = 90°$

$z_1 = 3\,\text{cis}90°$ in trigonometric form

$z_2 = -4i$ where $x = 0$, $y = -4$, $r = 4$ and $\theta = 270°$

$z_2 = 4\,\text{cis}(270°)$ in trigonometric form

$$z_1 z_2 = 3\,\text{cis}(90°) \cdot 4\,\text{cis}(270°)$$
$$= 3 \cdot 4\,\text{cis}(90° + 270°)$$
$$= 12\,\text{cis}(360°) = 12(1 + i \cdot 0) = 12$$

13.

$$z_1 z_2 = (1+i)(4i)$$
$$= 4i + 4i^2$$
$$= 4i + 4(-1) = -4 + 4i$$

$z_1 = 1 + i$ where $x = 1, y = 1$, and $r = \sqrt{1^2 + 1^2} = \sqrt{2}$.

Also $\tan\theta = 1$ and θ is in QI. Therefore, $\theta = 45°$.

$z_1 = \sqrt{2}\text{cis}(45°)$ in trigonometric form

$z_2 = 4i$ where $x = 0, y = 4, r = 4$ and $\theta = 90°$

$z_2 = 4\text{cis}(90°)$ in trigonometric form

$$z_1 z_2 = \sqrt{2}\text{cis}(45°) \cdot 4\text{cis}(90°)$$
$$= \sqrt{2} \cdot 4\text{cis}(45° + 90°)$$
$$= 4\sqrt{2}\text{cis}(135°) = 4\sqrt{2}\left(\frac{-\sqrt{2}}{2} + i\frac{\sqrt{2}}{2}\right) = -4 + 4i$$

15.

$$z_1 z_2 = -5(1 + i\sqrt{3})$$
$$= -5 - 5i\sqrt{3}$$

17.

$$\left[2(\cos 10° + i\sin 10°)\right]^6 = 2^6\left[\cos(6 \cdot 10°) + i\sin(6 \cdot 10°)\right]$$
$$= 64(\cos 60° + i\sin 60°)$$
$$= 64\left(\frac{1}{2} + i \cdot \frac{\sqrt{3}}{2}\right) = 32 + 32i\sqrt{3}$$

19.

$$\left(\cos 12° + i\sin 12°\right)^{10} = 1^{10}\left[\cos(10 \cdot 12°) + i\sin(10 \cdot 12°)\right]$$
$$= 1(\cos 120° + i\sin 120°)$$
$$= -\frac{1}{2} + \frac{\sqrt{3}}{2}i$$

21.

$$\left(3\text{cis}60°\right)^4 = 3^4\left[\text{cis}(4 \cdot 60°)\right]$$
$$= 81(\cos 240° + i\sin 240°)$$
$$= 81\left(-\frac{1}{2} + i \cdot -\frac{\sqrt{3}}{2}\right)$$
$$= -\frac{81}{2} - \frac{81\sqrt{3}}{2}i$$

23.
$$\left(\sqrt{2}\operatorname{cis}45^\circ\right)^{10} = \left(\sqrt{2}\right)^{10}\left[\operatorname{cis}(10\cdot 45^\circ)\right]$$
$$= 32(\cos 450^\circ + i\sin 450^\circ)$$
$$= 32(0 + i\cdot 1) = 32i$$

25. In problem 7, we found that $1 + i$ was $\sqrt{2}(\cos 45^\circ + i\sin 45^\circ)$ in trigonometric form.
$$z^4 = \left(\sqrt{2}\right)^4\left[\cos(4\cdot 45^\circ) + i\sin(4\cdot 45^\circ)\right]$$
$$= 4(\cos 180^\circ + i\sin 180^\circ)$$
$$= 4\left[-1 + i(0)\right] = -4$$

27. In problem 9, we found that $-\sqrt{3} + i$ was $2(\cos 150^\circ + i\sin 150^\circ)$ in trigonometric form.
$$z^4 = (2)^4\left[\cos(4\cdot 150^\circ) + i\sin(4\cdot 150^\circ)\right]$$
$$= 16(\cos 600^\circ + i\sin 600^\circ)$$
$$= 16\left[-\frac{1}{2} + i\left(-\frac{\sqrt{3}}{2}\right)\right] = -8 - 8i\sqrt{3}$$

29. In problem 7, we found that $1 + i$ was $\sqrt{2}(\cos 45^\circ + i\sin 45^\circ)$ in trigonometric form.
$$z^6 = \left(\sqrt{2}\right)^6\left[\cos(6\cdot 45^\circ) + i\sin(6\cdot 45^\circ)\right]$$
$$= 8(\cos 270^\circ + i\sin 270^\circ)$$
$$= 8\left[0 + i(-1)\right] = -8i$$

31. First we write $-2 + 2i$ in trigonometric form: $x = -2$, $y = 2$, and $r = \sqrt{(-2)^2 + 2^2} = \sqrt{8} = 2\sqrt{2}$

Also $\tan\theta = \dfrac{-2}{2} = -1$ and θ is in QII. Therefore, $\theta = 135^\circ$ and $z = 2\sqrt{2}\operatorname{cis}(135^\circ)$.

$$z^3 = \left(2\sqrt{2}\right)^3\left[\operatorname{cis}(3\cdot 135^\circ)\right]$$
$$= 16\sqrt{2}\left(\cos 405^\circ + i\sin 405^\circ\right)$$
$$= 16\sqrt{2}\left(\frac{\sqrt{2}}{2} + i\cdot\frac{\sqrt{2}}{2}\right) = 16 + 16i$$

33.
$$\frac{20\left(\cos 75^\circ + i\sin 75^\circ\right)}{5\left(\cos 40^\circ i\sin 40^\circ\right)} = \frac{20}{5}\left[\cos(75^\circ - 40^\circ) + i\sin(75^\circ - 40^\circ)\right]$$
$$= 4(\cos 35^\circ + i\sin 35^\circ)$$

35.
$$\frac{18\left(\cos 51° + i\sin 51°\right)}{12\left(\cos 32° \, i\sin 32°\right)} = \frac{18}{12}\left[\cos(51° - 32°) + i\sin(51° - 32°)\right]$$
$$= 1.5(\cos 19° + i\sin 19°)$$

37.
$$\frac{4\text{cis}90°}{8\text{cis}30°} = \frac{4}{8}\text{cis}(90° - 30°)$$
$$= 0.5\text{cis}60°$$

39.
$$\frac{z_1}{z_2} = \frac{2+2i}{1+i} \cdot \frac{1-i}{1-i}$$
$$= \frac{2-2i^2}{1-i^2} = \frac{2+2}{1+1} = \frac{4}{2} = 2$$

$z_1 = 2+2i$ where $x = 2$, $y = 2$, and $r = \sqrt{2^2 + 2^2} = 2\sqrt{2}$.

Also, $\tan\theta = \frac{2}{2} = 1$ and θ is in QI. Therefore, $\theta = 45°$ and $z_1 = 2\sqrt{2}\text{cis}45°$.

In problem 7, we found that $z_2 = \sqrt{2}(\cos 45° + i\sin 45°)$ in trigonometric form.

$$\frac{z_1}{z_2} = \frac{2\sqrt{2}}{\sqrt{2}}\text{cis}(45° - 45°)$$
$$= 2\text{cis}0° = 2(1 + i \cdot 0) = 2$$

41.
$$\frac{z_1}{z_2} = \frac{\sqrt{3}+i}{2i} \cdot \frac{i}{i}$$
$$= \frac{i\sqrt{3} + i^2}{2i^2} = \frac{-1 + i\sqrt{3}}{-2} = \frac{1}{2} - \frac{\sqrt{3}}{2}i$$

$z_1 = \sqrt{3} + i$ where $x = \sqrt{3}$, $y = 1$, and $r = \sqrt{\left(\sqrt{3}\right)^2 + 1^2} = 2$.

Also, $\tan\theta = \frac{1}{\sqrt{3}}$ and θ is in QI. Therefore, $\theta = 30°$ and $z_1 = 2\text{cis}30°$.

$z_2 = 2i$ where $x = 0$, $y = 2$, $r = 2$, and $\theta = 90°$. In trigonometric form, $z_2 = 2\text{cis}90°$.

$$\frac{z_1}{z_2} = \frac{2\,\text{cis}30°}{2\,\text{cis}90°} = \frac{2}{2}\text{cis}(30° - 90°)$$
$$= \text{cis}(-60°) = \frac{1}{2} + i\left(-\frac{\sqrt{3}}{2}\right) = \frac{1}{2} - \frac{\sqrt{3}}{2}i$$

43. $\dfrac{z_1}{z_2} = \dfrac{4+4i}{2-2i} \cdot \dfrac{2+2i}{2+2i}$

$$= \frac{8+16i+8i^2}{4-4i^2}$$

$$= \frac{8+16i-8}{4+4} = \frac{16i}{8} = 2i$$

$z_1 = 4+4i$ where $x=4$, $y=4$, and $r = \sqrt{4^2+4^2} = \sqrt{32} = 4\sqrt{2}$.

Also, $\tan\theta = \dfrac{4}{4} = 1$ and θ is in QI. Therefore, $\theta = 45°$ and $z_1 = 4\sqrt{2}\operatorname{cis}45°$.

$z_2 = 2-2i$ where $x=2$, $y=-2$, and $r = \sqrt{2^2+(-2)^2} = \sqrt{8} = 2\sqrt{2}$.

Also, $\tan\theta = \dfrac{2}{-2} = -1$ and θ is in QIV. Therefore, $\theta = 315°$ and $z_2 = 2\sqrt{2}\operatorname{cis}315°$.

$\dfrac{z_1}{z_2} = \dfrac{4\sqrt{2}\operatorname{cis}45°}{2\sqrt{2}\operatorname{cis}315°} = \dfrac{4\sqrt{2}}{2\sqrt{2}}\operatorname{cis}(45°-315°)$

$$= 2\operatorname{cis}(-270°)$$

$$= 2(\cos 90° + i\sin 90°) = 2(0 + i\cdot 1) = 2i$$

45. $\dfrac{z_1}{z_2} = \dfrac{8}{-4} = -2$

$z_1 = 8$ where $x=8$, $y=0$, $r=8$, and $\theta = 0°$. In trigonometric form, $z_1 = 8\operatorname{cis}0°$.

$z_2 = -4$ where $x=-4$, $y=0$, $r=4$, and $\theta = 180°$. In trigonometric form, $z_1 = 4\operatorname{cis}180°$.

$\dfrac{z_1}{z_2} \dfrac{8\operatorname{cis}0°}{4\operatorname{cis}180°} = \dfrac{8}{4}\operatorname{cis}(0°-180°)$

$$= 2\operatorname{cis}(-180°)$$

$$= 2(\cos 180° + i\sin 180°) = 2(-1 + i\cdot 0) = -2$$

47. In problem 7, we found that $z_1 = 1+i$ was $\sqrt{2}(\cos 45° + i\sin 45°)$ in trigonometric form.

In problem 41, we found that $z_2 = 2i$ was $2(\cos 90° + i\sin 90°)$ in trigonometric form.

In problem 31, we found that $z_3 = -2+2i$ was $2\sqrt{2}(\cos 135° + i\sin 135°)$ in trigonometric form.

$$\frac{(z_1)^4(z_2)^2}{z_3} = \frac{\left[(\sqrt{2})^4\operatorname{cis}(4\cdot 45°)\right]\left[(2)^2\operatorname{cis}(2\cdot 90°)\right]}{2\sqrt{2}\operatorname{cis}(135°)}$$

$$= \frac{4\cdot 4(\operatorname{cis}180°)(\operatorname{cis}180°)}{2\sqrt{2}\operatorname{cis}135°}$$

This problem is continued on the next page.

$$= \frac{16 \text{cis} 360°}{2\sqrt{2} \text{cis} 135°}$$

$$= \frac{8}{\sqrt{2}} \left[\text{cis}(360° - 135°) \right] = \frac{8}{\sqrt{2}} \text{cis} 225°$$

$$= \frac{8}{\sqrt{2}} \left(-\frac{\sqrt{2}}{2} + i \cdot -\frac{\sqrt{2}}{2} \right) = -4 - 4i$$

49. In problem 9, we found that $z_1 = 1 + i\sqrt{3}$ was $2(\cos 60° + i \sin 60°)$ in trigonometric form.

$z_2 = \sqrt{3} - i$ where $x = \sqrt{3}$, $y = -1$, and $r = \sqrt{\left(\sqrt{3}\right)^2 + (-1)^2} = 2$.

Also, $\tan \theta = -\frac{1}{\sqrt{3}}$ and θ is in QIV. Therefore, $\theta = 330°$ and $z_2 = 2(\cos 330° + i \sin 330°)$.

$z_3 = 1 - i\sqrt{3}$ where $x = 1$, $y = -\sqrt{3}$, and $r = \sqrt{1^2 + \left(\sqrt{3}\right)^2} = 2$.

Also, $\tan \theta = -\sqrt{3}$ and θ is in QIV. Therefore, $\theta = 300°$ and $z_3 = 2(\cos 300° + i \sin 300°)$.

$$\frac{(z_1)^4 (z_2)^2}{(z_3)^3} = \frac{\left[(2)^4 \text{cis}(4 \cdot 60°) \right] \left[(2)^2 \text{cis}(2 \cdot 330°) \right]}{2^3 \text{cis}(3 \cdot 300°)}$$

$$= \frac{16 \cdot 4 \left(\text{cis} 240° \right) \left(\text{cis} 660° \right)}{8 \text{cis} 900°}$$

$$= \frac{64 \text{cis}(240° + 660°)}{8 \text{cis} 900°}$$

$$= \frac{64 \text{cis} 900°}{8 \text{cis} 900°} = 8$$

51.
$$x^2 - 2x + 4 = (2\text{cis} 60°)^2 - 2(2\text{cis} 60°) + 4$$

$$= 4\text{cis}(120°) - 4\text{cis}(60°) + 4$$

$$= 4\left(-\frac{1}{2} + i \cdot \frac{\sqrt{3}}{2} \right) - 4\left(\frac{1}{2} + i \cdot \frac{\sqrt{3}}{2} \right) + 4$$

$$= -2 + 2i\sqrt{3} - 2 - 2i\sqrt{3} + 4 = 0$$

53.
$$w^4 = \left[2(\cos 15° + i \sin 15°) \right]^4$$

$$= 2^4 \left[\cos(4 \cdot 15°) + i \sin(4 \cdot 15°) \right]$$

$$= 16 \left[\cos(60°) + i \sin(60°) \right]$$

$$= 16\left(\frac{1}{2} + i \cdot \frac{\sqrt{3}}{2} \right) = 8 + 8i\sqrt{3}$$

55. In problem 7, we found that $1+i$ was $\sqrt{2}(\cos 45° + i\sin 45°)$ in trigonometric form.

$$(1+i)^{-1} = \left(\sqrt{2}\right)^{-1}\left[\cos(-1\cdot45°) + i\sin(-1\cdot45°)\right]$$

$$= \frac{1}{\sqrt{2}}\left[\frac{\sqrt{2}}{2} + i\left(-\frac{\sqrt{2}}{2}\right)\right] = \frac{1}{2} - \frac{1}{2}i$$

57. In problem 49, we found that $z = \sqrt{3} - i$ was $2(\cos 330° + i\sin 330°)$ in trigonometric form.

$$z^{-1} = 2^{-1}\left[\cos(-1\cdot330°) + i\sin(-1\cdot330°)\right]$$

$$= \frac{1}{2}\left[\cos(-330°) + i\sin(-330°)\right]$$

$$= \frac{1}{2}\left[\frac{\sqrt{3}}{2} + i\left(\frac{1}{2}\right)\right] = \frac{\sqrt{3}}{4} + \frac{1}{4}i$$

59.
$$\cos 2A = 2\cos^2 A - 1$$

$$= 2\left(-\frac{1}{3}\right)^2 - 1$$

$$= \frac{2}{9} - 1 = -\frac{7}{9}$$

61.
$$\sin\frac{A}{2} = \sqrt{\frac{1-\cos A}{2}}$$

$$= \sqrt{\frac{1-(-1/3)}{2}}$$

$$= \sqrt{\frac{2}{3}} = \frac{\sqrt{6}}{3}$$

63.
$$\csc\frac{A}{2} = \frac{1}{\sin(A/2)} = \frac{1}{\sqrt{6}/3} = \frac{3}{\sqrt{6}} = \frac{\sqrt{6}}{2}$$

65. If $\cos A = -\frac{1}{3}$ and A is in QII, then $\sin A = \sqrt{1-\cos^2 A} = \sqrt{1 - \frac{1}{9}} = \sqrt{\frac{8}{9}} = \frac{2\sqrt{2}}{3}$.

Also, $\tan A = \dfrac{\sin A}{\cos A} = \dfrac{2\sqrt{3}/3}{-1/3} = -2\sqrt{2}$

Therefore, $\tan 2A = \dfrac{2\tan A}{1-\tan^2 A} = \dfrac{2\left(-2\sqrt{2}\right)}{1-\left(-2\sqrt{2}\right)^2} = \dfrac{-4\sqrt{2}}{1-8} = \dfrac{4\sqrt{2}}{7}$

67.
$$\angle M = 180° - (27° + 35°) = 118°$$

$$x = \frac{9.2\sin 35°}{\sin 118°} = 6.0 \text{ miles}$$

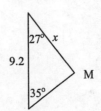

69. $|g| = \sqrt{170^2 + 28^2 - 2(170)(28)\cos 68^\circ}$

$\quad\quad = \sqrt{26{,}118} = 160$ mph (to 2 significant digits)

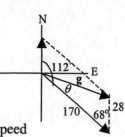

$\quad \sin\theta = \dfrac{28\sin 68^\circ}{160} = 0.1623$

$\quad\quad \theta = 9^\circ$ (to the nearest degree)

The true course is $112^\circ - \theta = 112^\circ - 9^\circ = 103^\circ$. The ground speed is 160 miles per hour.

Problem Set 8.4

1. The 2 square roots will be

$$w_k = 4^{1/2}\left(\text{cis}\,\frac{30^\circ + 360^\circ k}{2}\right) \text{ for } k = 0,1$$

$$= 2\text{cis}\left(15^\circ + 180^\circ k\right)$$

We replace k with 0 and 1:

$\quad$ when $k = 0$, $w_0 = 2(\cos 15^\circ + i\sin 15^\circ)$

$\quad$ when $k = 1$, $w_1 = 2(\cos 195^\circ + i\sin 195^\circ)$

3. The 2 square roots will be

$$w_k = 25^{1/2}\left(\text{cis}\,\frac{210^\circ + 360^\circ k}{2}\right) \text{ for } k = 0,1$$

$$= 5\text{cis}\left(105^\circ + 180^\circ k\right)$$

We replace k with 0 and 1:

$\quad$ when $k = 0$, $w_0 = 5(\cos 105^\circ + i\sin 105^\circ)$

$\quad$ when $k = 1$, $w_1 = 5(\cos 285^\circ + i\sin 285^\circ)$

5. The 2 square roots will be

$$w_k = 49^{1/2}\left(\text{cis}\,\frac{180^\circ + 360^\circ k}{2}\right) \text{ for } k = 0,1$$

$$= 7\text{cis}\left(90^\circ + 180^\circ k\right)$$

We replace k with 0 and 1:

$\quad$ when $k = 0$, $w_0 = 7(\cos 90^\circ + i\sin 90^\circ)$

$\quad$ when $k = 1$, $w_1 = 7(\cos 270^\circ + i\sin 270^\circ)$

7. First, we put $2 + 2i\sqrt{3}$ in trigonometric form: $x = 2$, $y = 2\sqrt{3}$, and $r = \sqrt{2^2 + (2\sqrt{3})^2} = 4$

Also, $\tan\theta = \dfrac{2\sqrt{3}}{2} = \sqrt{3}$ and θ is in QI. Therefore, $\theta = 60^\circ$.

In trigonometric form $2 + 2i\sqrt{3} = 4\left(\cos 60^\circ + i\sin 60^\circ\right)$

The 2 square roots will be $w_k = 4^{1/2}\left(\text{cis}\,\dfrac{60^\circ + 360^\circ k}{2}\right)$ for $k = 0,1$

$$= 2\text{cis}\left(30^\circ + 180^\circ k\right)$$

This problem is continued on the next page

We replace k with 0 and 1:

when $k = 0$, $w_0 = 2(\cos 30° + i \sin 30°) = 2\left(\dfrac{\sqrt{3}}{2} + i \cdot \dfrac{1}{2}\right) = \sqrt{3} + i$

when $k = 1$, $w_1 = 2(\cos 210° + i \sin 210°) = 2\left(-\dfrac{\sqrt{3}}{2} + i \cdot -\dfrac{1}{2}\right) = -\sqrt{3} - i$

9. In trigonometric form, $4i = 4(\cos 90° + i \sin 90°)$

The 2 square roots will be $w_k = 4^{1/2}\left(\text{cis}\dfrac{90° + 360°k}{2}\right)$ for $k = 0, 1$

$$= 2\text{cis}\left(45° + 180°k\right)$$

We replace k with 0 and 1:

when $k = 0$, $w_0 = 2(\cos 45° + i \sin 45°) = 2\left(\dfrac{\sqrt{2}}{2} + i \cdot \dfrac{\sqrt{2}}{2}\right) = \sqrt{2} + i\sqrt{2}$

when $k = 1$, $w_1 = 2(\cos 225° + i \sin 225°) = 2\left(-\dfrac{\sqrt{2}}{2} + i \cdot -\dfrac{\sqrt{2}}{2}\right) = -\sqrt{2} - i\sqrt{2}$

11. In trigonometric form, $-25 = 25(\cos 180° + i \sin 180°)$

The 2 square roots will be $w_k = 25^{1/2}\left(\text{cis}\dfrac{180° + 360°k}{2}\right)$ for $k = 0, 1$

$$= 5\text{cis}\left(90° + 180°k\right)$$

We replace k with 0 and 1:

when $k = 0$, $w_0 = 5(\cos 90° + i \sin 90°) = 5(0 + i \cdot 1) = 5i$

when $k = 1$, $w_1 = 5(\cos 270° + i \sin 270°) = 5(0 + i \cdot -1) = -5i$

13. In trigonometric form, $1 + i\sqrt{3} = 2(\cos 60° + i \sin 60°)$

The 2 square roots will be $w_k = 2^{1/2}\left(\text{cis}\dfrac{60° + 360°k}{2}\right)$ for $k = 0, 1$

$$= \sqrt{2}\text{cis}\left(30° + 180°k\right)$$

We replace k with 0 and 1:

when $k = 0$, $w_0 = \sqrt{2}(\cos 30° + i \sin 30°) = \sqrt{2}\left(\dfrac{\sqrt{3}}{2} + i \cdot \dfrac{1}{2}\right) = \dfrac{\sqrt{6}}{2} + \dfrac{\sqrt{2}}{2}i$

when $k = 1$, $w_1 = \sqrt{2}(\cos 210° + i \sin 210°) = \sqrt{2}\left(-\dfrac{\sqrt{3}}{2} + i \cdot -\dfrac{1}{2}\right) = -\dfrac{\sqrt{6}}{2} - \dfrac{\sqrt{2}}{2}i$

15. The 3 cube roots will be $w_k = 8^{1/3} \left(\text{cis} \frac{210° + 360°k}{3} \right)$ for $k = 0, 1, 2$

$$= 2\text{cis}\left(70° + 120°k \right)$$

We replace k with 0, 1, and 2:

when $k = 0$, $w_0 = 2(\cos 70° + i \sin 70°)$

when $k = 1$, $w_1 = 2(\cos 190° + i \sin 190°)$

when $k = 2$, $w_2 = 2(\cos 310° + i \sin 310°)$

17. In trigonometric form, $4\sqrt{3} + 4i = 8(\cos 30° + i \sin 30°)$

The 3 cube roots will be $w_k = 8^{1/3} \left(\text{cis} \frac{30° + 360°k}{3} \right)$ for $k = 0, 1, 2$

$$= 2\text{cis}\left(10° + 120°k \right)$$

We replace k with 0, 1, and 2:

when $k = 0$, $w_0 = 2(\cos 10° + i \sin 10°)$

when $k = 1$, $w_1 = 2(\cos 130° + i \sin 130°)$

when $k = 2$, $w_2 = 2(\cos 250° + i \sin 250°)$

19. In trigonometric form, $-27 = 27(\cos 180° + i \sin 180°)$

The 3 cube roots will be $w_k = 27^{1/3} \left(\text{cis} \frac{180° + 360°k}{3} \right)$ for $k = 0, 1, 2$

$$= 3\text{cis}\left(60° + 120°k \right)$$

We replace k with 0, 1, and 2:

when $k = 0$, $w_0 = 3(\cos 60° + i \sin 60°)$

when $k = 1$, $w_1 = 3(\cos 180° + i \sin 180°)$

when $k = 2$, $w_2 = 3(\cos 300° + i \sin 300°)$

21. In trigonometric form, $64i = 64(\cos 90° + i \sin 90°)$

The 3 cube roots will be $w_k = 64^{1/3} \left(\text{cis} \frac{90° + 360°k}{3} \right)$ for $k = 0, 1, 2$

$$= 4\text{cis}\left(30° + 120°k \right)$$

We replace k with 0, 1, and 2:

when $k = 0$, $w_0 = 4(\cos 30° + i \sin 30°)$

when $k = 1$, $w_1 = 4(\cos 150° + i \sin 150°)$

when $k = 2$, $w_2 = 4(\cos 270° + i \sin 270°)$

23. $x^3 - 27 = 0$

   $x^3 = 27$ The solutions are the 3 cube roots of 27.

In trigonometric form, $27 = 27(\cos 0° + i \sin 0°)$

The 3 cube roots will be $w_k = 27^{1/3}\left(\text{cis}\dfrac{0° + 360° k}{3}\right)$ for $k = 0, 1, 2$

$$= 3\text{cis}\left(120° k\right)$$

We replace k with 0, 1, and 2:

when $k = 0$, $w_0 = 3(\cos 0° + i \sin 0°) = 3(1 + i \cdot 0) = 3$

when $k = 1$, $w_1 = 3(\cos 120° + i \sin 120°) = 3\left(-\dfrac{1}{2} + i \cdot \dfrac{\sqrt{3}}{2}\right) = -\dfrac{3}{2} + \dfrac{3\sqrt{3}}{2}i$

when $k = 2$, $w_2 = 3(\cos 240° + i \sin 240°) = 3\left(-\dfrac{1}{2} + i \cdot -\dfrac{\sqrt{3}}{2}\right) = -\dfrac{3}{2} - \dfrac{3\sqrt{3}}{2}i$

25. $x^4 - 16 = 0$

   $x^4 = 16$ The solutions are the 4 fourth roots of 16.

In trigonometric form, $16 = 16(\cos 0° + i \sin 0°)$

The 4 fourth roots will be $w_k = 16^{1/4}\left(\text{cis}\dfrac{0° + 360° k}{4}\right)$ for $k = 0, 1, 2, 3$

$$= 2\text{cis}\left(90° k\right)$$

We replace k with 0, 1, 2, and 3:

when $k = 0$, $w_0 = 2(\cos 0° + i \sin 0°) = 2(1 + i \cdot 0) = 2$

when $k = 1$, $w_1 = 2(\cos 90° + i \sin 90°) = 2(0 + i \cdot 1) = 2i$

when $k = 2$, $w_2 = 2(\cos 180° + i \sin 180°) = 2(-1 + i \cdot 0) = -2$

when $k = 3$, $w_3 = 2(\cos 270° + i \sin 270°) = 2(0 + i \cdot -1) = -2i$

27.   The 4 fourth roots will be $w_k = 16^{1/4}\left(\text{cis}\dfrac{120° + 360° k}{4}\right)$ for $k = 0, 1, 2, 3$

$$= 2\text{cis}\left(30° + 90° k\right)$$

We replace k with 0, 1, 2, and 3:

when $k = 0$, $w_0 = 2(\cos 30° + i \sin 30°) = 2\left(\dfrac{\sqrt{3}}{2} + i \cdot \dfrac{1}{2}\right) = \sqrt{3} + i$

when $k = 1$, $w_1 = 2(\cos 120° + i \sin 120°) = 2\left(-\dfrac{1}{2} + i \cdot \dfrac{\sqrt{3}}{2}\right) = -1 + i\sqrt{3}$

This problem is continued on the next page

when $k = 2$, $w_2 = 2(\cos 210° + i \sin 210°) = 2\left(-\dfrac{\sqrt{3}}{2} + i \cdot -\dfrac{1}{2}\right) = -\sqrt{3} - i$

when $k = 3$, $w_3 = 2(\cos 300° + i \sin 300°) = 2\left(\dfrac{1}{2} + i \cdot -\dfrac{\sqrt{3}}{2}\right) = 1 - i\sqrt{3}$

29. The 5 fifth roots will be $w_k = \left(10^5\right)^{1/5}\left(\text{cis}\dfrac{15° + 360°k}{5}\right)$ for $k = 0, 1, 2, 3, 4$

$$= 10\text{cis}\left(3° + 72°k\right)$$

We replace k with 0, 1, 2, 3, and 4:

when $k = 0$, $w_0 = 10(\cos 3° + i \sin 3°) = 10[0.999 + i(0.052)] = 9.99 + 0.52i$

when $k = 1$, $w_1 = 10(\cos 75° + i \sin 75°) = 10[0.259 + i(0.966)] = 2.59 + 9.66i$

when $k = 2$, $w_2 = 10(\cos 147° + i \sin 147°) = 10[-0.839 + i(0.545)] = -8.39 + 5.45i$

when $k = 3$, $w_3 = 10(\cos 219° + i \sin 219°) = 10[-0.777 + i(-0.629)] = -7.77 - 6.29i$

when $k = 4$, $w_4 = 10(\cos 291° + i \sin 291°) = 10[0.358 + i(-0.934)] = 3.58 - 9.34i$

31. In trigonometric form, $-1 = 1(\cos 180° + i \sin 180°)$

The 6 sixth roots will be $w_k = 1^{1/6}\left(\text{cis}\dfrac{180° + 360°k}{6}\right)$ for $k = 0, 1, 2, 3, 4, 5$

$$= 1\text{cis}\left(30° + 60°k\right)$$

We replace k with 0, 1, 2, 3, 4, and 5:

when $k = 0$, $w_0 = \cos 30° + i \sin 30°$

when $k = 1$, $w_1 = \cos 90° + i \sin 90°$

when $k = 2$, $w_2 = \cos 150° + i \sin 150°$

when $k = 3$, $w_3 = \cos 210° + i \sin 210°$

when $k = 4$, $w_4 = \cos 270° + i \sin 270°$

when $k = 5$, $w_5 = \cos 330° + i \sin 330°$

33. Applying the quadratic formula, we have $x^2 = \dfrac{2 \pm \sqrt{(-2)^2 - 4(1)(4)}}{2(1)}$

$$= \dfrac{2 \pm \sqrt{-12}}{2} = \dfrac{2 \pm 2i\sqrt{3}}{2} = 1 \pm i\sqrt{3}$$

First, we find the 2 square roots of $1 + i\sqrt{3}$:

In trigonometric form, $1 + i\sqrt{3} = 2(\cos 60° + i \sin 60°)$

The 2 square roots will be $w_k = 2^{1/2}\left(\text{cis}\dfrac{60° + 360°k}{2}\right)$ for $k = 0, 1$

This problem is continued on the next page

$$= \sqrt{2}\operatorname{cis}\left(30° + 180° k\right)$$

We replace k with 0 and 1:

when $k = 0$, $w_0 = \sqrt{2}(\cos 30° + i \sin 30°)$

when $k = 1$, $w_1 = \sqrt{2}(\cos 210° + i \sin 210°)$

Next, we find the 2 square roots of $1 - i\sqrt{3}$:

In trigonometric form, $1 - i\sqrt{3} = 2(\cos 300° + i \sin 300°)$

The 2 square roots will be $w_k = 2^{1/2}\left(\operatorname{cis}\dfrac{300° + 360° k}{2}\right)$ for $k = 0,1$

$$= \sqrt{2}\operatorname{cis}\left(150° + 180° k\right)$$

We replace k with 0 and 1:

when $k = 0$, $w_0 = \sqrt{2}(\cos 150° + i \sin 150°)$

when $k = 1$, $w_1 = \sqrt{2}(\cos 330° + i \sin 330°)$

35. Applying the quadratic formula, we have $\qquad x^2 = \dfrac{-2 \pm \sqrt{2^2 - 4(1)(2)}}{2(1)}$

$$= \dfrac{-2 \pm \sqrt{-4}}{2} = \dfrac{-2 \pm 2i}{2} = -1 \pm i$$

First, we find the 2 square roots of $-1 + i$:

In trigonometric form, $-1 + i = \sqrt{2}(\cos 135° + i \sin 135°)$

The 2 square roots will be $w_k = \left(\sqrt{2}\right)^{1/2}\left(\operatorname{cis}\dfrac{135° + 360° k}{2}\right)$ for $k = 0,1$

$$= \sqrt[4]{2}\operatorname{cis}\left(67.5° + 180° k\right)$$

We replace k with 0 and 1:

when $k = 0$, $w_0 = \sqrt[4]{2}(\cos 67.5° + i \sin 67.5°)$

when $k = 1$, $w_1 = \sqrt[4]{2}(\cos 247.5° + i \sin 247.5°)$

Next, we find the 2 square roots of $-1 - i$:

In trigonometric form, $-1 - i = \sqrt{2}(\cos 225° + i \sin 225°)$

The 2 square roots will be $w_k = \left(\sqrt{2}\right)^{1/2}\left(\operatorname{cis}\dfrac{225° + 360° k}{2}\right)$ for $k = 0,1$

$$= \sqrt[4]{2}\operatorname{cis}\left(112.5° + 180° k\right)$$

We replace k with 0 and 1:

when $k = 0$, $w_0 = \sqrt[4]{2}(\cos 112.5° + i \sin 112.5°)$

when $k = 1$, $w_1 = \sqrt[4]{2}(\cos 292.5° + i \sin 292.5°)$

37. $y = -2\sin(-3x) = 2\sin 3x$ because sine is an odd function.

The graph is a sine curve with amplitude $= 2$ and period $= \dfrac{2\pi}{3}$.

39.   The graph is a cosine curve that has been reflected across the x-axis with:

Amplitude $= 1$

Period $= \dfrac{2\pi}{2} = \pi$

Phase shift $= \dfrac{-\pi/2}{2} = -\dfrac{\pi}{4}$

41.   The graph is a sine curve with:

Amplitude $= 3$

Period $= \dfrac{2\pi}{\pi/3} = 6$

Phase shift $= \dfrac{\pi/3}{\pi/3} = 1$

43. $S = \dfrac{1}{2}bc\sin A$ Formula for area of a triangle

$\quad = \dfrac{1}{2}(2.65)(3.84)\sin 56.2°$ Substitute known values

$\quad = 4.23\,\text{cm}^2$ Simplify

45. $s = \dfrac{1}{2}(a+b+c)$ Formula for half-perimeter

$\quad = \dfrac{1}{2}(2.3+3.4+4.5) = 5.1$ Substitute known values and simplify

$S = \sqrt{s(s-a)(s-b)(s-c)}$ Formula for area of a triangle

$\quad = \sqrt{5.1(5.1-2.3)(5.1-3.4)(5.1-4.5)}$ Substitute known values

$\quad = \sqrt{14.5656} = 3.8\,\text{ft}^2$ Simplify and round to 2 significant digits

Problem Set 8.5

13.   All points of the form $(2, 60° + 360°k)$, where k is an integer, will name the point $(2, 60°)$.

For example, if $k = -1$, we have $(2, -300°)$.

Also, all points of the form $(-2, 240° + 360°k)$, where k is an integer, will name the point $(2, 60°)$.

For example, if $k = 0$, we have $(-2, 240°)$ and if $k = -1$, we have $(-2, -120°)$.

15. All points of the form $(5,135° + 360°k)$, where k is an integer, will name the point $(5,135°)$.

For example, if $k = -1$, we have $(5,-225°)$.

Also, all points of the form $(-5,315° + 360°k)$, where k is an integer, will name the point $(5,135°)$.

For example, if $k = 0$, we have $(-5,315°)$ and if $k = -1$, we have $(-5,-45°)$.

17. All points of the form $(-3,30° + 360°k)$, where k is an integer, will name the point $(-3,30°)$.

For example, if $k = -1$, we have $(-3,-330°)$.

Also, all points of the form $(3,210° + 360°k)$, where k is an integer, will name the point $(-3,30°)$.

For example, if $k = 0$, we have $(3,210°)$ and if $k = -1$, we have $(3,-150°)$.

19.
$$x = r\cos\theta \quad \text{and} \quad y = r\sin\theta$$
$$= 2\cos 60° \qquad\qquad = 2\sin 60°$$
$$= 2\left(\frac{1}{2}\right) = 1 \qquad = 2\left(\frac{\sqrt{3}}{2}\right) = \sqrt{3}$$
$$(2,60°) = \left(1,\sqrt{3}\right)$$

21.
$$x = r\cos\theta \quad \text{and} \quad y = r\sin\theta$$
$$= 3\cos 270° \qquad\qquad = 3\sin 270°$$
$$= 3(0) = 0 \qquad = 3(-1) = -3$$
$$(3,270°) = (0,-3)$$

23.
$$x = r\cos\theta \quad \text{and} \quad y = r\sin\theta$$
$$= \sqrt{2}\cos\left(-135°\right) \qquad = \sqrt{2}\sin\left(-135°\right)$$
$$= \sqrt{2}\left(-\frac{1}{\sqrt{2}}\right) = -1 \qquad = \sqrt{2}\left(-\frac{1}{\sqrt{2}}\right) = -1$$
$$\left(\sqrt{2},-135°\right) = (-1,-1)$$

25.
$$x = r\cos\theta \quad \text{and} \quad y = r\sin\theta$$
$$= -4\sqrt{3}\cos 30° \qquad = -4\sqrt{3}\sin 30°$$
$$= -4\sqrt{3}\left(\frac{\sqrt{3}}{2}\right) = -6 \quad = -4\sqrt{3}\left(\frac{1}{2}\right)$$
$$= -6 \qquad\qquad = -2\sqrt{3}$$
$$\left(-4\sqrt{3},30°\right) = \left(-6,-2\sqrt{3}\right)$$

31.
$$r = \sqrt{x^2 + y^2} \quad \text{and} \quad \tan\theta = \frac{y}{x}$$
$$= \sqrt{(-3)^2 + 3^2} \qquad = \frac{3}{-3}$$
$$= \sqrt{18} = 3\sqrt{2} \qquad = -1$$
$$\theta = 135° \text{ or } 315°$$

Since $(-3,3)$ is in QII, one solution is $\left(3\sqrt{2},135°\right)$.

33.
$$r = \sqrt{x^2 + y^2} \quad \text{and} \quad \tan\theta = \frac{y}{x}$$
$$= \sqrt{\left(-2\sqrt{3}\right)^2 + 2^2} \qquad = \frac{2}{-2\sqrt{3}}$$
$$= \sqrt{16} = 4 \qquad = -\frac{1}{\sqrt{3}}$$
$$\theta = 150° \text{ or } 330°$$

Since $(-2\sqrt{3},2)$ is in QII, one solution is $\left(4,150°\right)$.

35. $r = \sqrt{x^2 + y^2}$ and $\tan\theta = \dfrac{y}{x}$

$\qquad = \sqrt{2^2 + 0^2}$ $\qquad\qquad = \dfrac{0}{2}$

$\qquad = \sqrt{4} = 2$ $\qquad\qquad = 0$

$\qquad\qquad\qquad\qquad\qquad\quad \theta = 0°$

One solution is $(2, 0°)$.

37. $r = \sqrt{x^2 + y^2}$ and $\tan\theta = \dfrac{y}{x}$

$\qquad = \sqrt{\left(-\sqrt{3}\right)^2 + (-1)^2}$ $\qquad = \dfrac{-1}{-\sqrt{3}}$

$\qquad = \sqrt{4} = 2$ $\qquad\qquad\qquad = \dfrac{1}{\sqrt{3}}$

$\qquad\qquad\qquad\qquad\qquad\quad \theta = 30°$ or $210°$

Since $\left(-\sqrt{3}, -1\right)$ is in QIII, one solution is $(2, 210°)$.

39. $r = \sqrt{x^2 + y^2}$ and $\tan\theta = \dfrac{y}{x}$

$\qquad = \sqrt{3^2 + 4^2}$ $\qquad\qquad = \dfrac{4}{3}$

$\qquad = \sqrt{25} = 5$ $\qquad\quad \theta = 53.1°$ or $233.1°$

Since $(3, 4)$ is in QI, one solution is $(5, 53.1°)$.

41. $r = \sqrt{x^2 + y^2}$ and $\tan\theta = \dfrac{y}{x}$

$\qquad = \sqrt{(-1)^2 + 2^2}$ $\qquad = \dfrac{2}{-1} = -2$

$\qquad = \sqrt{5}$ $\qquad\qquad \theta = 116.6°$ or $296.6°$

Since $(-1, 2)$ is in QII, one solution is

$\qquad \left(\sqrt{5}, 116.6°\right)$.

43. $r = \sqrt{x^2 + y^2}$ and $\tan\theta = \dfrac{y}{x}$

$\qquad = \sqrt{(-2)^2 + (-3)^2}$ $\qquad = \dfrac{-3}{-2} = 1.5$

$\qquad = \sqrt{13}$ $\qquad\quad \theta = 56.3°$ or $236.3°$

Since $(-2, -3)$ is in QIII, one solution is

$\qquad \left(\sqrt{13}, 236.3°\right)$.

49. $\qquad r^2 = 9$

$\qquad x^2 + y^2 = 9 \qquad$ Substitution

51. $\qquad\quad r = 6\sin\theta$

$\qquad\quad r^2 = 6r\sin\theta \qquad$ Multiply both sides by r

$\quad x^2 + y^2 = 6y \qquad$ Substitution

53. $\qquad r^2 = 4\sin 2\theta$

$\qquad\quad = 4\left(2\sin\theta\cos\theta\right) \qquad$ Double-angle identity

$\qquad\quad = 8\left(\dfrac{y}{r}\right)\left(\dfrac{x}{r}\right) \qquad$ Substitution

$\qquad\quad = \dfrac{8xy}{r^2} \qquad$ Multiply

This problem is continued on the next page

$$r^4 = 8xy \qquad \text{Multiply both sides by } r^2$$
$$\left(x^2 + y^2\right)^2 = 8xy \qquad \text{Substitution}$$

55.
$$r\left(\cos\theta + \sin\theta\right) = 3$$
$$r\cos\theta + r\sin\theta = 3 \qquad \text{Distributive property}$$
$$x + y = 3 \qquad \text{Substitution}$$

57.
$$x - y = 5$$
$$r\cos\theta - r\sin\theta = 5 \qquad \text{Substitution}$$
$$r\left(\cos\theta - \sin\theta\right) = 5 \qquad \text{Factor}$$

59.
$$x^2 + y^2 = 4$$
$$r^2 = 4 \qquad \text{Substitution}$$

61.
$$x^2 + y^2 = 6x$$
$$r^2 = 6r\cos\theta \qquad \text{Substitution}$$
$$r = 6\cos\theta \qquad \text{Divide both sides by r}$$

63.
$$y = x$$
$$r\sin\theta = r\cos\theta \qquad \text{Substitution}$$
$$\sin\theta = \cos\theta \text{ or} \qquad \text{Divide both sides by } r$$
$$\theta = 45°$$

65. The graph is a sine curve with amplitude of 6.

67. The graph is a sine curve with amplitude of 4 and period of π.

69. The graph is a sine curve with amplitude of 2, period of 2π and a vertical translation of 4.

Problem Set 8.6

1.

θ	$r = 6\cos\theta$	(r,θ)
$0°$	$r = 6\cos 0° = 6$	$(6, 0°)$
$45°$	$r = 6\cos 45° = 4.2$	$(4.2, 45°)$
$90°$	$r = 6\cos 90° = 0$	$(0, 90°)$
$135°$	$r = 6\cos 135° = -4.2$	$(-4.2, 135°)$
$180°$	$r = 6\cos 180° = -6$	$(-6, 180°)$
$225°$	$r = 6\cos 225° = -4.2$	$(-4.2, 225°)$
$270°$	$r = 6\cos 270° = 0$	$(0, 270°)$
$315°$	$r = 6\cos 315° = 4.2$	$(4.2, 315°)$

3.

θ	$r = \sin 3\theta$	(r,θ)
$0°$	$r = \sin 3(0°) = 0$	$(0, 0°)$
$45°$	$r = \sin 3(45°) = 0.71$	$(0.71, 45°)$
$90°$	$r = \sin 3(90°) = -1$	$(-1, 90°)$
$135°$	$r = \sin 3(135°) = 0.71$	$(0.71, 135°)$
$180°$	$r = \sin 3(180°) = 0$	$(0, 180°)$
$225°$	$r = \sin 3(225°) = -0.71$	$(-0.71, 225°)$
$270°$	$r = \sin 3(270°) = 1$	$(1, 270°)$
$315°$	$r = \sin 3(315°) = -0.71$	$(-0.71, 315°)$

9.

$$r = 3$$

$$\pm\sqrt{x^2 + y^2} = 3 \qquad \text{Substitution}$$

$$x^2 + y^2 = 9 \qquad \text{Square both sides}$$

The graph is a circle with center at the origin or pole and radius of 3.

11.

$$\theta = 45°$$

$$\tan\theta = 1 \qquad \text{Find tangent of } 45°$$

$$\frac{y}{x} = 1 \qquad \text{Substitution}$$

$$y = x \qquad \text{Multiply both sides by } x$$

The graph is the line $y = x$.

13. To graph by hand, first we sketch $y = 3\sin x$. We note the relationship between the variations in x or θ and the corresponding variations in y or r.

Variations in x or θ	Corresponding Variations in y or r
0° to 90°	Increases from 0 to 3
90° to 180°	Decreases from 3 to 0
180° to 270°	Decreases from 0 to –3
270° to 360°	Increases from –3 to 0

Then we sketch the graph using this relationship.

15. To graph by hand, first we sketch $y = 4 + 2\sin x$. We note the relationship between the variations in x or θ and the corresponding variations in y or r.

Variations in x or θ	Corresponding Variations in y or r
0° to 90°	Increases from 4 to 6
90° to 180°	Decreases from 6 to 4
180° to 270°	Decreases from 4 to 2
270° to 360°	Increases from 2 to 4

Then we sketch the graph using this relationship.

17. To graph by hand, first we sketch $y = 2 + 4\cos x$. We note the relationship between the variations in x or θ and the corresponding variations in y or r.

Variations in x or θ	Corresponding Variations in y or r
0° to 90°	Decreases from 6 to 2
90° to 180°	Decreases from 2 to –2
180° to 270°	Increases from –2 to 2
270° to 360°	Increases from 2 to 6

Then we sketch the graph using this relationship.

19. To graph by hand, first we sketch $y = 2 + 2\sin x$. We note the relationship between the variations in x or θ and the corresponding variations in y or r.

Variations in x or θ	Corresponding Variations in y or r
$0°$ to $90°$	Increases from 2 to 4
$90°$ to $180°$	Decreases from 4 to 2
$180°$ to $270°$	Decreases from 2 to 0
$270°$ to $360°$	Increases from 0 to 2

Then we sketch the graph using this relationship.

21. $r^2 = 9\sin 2\theta$ is a lemniscate or two-leaved rose where $a = 9$. (See figure 19 in textbook.) The endpoints of the leaves are $\left(\sqrt{a}, 45°\right) = \left(3, 45°\right)$ and $\left(-\sqrt{a}, 45°\right) = \left(-3, 45°\right)$.

23. $r = 2\sin 2\theta$ is a four-leaved rose where $a = 2$. (See figure 17 in textbook.) The endpoints of the leaves are $\left(2, 45°\right), \left(2, 135°\right), \left(2, 225°\right)$, and $\left(2, 315°\right)$.

25. $r = 4\cos 3\theta$ is a three-leaved rose where $a = 4$. (See figure 18 in textbook.) The endpoints of the leaves are $\left(4, 0°\right), \left(4, 120°\right)$ and $\left(4, 240°\right)$.

39. $x^2 + y^2 = 16$
$\qquad r^2 = 16$
The graph is a circle with center at the pole and radius of 4.

41. $x^2 + y^2 = 6x$
$\qquad r^2 = 6r\cos\theta$
$\qquad r = 6\cos\theta$
The graph is a circle with center at $(3, 0°)$ and radius of 3.

43. $\left(x^2 + y^2\right)^2 = 2xy$

$\quad \left(r^2\right)^2 = 2(r\cos\theta)(r\sin\theta)$ Substitution

$\quad r^4 = r^2\left(2\sin\theta\cos\theta\right)$ Simplify

$\quad r^2 = \sin 2\theta$ Divide both sides by r^2 and use the double-angle identity

The graph is a lemniscate with $a = 1$. (See figure 19 in textbook). The endpoints of the leaves are at $\left(1, 45°\right)$ and $\left(1, 225°\right)$.

45. $r(2\cos\theta + 3\sin\theta) = 6$

$\quad\quad 2r\cos\theta + 3r\sin\theta = 6$ Multiply

$\quad\quad\quad\quad\quad 2x + 3y = 6$ Substitution

The graph is a line through (0, 2) and (3, 0) in rectangular coordinates.

47. $r(1 - \cos\theta) = 1$

$\quad\quad\quad r - r\cos\theta = 1$ Multiply

$\quad\quad\quad\quad\quad r - x = 1$ Substitution

$\quad\quad\quad\quad\quad\quad r = x + 1$ Add x to both sides

$\quad\quad\quad\quad\quad r^2 = (x + 1)^2$ Square both sides

$\quad x^2 + y^2 = x^2 + 2x + 1$ Substitution on left side and multiply right side

$\quad\quad\quad\quad\quad y^2 = 2x + 1$ Subtract x^2 from both sides

$\quad\quad\quad\quad\quad x = \dfrac{1}{2}y^2 - \dfrac{1}{2}$ Solve for x

The graph is a parabola that opens to the right with vertex at $(-½, 0)$ in rectangular coordinates.

49. $r = 4\sin\theta$

$\quad\quad\quad\quad r^2 = 4r\sin\theta$ Multiply both sides by r

$\quad\quad x^2 + y^2 = 4y$ Substitution

$\quad x^2 + (y - 2)^2 = 4$ Complete the square and put in standard form

The graph is a circle with center at (0, 2) and radius of 2 in rectangular coordinates.

51.   We graph the two equations. The equation $r_1 = 2\sin\theta$ is a circle with center at (0, 1) and radius of 1. The equation $r_2 = 2\cos\theta$ is a circle with center at (1, 0) and radius of 1.

Next, we solve the equations algebraically by substitution:

$\quad\quad 2\sin\theta = 2\cos\theta$ $r = r$

$\quad\quad \dfrac{2\sin\theta}{2\cos\theta} = 1$ Divide both sides by $2\cos\theta$

$\quad\quad\quad \tan\theta = 1$ Ratio identity

$\quad\quad \theta = 45°$ or $\theta = 225°$

Then we solve for r in each case:

$\quad\quad \theta = 45°,\quad r = 2\sin 45°$ $\quad\quad\quad\quad\quad\quad \theta = 225°,\quad r = 2\sin 225°$

$\quad\quad\quad\quad\quad\quad = 2\left(\dfrac{\sqrt{2}}{2}\right) = \sqrt{2}$ $\quad\quad\quad\quad\quad\quad\quad\quad = 2\left(-\dfrac{\sqrt{2}}{2}\right) = -\sqrt{2}$

Therefore, the points of intersection are $\left(\sqrt{2}, 45°\right)$ and $\left(-\sqrt{2}, 225°\right)$. The graphs have the point (0, 0) in common. On one graph it is the point $\left(0, 0°\right)$ and on the other it is $\left(0, 90°\right)$.

53. Let $y_1 = \sin x$ (the basic sine curve) and $y_2 = -\cos x$ (a cosine curve reflected across the x-axis). Graph y_1, y_2, and $y = y_1 + y_2$ on the same coordinate axes.

55. Let $y_1 = x$ (a line through the origin with a slope of 1) and $y_2 = \sin \pi x$ (a sine curve with period $\dfrac{2\pi}{\pi} = 2$). Graph y_1, y_2, and $y = y_1 + y_2$ on the same coordinate axes.

57. Let $y_1 = 3\sin x$ (a sine curve with amplitude of 3) and $y_2 = \cos 2x$ (a cosine curve with period of $\dfrac{2\pi}{2} = \pi$). Graph y_1, y_2, and $y = y_1 + y_2$ on the same coordinate axes.

Chapter 8 Test

1. $\sqrt{-25} = i\sqrt{25} = 5i$

2. $\sqrt{-12} = i\sqrt{12} = 2i\sqrt{3}$

3. $7x = 14$ and $-6 = -3y$
$x = 2$ $\qquad$ $y = 2$

4.
$$x^2 - 3x = 10 \quad \text{and} \quad 16 = 8y$$
$$x^2 - 3x - 10 = 0 \qquad\qquad y = 2$$
$$(x-5)(x+2) = 0$$
$$x = 5 \text{ or } x = -2$$

5.
$$\begin{aligned}
(6 - 3i) + \left[(4 - 2i) - (3 + i)\right] &= 6 - 3i + 4 - 2i - 3 - i \\
&= (6 + 4 - 3) + (-3i - 2i - i) \\
&= 7 - 6i
\end{aligned}$$

6.
$$\begin{aligned}
(7 + 3i) - \left[(2 + i) - (3 - 4i)\right] &= 7 + 3i - \left[2 + i - 3 + 4i\right] \\
&= 7 + 3i - (-1 + 5i) \\
&= 7 + 3i + 1 - 5i \\
&= 8 - 2i
\end{aligned}$$

7. $i^{16} = \left(i^4\right)^4 = 1^4 = 1$

8. $i^{17} = \left(i^4\right)^4 (i) = 1^4 i = i$

9.
$$\begin{aligned}
(8 + 5i)(8 - 5i) &= 64 - 25i^2 \\
&= 64 - 25(-1) \\
&= 64 + 25 = 89
\end{aligned}$$

10.
$$\begin{aligned}
(3 + 5i)^2 &= 9 + 30i + 25i^2 \\
&= 9 + 30i + 25(-1) \\
&= 9 + 30i - 25 = -16 + 30i
\end{aligned}$$

11.
$$\begin{aligned}
\frac{5 - 4i}{2i} \cdot \frac{i}{i} &= \frac{5i - 4i^2}{2i^2} \\
&= \frac{5i - 4(-1)}{2(-1)} \\
&= \frac{4 + 5i}{-2} = -2 - \frac{5}{2}i
\end{aligned}$$

12.
$$\begin{aligned}
\frac{6 + 5i}{6 - 5i} \cdot \frac{6 + 5i}{6 + 5i} &= \frac{36 + 60i + 25i^2}{36 - 25i^2} \\
&= \frac{36 + 60i + 25(-1)}{36 - 25(-1)} \\
&= \frac{36 + 60i - 25}{36 + 25} \\
&= \frac{11 + 60i}{61} = \frac{11}{61} + \frac{60}{61}i
\end{aligned}$$

13. a. $|3 + 4i| = \sqrt{3^2 + 4^2} = \sqrt{25} = 5$ $\qquad$ b. $-(3 + 4i) = -3 - 4i$ $\qquad$ c. The conjugate of $3 + 4i$ is $3 - 4i$.

14. a. $|3-4i| = \sqrt{3^2 + (-4)^2} = \sqrt{25} = 5$ b. $-(3-4i) = -3+4i$ c. The conjugate of $3-4i$ is $3+4i$.

15. a. $|8i| = \sqrt{0^2 + 8^2} = \sqrt{64} = 8$ b. $-(8i) = -8i$ c. The conjugate of $0+8i$ is $0-8i$ or $-8i$.

16. a. $|-4| = \sqrt{(-4)^2 + 0^2} = \sqrt{16} = 4$ b. $-(-4) = 4$ c. The conjugate of $-4+0i$ is $-4-0i$ or -4.

17. $8(\cos 330^\circ + i\sin 330^\circ) = 8\left[\dfrac{\sqrt{3}}{2} + i\left(-\dfrac{1}{2}\right)\right]$
$$= 4\sqrt{3} - 4i$$

18. $2(\cos 135^\circ + i\sin 135^\circ) = 2\left[-\dfrac{\sqrt{2}}{2} + i\left(\dfrac{\sqrt{2}}{2}\right)\right]$
$$= -\sqrt{2} + i\sqrt{2}$$

19. $x = 2$ and $y = 2$ $\tan\theta = \dfrac{2}{2}$ and θ is in QII

$r = \sqrt{2^2 + 2^2}$ $= 1$
$= \sqrt{8} = 2\sqrt{2}$ $\theta = 45^\circ$

Therefore, $2 + 2i = 2\sqrt{2}(\cos 45^\circ + i\sin 45^\circ)$

20. $x = -\sqrt{3}$ and $y = 1$ $\tan\theta = \dfrac{1}{-\sqrt{3}}$ and θ is in QII

$r = \sqrt{\left(-\sqrt{3}\right)^2 + 1^2}$ $\theta = 150^\circ$
$= \sqrt{4} = 2$

Therefore, $-\sqrt{3} + i = 2(\cos 150^\circ + i\sin 150^\circ)$

21. $5i$ lies on the positive y-axis. Therefore, $r = 5$ and $\theta = 90^\circ$.
$5i = 5(\cos 90^\circ + i\sin 90^\circ)$

22. -3 lies on the negative x-axis. Therefore $r = 3$ and $\theta = 180^\circ$. $-3 = 3\left(\cos 180^\circ + i\sin 180^\circ\right)$.

23. $5\left(\cos 25^\circ + i\sin 25^\circ\right) \cdot 3\left(\cos 40^\circ + i\sin 40^\circ\right) = 5\cdot 3\left[5\cos(25^\circ + 40^\circ) + i\sin(25^\circ + 40^\circ)\right]$
$$= 15\left(\cos 65^\circ + i\sin 65^\circ\right)$$

24.
$$\frac{10\left(\cos 50° + i\sin 50°\right)}{2\left(\cos 20° + i\sin 20°\right)} = \frac{10}{2}\left[\cos(50° - 20°) + i\sin(50° - 20°)\right]$$
$$= 5(\cos 30° + i\sin 30°)$$

25.
$$\left[2\left(\cos 10° + i\sin 10°\right)\right]^5 = 2^5\left[\cos(5 \cdot 10°) + i\sin(5 \cdot 10°)\right]$$
$$= 32(\cos 50° + i\sin 50°)$$

26.
$$\left[3\left(\cos 20° + i\sin 20°\right)\right]^4 = 3^4\left[\cos(4 \cdot 20°) + i\sin(4 \cdot 20°)\right]$$
$$= 81(\cos 80° + i\sin 80°)$$

27. The 2 square roots will be
$$w_k = 49^{1/2}\left(\operatorname{cis}\frac{50° + 360°k}{2}\right) \text{ for } k = 0,1$$
$$= 7\operatorname{cis}\left(25° + 180°k\right)$$

We replace k with 0 and 1:

when $k = 0$, $w_0 = 7(\cos 25° + i\sin 25°)$

when $k = 1$, $w_1 = 7(\cos 205° + i\sin 205°)$

28. First, we put $2 + 2i\sqrt{3}$ in trigonometric form: $x = 2$, $y = 2\sqrt{3}$, and $r = \sqrt{2^2 + (2\sqrt{3})^2} = 4$

Also, $\tan\theta = \dfrac{2\sqrt{3}}{2} = \sqrt{3}$ and θ is in QI. Therefore, $\theta = 60°$.

In trigonometric form $2 + 2i\sqrt{3} = 4\left(\cos 60° + i\sin 60°\right)$

The 4 fourth roots will be $w_k = 4^{1/4}\left(\operatorname{cis}\dfrac{60° + 360°k}{4}\right)$ for $k = 0,1,2,3$
$$= \sqrt{2}\operatorname{cis}\left(15° + 90°k\right)$$

We replace k with 0, 1, 2, and 3:

when $k = 0$, $w_0 = \sqrt{2}(\cos 15° + i\sin 15°)$

when $k = 1$, $w_1 = \sqrt{2}(\cos 105° + i\sin 105°)$

when $k = 2$, $w_2 = \sqrt{2}(\cos 195° + i\sin 195°)$

when $k = 3$, $w_3 = \sqrt{2}(\cos 285° + i\sin 285°)$

29. Using the quadratic formula, we have $x^2 = \dfrac{2\sqrt{3} \pm \sqrt{12 - 4(1)(4)}}{2(1)}$

$$= \dfrac{2\sqrt{3} \pm \sqrt{-4}}{2}$$

$$= \dfrac{2\sqrt{3} \pm 2i}{2} = \sqrt{3} \pm i$$

First, we will find the 2 square root of $\sqrt{3} + i$:

$\sqrt{3} + i = 2(\cos 30° + i \sin 30°)$ in trigonometric form

The 2 square roots will be $w_k = 2^{1/2}\left(\operatorname{cis}\dfrac{30° + 360°k}{2}\right)$ for $k = 0, 1$

$$= \sqrt{2}\operatorname{cis}(15° + 180°k)$$

For $k = 0$, $w_0 = \sqrt{2}(\cos 15° + i \sin 15°)$

For $k = 1$, $w_1 = \sqrt{2}(\cos 195° + i \sin 195°)$

Next, we will find the 2 square root of $\sqrt{3} - i$:

$\sqrt{3} - i = 2(\cos 330° + i \sin 330°)$ in trigonometric form

The 2 square roots will be $w_k = 2^{1/2}\left(\operatorname{cis}\dfrac{330° + 360°k}{2}\right)$ for $k = 0, 1$

$$= \sqrt{2}\operatorname{cis}(165° + 180°k)$$

For $k = 0$, $w_0 = \sqrt{2}(\cos 165° + i \sin 165°)$

For $k = 1$, $w_1 = \sqrt{2}(\cos 345° + i \sin 345°)$

30. In trigonometric form, $-1 = 1(\cos 180° + i \sin 180°)$

The 3 cube roots will be $w_k = 1^{1/3}\left(\operatorname{cis}\dfrac{180° + 360°k}{3}\right)$ for $k = 0, 1, 2$

$$= 1\operatorname{cis}\left(60° + 120°k\right)$$

For $k = 0$, $w_0 = \cos 60° + i \sin 60°$

For $k = 1$, $w_1 = \cos 180° + i \sin 180°$

For $k = 2$, $w_2 = \cos 300° + i \sin 300°$

31. All points of the form $(4, 225° + 360°k)$, where k is an integer, will name the point $(4, 225°)$. For example, if $k = -1$, we have $(4, -135°)$.

Also, all points of the form $(-4, 45° + 360°k)$, where k is an integer, will name the point $(4, 225°)$. For example, if $k = 0$, we have $(-4, 45°)$.

This problem is continued on the next page.

Since $r = 4$ and $\theta = 225°$, $\quad x = 4\cos 225°\quad$ and $\quad y = 4\sin 225°$

$$= 4\left(-\frac{\sqrt{2}}{2}\right) = -2\sqrt{2} \qquad\qquad = 4\left(-\frac{\sqrt{2}}{2}\right) = -2\sqrt{2}$$

Therefore, $(4, 225°) = (-2\sqrt{2}, -2\sqrt{2})$.

32. All points of the form $(-6, 60° + 360° k)$, where k is an integer, will name the point $(-6, 60°)$. For example, if $k = -1$, we have $(-6, -300°)$.

Also, all points of the form $(6, 240° + 360° k)$, where k is an integer, will name the point $(-6, 60°)$. For example, if $k = 0$, we have $(6, 240°)$ and if $k = -1$, we have $(6, -120°)$.

Since $r = -6$ and $\theta = 60°$, $\quad x = -6\cos 60°\quad$ and $\quad y = -6\sin 60°$

$$= -6\left(\frac{1}{2}\right) = -3 \qquad\qquad = -6\left(\frac{\sqrt{3}}{2}\right) = -3\sqrt{3}$$

Therefore, $(-6, 60°) = (-3, -3\sqrt{3})$.

33. $\quad r = \sqrt{x^2 + y^2}\qquad$ and $\qquad \tan\theta = \dfrac{y}{x}$

$$= \sqrt{(-3)^2 + 3^2} \qquad\qquad = \frac{3}{-3} = -1$$

$$= \sqrt{18} = 3\sqrt{2} \qquad\qquad \theta = 135° \text{ or } 315°$$

Since $(-3, 3)$ is in QII, one solution is $\left(3\sqrt{2}, 135°\right)$.

34. Since $(0, 5)$ lies on the positive y-axis, $r = 5$ and $\theta = 90°$. Therefore, $(0, 5) = (5, 90°)$.

35.
$$r = 6\sin\theta$$
$$r^2 = 6r\sin\theta \qquad \text{Multiply both sides by } r$$
$$x^2 + y^2 = 6y \qquad \text{Substitution}$$

36.
$$r = \sin 2\theta$$
$$r = 2\sin\theta\cos\theta \qquad \text{Double-angle identity}$$
$$r = 2\left(\frac{y}{r}\right)\left(\frac{x}{r}\right) \qquad \text{Substitution}$$
$$r = \frac{2xy}{r^2} \qquad \text{Simplify}$$

This problem is continued on the next page.

$$r^3 = 2xy \qquad \text{Multiply both sides by } r^2$$
$$\left(r^2\right)^{3/2} = 2xy \qquad \text{Rewrite in terms of } r^2$$
$$\left(x^2 + y^2\right)^{3/2} = 2xy \qquad \text{Substitution}$$

37.
$$x + y = 2$$
$$r\cos\theta + r\sin\theta = 2 \qquad \text{Substitution}$$
$$r(\cos\theta + \sin\theta) = 2 \qquad \text{Factor}$$

38.
$$x^2 + y^2 = 8y$$
$$r^2 = 8r\sin\theta \qquad \text{Substitution}$$
$$r = 8\sin\theta \qquad \text{Divide both sides by } r$$

39. The graph of $r = 4$ is a circle with center at the pole and radius of 4.

40. The graph of $\theta = 45°$ is a straight line through the pole where θ is $45°$. This is the line $y = x$ in rectangular coordinates.

41. First, we sketch the graph $y = 4 + 2\cos x$. Next, we note the relationship between variations in x or θ, and the corresponding variations in y or r.

Variations in x or θ	Corresponding Variations in y or r
$0°$ to $90°$	Decreases from 6 to 4
$90°$ to $180°$	Decreases from 4 to 2
$180°$ to $270°$	Increases from 2 to 4
$270°$ to $360°$	Increases from 4 to 6

Then we sketch the graph using this relationship.

42. The graph $r = \sin 2\theta$ is a 4-leaved rose where $a = 1$. (See figure 17 in section 8.6 in textbook). The endpoints of the leaves are $(1, 45°)$, $(1, 135°)$, $(1, 225°)$, and $(1, 315°)$.

APPENDIX A – REVIEW OF FUNCTIONS

Problem Set A.1

1. The domain is the x-values of the ordered pairs: Domain = $\{1,2,4\}$

   The range is the y-values of the ordered pairs: Range = $\{3,5,1\}$

   It is a function because for each x there is only one y.

3. The domain is the x-values of the ordered pairs: Domain = $\{-1,1,2\}$

   The range is the y-values of the ordered pairs: Range = $\{3,-5\}$

   It is a function because for each x there is only one y.

5. The domain is the x-values of the ordered pairs: Domain = $\{7,3\}$

   The range is the y-values of the ordered pairs: Range = $\{-1,4\}$

   It is not a function because 7 corresponds to -1 and 4.

7. It is a function because any vertical line will pass through the curve only once which means that for each x there is only one y.

9. It is not a function because the y-axis passes through the curve 3 times which means that for some x-values there is more than one y.

11. It is not a function because there are many lines that pass through the curve 2 times which means that for some x-values there is more than one y.

13. It is a function because any vertical line will pass through the curve only once which means that for each x there is only one y.

15. The graph is a parabola that opens upwards and has been shifted down one unit. The domain is all real numbers and the range is all real numbers greater than or equal to negative one. It is a function because it passes the vertical line test.

17. The graph is a parabola that opens to the right and has been shifted right four units. The domain is all real numbers greater than or equal to four and the range is all real numbers. It is not a function because it does not pass the vertical line test.

19. The graph is a "V" that opens upwards and has been shifted right two units. The domain is all real numbers and the range is all real numbers greater than or equal to zero. It is a function because it passes the vertical line test.

21. The graph is a "V" that opens upwards and has been shifted down two units. The domain is all real numbers and the range is all real numbers greater than or equal to negative two. It is a function because it passes the vertical line test.

23. (a) Let x = the number of hours worked in a week
The amount of money you earn is equal to the rate per hour times the number of hours.
Therefore the equation is $y = 8.50x$.
You work anywhere between 10 and 40 hours. Therefore, the restriction is $10 \le x \le 40$.

(d) Since x is restricted between 10 and 40, the domain is $\{x | 10 \le x \le 40\}$.
The y is also restricted because of x; the range is $\{y | 85 \le y \le 340\}$.

25. (b) Since t is restricted between 0 and 1, the domain is $\{t | 0 \le t \le 1\}$.
The h is also restricted because of t; the range is $\{h | 0 \le h \le 4\}$.

27. (b) Since r is restricted between 0 and 3, the domain is $\{r | 0 \le t \le 3\}$.
The area, A, is also restricted because of r; the range is $\{A | 0 \le A \le 9\pi\}$.

29. $f(2) = 2 \cdot 2 - 5 = 4 - 5 = -1$

31. $f(-3) = 2(-3) - 5 = -6 - 5 = -11$

33. $g(-1) = (-1)^2 + 3(-1) + 4 = 1 - 3 + 4 = 2$

35. $g(-3) = (-3)^2 + 3(-3) + 4 = 9 - 9 + 4 = 4$

37. $g(4) = (4)^2 + 3(4) + 4 = 16 + 12 + 4 = 32$
$f(4) = 2(4) - 5 = 8 - 5 = 3$
$g(4) + f(4) = 32 + 3 = 35$

39. $f(3) = 2(3) - 5 = 6 - 5 = 1$
$g(2) = (2)^2 + 3(2) + 4 = 4 + 6 + 4 = 14$
$f(3) - g(2) = 1 - 14 = -13$

41. $f(0) = 3(0)^2 - 4(0) + 1 = 0 - 0 + 1 = 1$

43. $g(-4) = 2(-4) - 1 = -8 - 1 = -9$

45. $f(-1) = 3(-1)^2 - 4(-1) + 1 = 3 + 4 + 1 = 8$

47. $g(10) = 2(10) - 1 = 20 - 1 = 19$

49. $f(3) = 3(3)^2 - 4(3) + 1 = 27 - 12 + 1 = 16$

51. $g(1/2) = 2(1/2) - 1 = 1 - 1 = 0$

53. $f(a) = 3(a)^2 - 4(a) + 1 = 3a^2 - 4a + 1$

55. In the function f, when $x = 1, y = 4$.
Therefore, $f(1) = 4$.

57. In the function g, when $x = \dfrac{1}{2}$, the second coordinate is 0.

Therefore $g\left(\dfrac{1}{2}\right) = 0$

59. In the function f, when $x = -2$, the second coordinate is 2.

Therefore, $f(-2) = 2$

63. $V(3) = 150 \cdot 2^{3/3} = 150 \cdot 2 = 300$ This means that the painting is worth \$300 in 3 years.

$V(6) = 150 \cdot 2^{6/3} = 150 \cdot 2^2 = 600$ This means that the painting is worth \$600 in 6 years.

65. $A(2) = \pi(2)^2 = 4\pi = 12.56$ $A(5) = \pi(5)^2 = 25\pi = 78.53$ $A(10) = \pi(10)^2 = 100\pi = 314.16$

Problem Set A.2

1.
$$y = 3x - 1$$
$$x = 3y - 1$$
$$3y = x + 1$$
$$y = \frac{x+1}{3}$$
$$f^{-1}(x) = \frac{x+1}{3}$$

3.
$$y = x^3$$
$$x = y^3$$
$$y = \sqrt[3]{x}$$
$$f^{-1}(x) = \sqrt[3]{x}$$

5.
$$y = \frac{x-3}{x-1}$$
$$x = \frac{y-3}{y-1}$$
$$x(y-1) = y-3$$
$$xy - x = y - 3$$

$$xy - y = x - 3$$
$$y(x-1) = x - 3$$
$$y = \frac{x-3}{x-1}$$
$$f^{-1}(x) = \frac{x-3}{x-1}$$

7.
$$y = \frac{x-3}{4}$$
$$x = \frac{y-3}{4}$$
$$4x = y - 3$$
$$y = 4x + 3$$
$$f^{-1}(x) = 4x + 3$$

9.
$$y = \frac{1}{2}x - 3$$
$$x = \frac{1}{2}y - 3$$
$$2x = y - 6$$
$$y = 2x + 6$$
$$f^{-1}(x) = 2x + 6$$

11.
$$y = \frac{2x+1}{3x+1}$$
$$x = \frac{2y+1}{3y+1}$$
$$x(3y+1) = 2y+1$$
$$3xy + x = 2y + 1$$

$$3xy - 2y = 1 - x$$
$$y(3x-2) = 1 - x$$
$$y = \frac{1-x}{3x-2}$$
$$f^{-1}(x) = \frac{1-x}{3x-2}$$

27. (a) The function is one-to-one because it passes the horizontal line test.

(b) The function is not one-to-one because it does not pass the horizontal line test.

29. (a) $f(2) = 3(2) - 2 = 4$ (b) $f^{-1}(2) = \dfrac{2+2}{3} = \dfrac{4}{3}$

(c) $f\left[f^{-1}(2)\right] = f\left(\dfrac{4}{3}\right) = 3\left(\dfrac{4}{3}\right) - 2 = 2$ (d) $f^{-1}\left[f(2)\right] = f^{-1}(4) = \dfrac{4+2}{3} = 2$

31.
$$y = \frac{1}{x}$$
$$x = \frac{1}{y}$$
$$xy = 1$$
$$y = \frac{1}{x}$$
$$f^{-1}(x) = \frac{1}{x}$$

33. (a) $f\left[g(-3)\right] = f(2) = -3$

(b) $g\left[f(-6)\right] = g(3) = -6$

(c) $g\left[f(2)\right] = g(-3) = 2$

(d) $f\left[g(3)\right] = f(-6) = 3$

(e) $f\left[g(-2)\right] = f(3) = -2$

(f) $g\left[f(3)\right] = g(-2) = 3$

(g) They are inverses of each other.

35.
$$y = 3x + 5$$
$$x = 3y + 5$$
$$3y = x - 5$$
$$y = \frac{x-5}{3}$$
$$f^{-1}(x) = \frac{x-5}{3}$$

$f\left[f^{-1}(x)\right] = f\left(\dfrac{x-5}{3}\right)$

$= 3\left(\dfrac{x-5}{3}\right) + 5$

$= x - 5 + 5 = x$

37.
$$y = x^3 + 1$$
$$x = y^3 + 1$$
$$y^3 = x - 1$$
$$y = \sqrt[3]{x-1}$$
$$f^{-1}(x) = \sqrt[3]{x-1}$$

$f\left[f^{-1}(x)\right] = f\left(\sqrt[3]{x-1}\right)$

$= \left(\sqrt[3]{x-1}\right)^3 + 1$

$= x - 1 + 1 = x$

APPENDIX B – EXPONENTIAL AND LOGARITHMIC FUNCTIONS

Problem Set B.1

1. $g(0) = \left(\dfrac{1}{2}\right)^0 = 1$

3. $g(-1) = \left(\dfrac{1}{2}\right)^{-1} = 2$

5. $f(-3) = (3)^{-3} = \dfrac{1}{3^3} = \dfrac{1}{27}$

7. $f(2) = 3^2 = 9$ and $g(-2) = \left(\dfrac{1}{2}\right)^{-2} = 2^2 = 4$

$f(2) + g(-2) = 9 + 4 = 13$

9. If $x = -1$, then $y = 4^{-1} = \dfrac{1}{4}$.   This gives us $\left(-1, \dfrac{1}{4}\right)$ as a point on the curve.

If $x = 0$, then $y = 4^0 = 1$.   This gives us $(0, 1)$ as a point on the curve.

If $x = 1$, then $y = 4^1 = 4$.   This gives us $(1, 4)$ as a point on the curve.

Graph these three points and then draw a smooth curve through them. Also remember that y can never be equal to zero and therefore, the x-axis is a horizontal asymptote.

11. If $x = -1$, then $y = 3^{-(-1)} = 3$.   This gives us $(-1, 3)$ as a point on the curve.

If $x = 0$, then $y = 3^0 = 1$.   This gives us $(0, 1)$ as a point on the curve.

If $x = 1$, then $y = 3^{-1} = \dfrac{1}{3}$.   This gives us $\left(1, \dfrac{1}{3}\right)$ as a point on the curve.

Graph these three points and then draw a smooth curve through them. Also remember that y can never be equal to zero and therefore, the x-axis is a horizontal asymptote.

13. If $x = -1$, then $y = 2^{-1+1} = 2^0 = 1$.   This gives us $(-1, 1)$ as a point on the curve.

If $x = 0$, then $y = 2^{0+1} = 2^1 = 2$.   This gives us $(0, 2)$ as a point on the curve.

If $x = 1$, then $y = 2^{1+1} = 2^2 = 4$.   This gives us $(1, 4)$ as a point on the curve.

Graph these three points and then draw a smooth curve through them. Also remember that y can never be equal to zero and therefore, the x-axis is a horizontal asymptote.

15.   If $x = -1$, then $y = e^{-1} = \dfrac{1}{e} = 0.37$.   This gives us $(-1, 0.37)$ as a point on the curve.

If $x = 0$, then $y = e^0 = 1$.   This gives us $(0, 1)$ as a point on the curve.

If $x = 1$, then $y = e^1 = 2.72$.   This gives us $(1, 2.72)$ as a point on the curve.

Graph these three points and then draw a smooth curve through them. Also remember that y can never be equal to zero and therefore, the x-axis is a horizontal asymptote.

17.   The graph of $y = 2x$ is a straight line with a slope of 2 and a y-intercept of 0. We can draw the straight line through (0, 0) and (2, 4).

The graph of $y = x^2$ is a parabola: If $x = 0$, then $y = 0^2 = 0$.

If $x = 1$, then $y = 1^2 = 1$.

If $x = 2$, then $y = 2^2 = 4$.

We draw the curve through (0, 0), (1, 1), and (2, 4).

The graph of $y = 2^x$ is an exponential function: If $x = 0$, then $y = 2^0 = 1$.

If $x = 1$, then $y = 2^1 = 2$.

If $x = 2$, then $y = 2^2 = 4$

We draw the curve through (0, 1), (1, 2), and (2, 4).

21. $h = 6\left(\dfrac{2}{3}\right)^n$ When n is 5, $h = 6\left(\dfrac{2}{3}\right)^5 = 0.79$ feet.

23. $A(t) = 1400 \cdot 2^{-t/8}$ If $t = 8$, then $A(8) = 1400 \cdot 2^{-8/8} = 1400 \cdot 2^{-1} = 1400\left(\dfrac{1}{2}\right) = 700$ micrograms.

If $t = 11$, then $A(11) = 1400 \cdot 2^{-11/8} = \dfrac{1400}{2^{11/8}} = 539.8$ micrograms.

25.   (a) $A(t) = P\left(1 + \dfrac{r}{n}\right)^{nt}$ where $P = 1200$, $r = 0.06$, and $n = 4$

$A(t) = 1200\left(1 + \dfrac{0.06}{4}\right)^{4t}$

(b) $A(8) = 1200\left(1 + \dfrac{0.06}{4}\right)^{4(8)} = \$1932.39$

(c) $A = Pe^{rt}$ where $P = 1200$ and $r = 0.06$
$A = 1200e^{(0.06)(8)} = \$1939.29$

27. $A(t) = P\left(1 + \dfrac{r}{n}\right)^{nt}$ where $P = 5000$, $r = 0.12$, and $n = 4$

$A(10) = 5000\left(1 + \dfrac{0.12}{4}\right)^{4(10)} = \$16,310.19$

29. $f(1) = 100 \cdot 2^1 = 200$ bacteria

$f(2) = 100 \cdot 2^2 = 400$ bacteria

$f(3) = 100 \cdot 2^3 = 800$ bacteria

$f(4) = 100 \cdot 2^4 = 1600$ bacteria

31. (a) In 1990, $t = 20$: $E(20) = 78.16(1.11)^{20} = \$630$ billion

$699 - 630 = 69$ billion   The function underestimated the expenditures by \$69 billion.

(b) In 2005, $t = 35$: $E(35) = 78.16(1.11)^{35} = \$3015$ billion

In 2006, $t = 36$: $E(36) = 78.16(1.11)^{36} = \$3347$ billion

In 2007, $t = 37$ $E(37) = 78.16(1.11)^{37} = \$3715$ billion

33. If $t = 0$, then $V(0) = 150 \cdot 2^{0/3} = 150 \cdot 1 = 150$.   This gives us $(0, 150)$ as a point on the curve.

If $t = 1$, then $V(1) = 150 \cdot 2^{1/3} = 189$.   This gives us $(1, 189)$ as a point on the curve.

If $t = 2$, then $V(2) = 150 \cdot 2^{2/3} = 238$.   This gives us $(2, 238)$ as a point on the curve.

If $t = 4$, then $V(4) = 150 \cdot 2^{4/3} = 378$.   This gives us $(4, 378)$ as a point on the curve.

If $t = 6$, then $V(6) = 150 \cdot 2^{6/3} = 150 \cdot 2^2 = 600$.   This gives us $(6, 600)$ as a point on the curve.

Graph these 5 points and then draw a smooth curve through them.

35. (a) Three years and 6 months is equivalent to 3.5 years:
$$V(3.5) = 450,000(1 - 0.30)^{3.5} = \$129,138.48$$

(b) The function is good for the first 6 years of use. Therefore, the domain is $\{t \,|\, 0 \le t \le 6\}$.

Problem Set B.2

25. $x = 3^2 = 9$

27. $x = 5^{-3} = \dfrac{1}{5^3} = \dfrac{1}{125}$

29. $16 = 2^x$

$x = 4$

31. $2 = 8^x$

$2^1 = 2^{3x}$

$1 = 3x$

$x = \dfrac{1}{3}$

33. $4 = x^2$

$x = \pm 2$

$x = 2$ because the base must be positive

35. $5 = x^3$

$x = \sqrt[3]{5}$

37. If we change this function to exponential form, we get $x = 3^y$. Now we can substitute values for y and find the corresponding x-values:

If $y = -1$, then $x = 3^{-1} = \dfrac{1}{3}$.
This gives us $\left(\dfrac{1}{3}, -1\right)$ as a point on the curve.

If $y = 0$, then $x = 3^0 = 1$.
This gives us $(1, 0)$ as a point on the curve.

If $y = 1$, then $x = 3^1 = 3$.
This gives us $(3, 1)$ as a point on the curve.

Graph these 3 points and then draw a smooth curve through them. Remember there is a vertical asymptote at the y-axis.

39. If we change this function to exponential form, we get $x = \left(\dfrac{1}{3}\right)^y$. Now we can substitute values for y and find the corresponding x-values:

If $y = -1$, then $x = \left(\dfrac{1}{3}\right)^{-1} = 3$.
This gives us $(3, -1)$ as a point on the curve.

If $y = 0$, then $x = \left(\dfrac{1}{3}\right)^0 = 1$.
This gives us $(1, 0)$ as a point on the curve.

If $y = 1$, then $x = \left(\dfrac{1}{3}\right)^1 = \dfrac{1}{3}$.
This gives us $\left(\dfrac{1}{3}, 1\right)$ as a point on the curve.

Graph these 3 points and then draw a smooth curve through them. Remember there is a vertical asymptote at the y-axis.

45. If $x = \log_2 16$, then $2^x = 16$ and $x = 4$.

47. If $x = \log_{25} 125$, then
$$25^x = 125$$
$$5^{2x} = 5^3$$
$$2x = 3$$
$$x = \frac{3}{2}$$

49. If $x = \log_{10} 1000$, then $10^x = 1000$ and $x = 3$.

51. If $x = \log_3 3$, then $3^x = 3$ and $x = 1$.

53. If $x = \log_5 1$, then $5^x = 1$ and $x = 0$.

55. We work from the inside outwards. First let $x = \log_6 6$. Then $6^x = 6$ and $x = 1$.
Next we must find $\log_3 1$. We'll let $y = \log_3 1$, then $3^y = 1$ and $y = 0$.
Therefore, $\log_3(\log_6 6) = 0$

57. We begin with the innermost grouping symbol. First let $x = \log_2 16$. Then $2^x = 16$ and $x = 4$.
Next we must find $\log_2 4$. We'll let $y = \log_2 4$, then $2^y = 4$ and $y = 2$.
Last we must find $\log_4 2$. We'll let $z = \log_4 2$, then $4^z = 2$
$$2^{2z} = 2^1$$
$$2z = 1 \quad \text{or} \quad z = \frac{1}{2}$$
Therefore, $\log_4 \left[\log_2(\log_2 16)\right] = \frac{1}{2}$.

59.
$$\text{pH} = -\log_{10}\left[10^{-7}\right]$$
$$= -(-7) = 7$$

61.
$$6 = -\log_{10}\left[\text{H}^+\right]$$
$$\log_{10}\left[\text{H}^+\right] = -6$$
$$\left[\text{H}^+\right] = 10^{-6} = 0.000001$$

63.
$$M = \log_{10} 100$$
$$10^M = 100$$
$$M = 2$$

65.
$$8 = \log_{10} T$$
$$T = 10^8$$
It is 10^8 times as large.

Problem Set B.3

7. $\log_9 \sqrt[3]{z} = \log_9 z^{1/3} = \dfrac{1}{3}\log_9 z$

9. $\log_6 x^2 y^4 = \log_6 x^2 + \log_6 y^4$

$$= 2\log_6 x + 4\log_6 y$$

11. $\log_5\left(\sqrt{x}\cdot y^4\right) = \log_5 x^{1/2} + \log_5 y^4$

$$= \dfrac{1}{2}\log_5 x + 4\log_5 y$$

13. $\log_b \dfrac{xy}{z} = \log_b xy - \log_b z$

$$= \log_b x + \log_b y - \log_b z$$

15. $\log_{10} \dfrac{4}{xy} = \log_{10} 4 - \log_{10} xy$

$$= \log_{10} 4 - \left(\log_{10} x + \log_{10} y\right)$$

$$= \log_{10} 4 - \log_{10} x - \log_{10} y$$

17. $\log_{10} \dfrac{x^2 y}{\sqrt{z}} = \log_{10} x^2 y - \log_{10} z^{1/2}$

$$= \log_{10} x^2 + \log_{10} y - \log_{10} z^{1/2}$$

$$= 2\log_{10} x + \log_{10} y - \dfrac{1}{2}\log_{10} z$$

19. $\log_{10} \dfrac{x^3 \sqrt{y}}{z^4} = \log_{10} x^3 y^{1/2} - \log_{10} z^4$

$$= \log_{10} x^3 + \log_{10} y^{1/2} - \log_{10} z^4$$

$$= 3\log_{10} x + \dfrac{1}{2}\log_{10} y - 4\log_{10} z$$

21. $\log_b \sqrt[3]{\dfrac{x^2 y}{z^4}} = \log_b \left(\dfrac{x^2 y}{z^4}\right)^{1/3}$

$$= \dfrac{1}{3}\left(\log_b \dfrac{x^2 y}{z^4}\right)$$

$$= \dfrac{1}{3}\left(\log_b x^2 y - \log_b z^4\right)$$

$$= \dfrac{1}{3}\left(\log_b x^2 + \log_b y - \log_b z^4\right)$$

$$= \dfrac{1}{3}\left(2\log_b x + \log_b y - 4\log_b z\right)$$

$$= \dfrac{2}{3}\log_b x + \dfrac{1}{3}\log_b y - \dfrac{4}{3}\log_b z$$

23. $\log_b x + \log_b y = \log_b (xy)$

25. $2\log_3 x - 3\log_3 y = \log_3 x^2 - \log_3 y^3$

$$= \log_3 \dfrac{x^2}{y^3}$$

27. $\dfrac{1}{2}\log_{10} x + \dfrac{1}{3}\log_{10} y = \log_{10} x^{1/2} + \log_{10} y^{1/3}$

$$= \log_{10}\left(x^{1/2} y^{1/3}\right)$$

$$= \log_{10} \sqrt{x}\sqrt[3]{y}$$

29. $3\log_2 x + \dfrac{1}{2}\log_2 y - \log_2 z = \log_2 x^3 + \log_2 y^{1/2} - \log_2 z$

$$= \log_2 x^3 y^{1/2} - \log_2 z$$

$$= \log_2 \frac{x^3 \sqrt{y}}{z}$$

31. $\dfrac{1}{2}\log_2 x - 3\log_2 y - 4\log_2 z = \log_2 x^{1/2} - \left(\log_2 y^3 + \log_2 z^4\right)$

$$= \log_2 \sqrt{x} - \log_2 y^3 z^4$$

$$= \log_2 \frac{\sqrt{x}}{y^3 z^4}$$

33. $\dfrac{3}{2}\log_{10} x - \dfrac{3}{4}\log_{10} y - \dfrac{4}{5}\log_{10} z = \log_{10} x^{3/2} - \left(\log_{10} y^{3/4} + \log_{10} z^{4/5}\right)$

$$= \log_{10} x^{3/2} - \log_{10} y^{3/4} z^{4/5}$$

$$= \log_{10} \frac{x^{3/2}}{y^{3/4} z^{4/5}}$$

35. $\log_2 x + \log_2 3 = 1$

$\log_2 3x = 1$

$3x = 2^1$

$x = \dfrac{2}{3}$ (It checks.)

37. $\log_3 x - \log_3 2 = 2$

$\log_3 \dfrac{x}{2} = 2$

$\dfrac{x}{2} = 3^2$

$x = 18$ (It checks.)

39. $\log_3 x + \log_3 (x-2) = 1$

$\log_3 x(x-2) = 1$

$x(x-2) = 3^1$

$x^2 - 2x - 3 = 0$

$(x-3)(x+1) = 0$

$x = 3$ or $x = -1$

$x = 3$ is the only solution because

$x = -1$ does not check

41. $\log_3 (x+3) - \log_3 (x-1) = 1$

$\log_3 \dfrac{x+3}{x-1} = 1$

$\dfrac{x+3}{x-1} = 3^1$

$x + 3 = 3x - 3$

$-2x = -6$

$x = 3$ (It checks)

43. $\log_2 x + \log_2 (x-2) = 3$

$\log_2 x(x-2) = 3$

$x(x-2) = 2^3$

$x^2 - 2x - 8 = 0$

$(x-4)(x+2) = 0$

$x = 4 \quad \text{or} \quad x = -2$

$x = 4$ is the only solution because
$x = -2$ does not check

45. $\log_8 x + \log_8 (x-3) = \dfrac{2}{3}$

$\log_8 x(x-3) = \dfrac{2}{3}$

$x(x-3) = 8^{2/3}$

$x^2 - 3x - 4 = 0$

$(x-4)(x+1) = 0$

$x = 4 \quad \text{or} \quad x = -1$

$x = 4$ is the only solution because $x = -1$
does not check

47. $\log_5 \sqrt{x} + \log_5 \sqrt{6x+5} = 1$

$\log_5 \sqrt{x(6x+5)} = 1$

$\sqrt{x(6x+5)} = 5^1$

$x(6x+5) = 5^2$

$6x^2 + 5x - 25 = 0$

$(3x-5)(2x+5) = 0$

$x = \dfrac{5}{3} \quad \text{or} \quad x = -\dfrac{5}{2}$

$x = \dfrac{5}{3}$ is the only solution because $x = -\dfrac{5}{2}$

does not check.

49. $M = 0.21\left(\log_{10} 1 - \log_{10} 10^{-12}\right)$

$= 0.21[0 - (-12)]$

$= 0.21(12) = 2.52 \text{ min}$

$M = 0.21\log_{10} \dfrac{1}{10^{-12}}$

$= 0.21\log_{10} 10^{12}$

$= 0.21(12) = 2.52 \text{ min}$

51. $\text{pH} = 6.1 + \log_{10}\left(\dfrac{x}{y}\right)$

$= 6.1 + \log_{10} x - \log_{10} y$

53. $D = 10\log_{10}\left(\dfrac{I}{I_0}\right)$

$= 10(\log_{10} I - \log_{10} I_0)$

Problem Set B.4

41. $\ln e = \log_e e = 1$

43. $\ln e^5 = 5\ln e = 5(1) = 5$

45. $\ln e^x = x\ln e = x(1) = x$

47. $\ln 10e^{3t} = \ln 10 + \ln e^{3t}$

$= \ln 10 + 3t\ln e$

$= \ln 10 + 3t(1) = \ln 10 + 3t$

49.
$$\ln Ae^{-2t} = \ln A + \ln e^{-2t}$$
$$= \ln A + (-2t)\ln e$$
$$= \ln A - 2t$$

51.
$$\ln 15 = \ln(5 \cdot 3)$$
$$= \ln 5 + \ln 3$$
$$= 1.6094 + 1.0986 = 2.7080$$

53.
$$\ln \frac{1}{3} = \ln 1 - \ln 3$$
$$= 0 - 1.0986 = -1.0986$$

55.
$$\ln 9 = \ln 3^2$$
$$= 2\ln 3$$
$$= 2(1.0986) = 2.1972$$

57.
$$\ln 16 = \ln 2^4$$
$$= 4\ln 2$$
$$= 4(0.6931) = 2.7724$$

59.
$$pH = -\log(6.50 \cdot 10^{-4})$$
$$= 3.19$$

61.
$$4.75 = -\log\left[H^+\right]$$
$$\log_{10}\left[H^+\right] = -4.75$$
$$\left[H^+\right] = 10^{-4.75} = 1.78 \times 10^{-5}$$

63.
$$5.5 = \log T$$
$$T = 10^{5.5} = 3.16 \times 10^5$$

65.
$$8.3 = \log T$$
$$T = 10^{8.3} = 2.00 \times 10^8$$

67.
$$6.5 = \log T$$
$$T = 10^{6.5} = 3.16 \times 10^6$$
$$5.5 = \log T$$
$$T = 10^{5.5} = 3.16 \times 10^5$$
It is 10 times larger.

69.

Moresby Island:	$T = 10^{4.0} = 1.00 \times 10^4$
Vancouver Island	$M = \log(1.99 \times 10^5) = 5.3$
Quebec City	$T = 10^{3.2} = 1.58 \times 10^3$
Mould Bay	$T = 10^{5.2} = 1.58 \times 10^5$
St. Lawrence	$M = \log(5.01 \times 10^3) = 3.7$

71.
$$\log(1-r) = \frac{1}{5}\log\frac{4500}{9000}$$
$$\log(1-r) = 0.2\log(0.5)$$
$$\log(1-r) = -0.0602$$
$$1-r = 10^{-0.0602}$$
$$1-r = 0.871$$
$$r = 0.129 = 12.9\%$$

73.
$$\log(1-r) = \frac{1}{5}\log\frac{5750}{7550}$$
$$\log(1-r) = 0.2\log(0.7616)$$
$$\log(1-r) = -0.0237$$
$$1-r = 10^{-0.0237}$$
$$1-r = 0.947$$
$$r = 0.053 = 5.3\%$$

Problem Set B.5

1.
$$3^x = 5$$
$$\log 3^x = \log 5$$
$$x \log 3 = \log 5$$
$$x = \frac{\log 5}{\log 3} = 1.4650$$

3.
$$5^x = 3$$
$$\log 5^x = \log 5$$
$$x \log 5 = \log 3$$
$$x = \frac{\log 3}{\log 5} = 0.6826$$

5.
$$5^{-x} = 12$$
$$\log 5^{-x} = \log 12$$
$$-x \log 5 = \log 12$$
$$x = \frac{\log 12}{-\log 5} = -1.5440$$

7.
$$12^{-x} = 5$$
$$\log 12^{-x} = \log 5$$
$$-x \log 12 = \log 5$$
$$x = \frac{\log 5}{-\log 12} = -0.6477$$

9.
$$8^{x+1} = 4$$
$$2^{3(x+1)} = 2^2$$
$$3x + 3 = 2$$
$$3x = -1$$
$$x = -\frac{1}{3}$$

11.
$$4^{x-1} = 4^1$$
$$x - 1 = 1$$
$$x = 2$$

13.
$$3^{2x+1} = 2$$
$$\log 3^{2x+1} = \log 2$$
$$(2x+1) \log 3 = \log 2$$
$$2x + 1 = \frac{\log 2}{\log 3}$$
$$2x = \frac{\log 2}{\log 3} - 1$$
$$x = \frac{1}{2}\left(\frac{\log 2}{\log 3} - 1\right) = -0.1845$$

15.
$$3^{1-2x} = 2$$
$$\log 3^{1-2x} = \log 2$$
$$(1-2x) \log 3 = \log 2$$
$$1 - 2x = \frac{\log 2}{\log 3}$$
$$-2x = \frac{\log 2}{\log 3} - 1$$
$$x = -\frac{1}{2}\left(\frac{\log 2}{\log 3} - 1\right) = 0.1845$$

17.
$$15^{3x-4} = 10$$
$$(3x-4) \log 15 = \log 10 \quad (\log 10 = 1)$$
$$3x - 4 = \frac{1}{\log 15}$$
$$3x - 4 = 0.8503$$
$$3x = 4.8503$$
$$x = 1.6168$$

19.
$$6^{5-2x} = 4$$
$$\log 6^{5-2x} = \log 4$$
$$(5-2x)\log 6 = \log 4$$
$$5 - 2x = \frac{\log 4}{\log 6}$$
$$-2x = \frac{\log 4}{\log 6} - 5$$
$$x = -\frac{1}{2}\left(\frac{\log 4}{\log 6} - 5\right) = 2.1131$$

21. $\log_8 16 = \dfrac{\log 16}{\log 8} = 1.3333$

23. $\log_{16} 8 = \dfrac{\log 8}{\log 16} = 0.7500$

25. $\log_7 15 = \dfrac{\log 15}{\log 7} = 1.3917$

27. $\log_{15} 7 = \dfrac{\log 7}{\log 15} = 0.7186$

29. $\log_8 240 = \dfrac{\log 240}{\log 8} = 2.6356$

31. $\log_4 321 = \dfrac{\log 321}{\log 4} = 4.1632$

41.
$$1000 = 500\left(1 + \frac{0.06}{2}\right)^{2t}$$
$$2 = (1.03)^{2t}$$
$$\log 2 = \log(1.03)^{2t}$$
$$\log 2 = 2t\log(1.03)$$
$$t = \frac{\log 2}{2\log 1.03} = 11.7 \text{ years}$$

43.
$$3000 = 1000\left(1 + \frac{0.12}{6}\right)^{6t}$$
$$3 = (1.02)^{6t}$$
$$\log 3 = \log(1.02)^{6t}$$
$$\log 3 = 6t\log(1.02)$$
$$t = \frac{\log 3}{6\log 1.02} = 9.25 \text{ years}$$

45.
$$2P = P\left(1 + \frac{0.08}{4}\right)^{4t}$$
$$2 = (1.02)^{4t}$$
$$\log 2 = \log(1.02)^{4t}$$
$$\log 2 = 4t\log(1.02)$$
$$t = \frac{\log 2}{4\log(1.02)} = 8.75 \text{ years}$$

47.
$$75 = 25\left(1 + \frac{0.06}{2}\right)^{2t}$$
$$3 = (1.03)^{2t}$$
$$\log 3 = \log(1.03)^{2t}$$
$$\log 3 = 2t\log(1.03)$$
$$t = \frac{\log 3}{2\log 1.03} = 18.6 \text{ years}$$

49.
$$1000 = 500e^{0.06t}$$
$$2 = e^{0.06t}$$
$$0.06t = \ln 2$$
$$t = \frac{\ln 2}{0.06} = 11.6 \text{ years}$$

51.
$$1500 = 500e^{0.06t}$$
$$3 = e^{0.06t}$$
$$0.06t = \ln 3$$
$$t = \frac{\ln 3}{0.06} = 18.3 \text{ years}$$

53.
$$64,000 = 32,000e^{0.05t}$$
$$2 = e^{0.05t}$$
$$0.05t = \ln 2$$
$$t = \frac{\ln 2}{0.05} = 13.9 \text{ years later}$$
$$\text{or near the end of 2007}$$

55.
$$45,000 = 15,000e^{0.04t}$$
$$3 = e^{0.04t}$$
$$0.04t = \ln 3$$
$$t = \frac{\ln 3}{0.04} = 27.5 \text{ years}$$